INTERREGIONAL UNCONFORMITIES AND HYDROCARBON ACCUMULATION

AAPG MEMOIR 36

INTERREGIONAL UNCONFORMITIES AND HYDROCARBON ACCUMULATION

EDITED BY JOHN S. SCHLEE

Published by
The American Association of Petroleum Geologists
Tulsa, Oklahoma 74101, U.S.A.

Association Editor: Richard Steinmetz
Science Director: Edward A. Beaumont
Project Editor: Douglas A. White
Design and Production: S. Wally Powell
Typographers: Tricia Kinion and Eula Matheny

CONTENTS

FOREWORD

In 1956, as a new employee of the U.S. Geological Survey (USGS), I was assigned to map Mesozoic rocks of the southern San Juan basin of northwestern New Mexico. In the "layer-cake" geology of the area, sedimentary formations spread colorfully across the area with seemingly little change and with apparent conformity to older and younger units. Yet, according to the regional stratigraphers whose job it was to correlate these formations over the entire Colorado Plateau, major low-angle unconformities of regional extent could be seen to truncate units over hundreds of miles lateral distance (see USGS Professional Paper 1035-A by Pipiringos and O'Sullivan). It was hard for those of us who mapped on a scale of $7 \frac{1}{2}$ minute quadrangles to see evidence for these great erosional gaps.

At the Dallas AAPG meeting in 1975, I was reintroduced to the idea of widespread unconformities, this time by Peter Vail and his colleagues at Exxon. The exciting part of Vail's presentation was that he showed a worldwide relative sea-level curve punctuated by sudden worldwide periodic drops or regression, and introduced a total approach to mapping depositional sequences based on the correlation of the bounding interregional unconformities. Vail and some of his colleagues were products of Northwestern University where they had been ably trained by Larry Sloss to think in terms of widespread interregional unconformities — the kind he had talked and written about for years. The important aspect of their activities at Exxon is that they worked 1) with state of the art multichannel seismic reflection profiles over the continental margin, an area where the sedimentary section is the thickest and most

extensive; and 2) they had drill-hole information from all over the world with which to check their seismic picks.

Their approach whet my appetite, so I enrolled in the second AAPG short course on seismic stratigraphy at which Vail and his associates attempted to convince us of the efficacy of their method. It was a stimulating and boisterous series of sessions duplicated when Vail gave a series of lectures to the students and faculty at Woods Hole Oceanographic Institution (WHOI) in the summer of 1977. Late in 1977, the AAPG published Memoir 26 which discussed in detail the Exxon approach to seismic stratigraphy — an all-time best-seller from the AAPG.

The perturbations in Vail's sea-level curve sparked derisive comments from stratigraphers, paleontologists, global tectonists and marine geologists alike. The suite of Exxon papers was long on technique and implications, and short on paleontologic documentation. Those of us who were trying to apply their method to the Atlantic continental margin were hampered both by the lack of deep stratigraphic holes and of fossils in the oldest part of the offshore section. Utilization of Vail's technique in the marine realm represented a major change of emphasis from the mapping of key horizons to the mapping of unconformities in the deep sea. Vail et al (1980) worked to stress the importance of unconformities in the stratigraphy of deep oceanic basins and their equivalence to some key seismic horizons.

Partly to air the growing differences between stratigraphers working on the continent or continental margin from those working in the deep sea, and partly to view how other marine stratigraphers were using interregional unconformities, I proposed to the Marine Geology Committee that a research conference on interregional unconformities be held in Woods Hole, Massachusetts, in April 1980. To my surprise, they endorsed it. I hoped the conference would comment on the causes of short-term shifts of sea level, the relationship between onland unconformities and those delineated in the margins and deep sea, and the evidences for Phanerozoic glaciations.

By April 28, 1980, when the conference began, we had scheduled 17 speakers for two and one-half days of talks and discussion. Five speakers were from oil companies and the remainder from university and research labs. Nearly 80 participants attended: 59% from oil companies, 24% from academe, and 7% from government labs.

Response to the conference was so favorable that the AAPG asked that I organize a half-day session for the 1981 AAPG Annual Meeting in San Francisco. This was accomplished in an overcrowed session room the first day of the meeting, and as a result it was decided to publish most of the papers presented there, plus some other papers not presented (Kidwell; Tucholke and Embley; Harris, Frost, Seiglie, and Schneidermann; Seiglie and Baker; Hazel, Edwards, and Bybell; Seiglie and Moussa).

Many people helped in organizing the conference and later session, among them, Dr. W. A. Berggren (WHOI), Hollis Hedberg (Princeton University), Mahlon Ball (USGS), Gary Howell, and Sondra Biggs (AAPG).

In 1980 Richard A. Kerr (Science wrote):

"Five years after it was first presented the Exxon sea-level curve continues to be debated but its basic integrity remains intact. . . . Support continues to increase for its usefulness as a general approach to deciphering the geologic record of the continental margin."

What is needed now is more documentation concerning the age and magnitude of sea level shifts, and ultimately, the cause of the short term shifts. Through the papers that follow, this memoir attempts to begin these tasks.

John S. Schlee
U.S. Geological Survey
Woods Hole, Massachusetts
August 1984

REFERENCES

Kerr, R. A., 1980, Changing global sea levels as a geologic index: Science, v. 209, p. 483-486.

Pipiringos, G. N., and R. B. O'Sullivan, 1978, Principal unconformities in Triassic and Jurassic rocks, Western Interior United States—a preliminary survey: U.S. Geological Survey Professional Paper 1035-A, 29 p.

Vail, P. R., et al, 1980, Unconformities of the North Atlantic: Philosophical Transactions of the Royal Society of London, Series A, v. 294, p. 137-155.

Comparative Anatomy of Cratonic Unconformities

L.L. Sloss
Northwestern University
Evanston, Illinois

Cratonic unconformities represent (1) coincidence of surfaces of sedimentary accumulation with depositional base level, or elevation of depositional surfaces above erosional base level, and (2) renewed deposition covering surfaces of nondeposition or erosion. The chronostratigraphic record of unconformities is best displayed on Wheeler diagrams on which geographic distances are plotted against chronostratigraphic intervals or absolute time. Assessing the lithostratigraphic significance of unconformities requires reconstruction of the pre-unconformity stratigraphy and estimation of the thickness (or volume) of the strata eroded and of their lithologic character.

Interregional cratonic unconformities fall into two major types: (1) those marked by subequal values of nondeposition and erosion, commonly involving 5 to 30 m.y. and the stripping of as much as 1 km (.6 mi) over very broad areas (for example, the sub-Kaskaskia unconformity-Early Devonian to Early Carboniferous); and, (2) unconformities characterized by short-term nondeposition (< 5 m.y.) and extremes of erosional vacuity (for example, the sub-Absaroka surface-Late Carboniferous).

Conventional wisdom suggests that episodes of cratonic nondeposition and erosion should equate with accelerated detrital deposition at continental margins and with perturbations of marine chemistry. Evidence is accumulating to indicate a degree of concomitance between cratonic events and oceanic geochemistry but no complementary pattern is clear in terms of slope/rise depositional rates. Indeed, certain major unconformities identified on continental slopes appear to have equivalents on cratons. These and related questions demand increased communication between land-based and seagoing stratigraphic and tectonic specialists.

INTRODUCTION

There are at least three classes of unconformities of regional and interregional scale. One of these is clearly the product of uplift and erosion accompanying deformation in orogenic belts; extreme contrast in the degree of deformation, of metamorphism, and of plutonism above and below the erosion surface is common. A second class is illustrated by the stratigraphy of ocean basins and of continental slopes and rises. These result from submarine scour by geostrophically driven contour currents, by thermohaline currents, and in the course of canyon cutting by turbidity currents, or by carbonate dissolution related to changes in the depth of the sea floor with respect to the lysocline. Many oceanic and continent-margin unconformities are obscure in the absence of detailed biostratigraphic and/or high-resolution seismic-profile control.

This paper considers a third class of unconformities which are confined to the stable cratonic interiors of continents. Here, the cessation of deposition, the initiation of erosion, and the reestablishment of depositional regimes responsible for regional and interregional unconformities are most commonly related to epeirogenic motions of continental interiors, to changes in sea level, or to combinations of the two. Marked structural and petrologic discordance across the surface of unconformity is not common except where basement rocks are involved.

CRATONIC UNCONFORMITY VARIABLES

Geographic Variables

The fundamental control on the accumulation and preservation of cratonic sediment is base level, the equilibrium surface separating erosional and depositional regimes. In general, base level approximates sea level, rising a few meters above sea level landward of the strandline and descending a few meters below sea level where depositional base level is commonly defined by wave base. To enter the stratigraphic record, sediments must be carried below base level by a rise in sea level or by subsidence of the depositional site. During long periods of cratonic history the supply of sediment significantly exceeds the capacity of cratonic sediment traps. Large volumes of sediment in transport are bypassed to subsiding basins and cratonic margins. Given a steady-state condition of slowly rising sea level or slowly subsiding crust and given a steady state climatic regime and an absence of tectonic, seismic, or other disturbing events, cratonic sediment accumulates at steady rates equal to sea-level rise or tectonic subsidence.

Such steady-state conditions, of course, do not exist over geologically significant time spans. Cratonic stratigraphy is therefore punctuated by repeated perturbations representing, among other non-steady-state departures, climatic events from seasonal storms to aperiodic hurricanes such that a particular cratonic succession has a high probability of recording in preserved sediment no more than a small fraction of the total time span identified. Each interruption is, in theory, represented by an unconformity of dimensions that defy detection, measurement, and mapping except, perhaps, at roadcut and quarry scales of observation.

Above the microscale just discussed, unconformities can be classified by the scope of the geographic areas over which they can be traced. Local unconformities are those confined to a few hundreds or thousands of square kilometers. They commonly represent syndepositional uplift of local tectonic elements such as anticlines or fault blocks, channeling episodes caused by drainage-pattern changes on delta platforms, locally effective catastrophic storms, etc.

Regional unconformities are typically mappable over entire sedimentary basins or adjacent basins but cannot be traced to other regions of the same craton. Unconformities of regional scope are often attributable to tectonic uplift of discrete segments of a craton. Interregional unconformities are virtually craton wide, traceable to variable distances outboard from cratonic interiors to continental shelf prisms, and are not confined to specific portions of cratons.

Evidence accumulates (Sloss 1972; Sloss and Speed 1974; Vail, Mitchum, and Thompson, 1977b) that a number of interregional unconformities, particularly those identified as sequence-bounding surfaces by Sloss (1963) and, in greater detail, by Vail, Mitchum and Thompson (1977b), are recognizable on all cratons and may be considered global cratonic unconformities.

Geometric Variables

There are many geometric attributes of cratonic unconformities; among these are structural discordance across the unconformity surface, geomorphology of the buried erosion surface, and the vertical amplitude of erosion.

Structural discordance

Most cratonic unconformities would be classed as disconformities; that is, there is apparent parallelism of bedding on either side of the unconformity surface at outcrop or oil-field scale. In actuality, of course, all unconformities of regional scope display discordant relationships if traced for significant distances. Some cratonic unconformities, especially those formed during episodes of rapid, fault-accompanied cratonic uplift, may exhibit all degrees of structural discordance.

Geomorphology

The character of topographic relief at ancient erosion sur-

faces has attracted attention for decades because of the importance of unconformity traps for oil and gas, both through truncation of underlying reservoir units and through development of trapping potentials controlled by the geomorphology of the surface undergoing transgression. This subject has been so thoroughly explored that it requires no further embellishment here.

Vertical amplitude

Ordinarily, it is not difficult to estimate the time value of rocks stripped in the creation of a cratonic unconformity; determining the thickness of rocks removed, leading to estimating the vertical amplitude of uplift, is more difficult. Pennsylvanian strata rest on Ordovician beds on the Ozark Dome, for example. What was the amplitude of uplift prior to Pennsylvanian onlap? Depending on whether one extrapolates pre-Pennsylvanian thickness over the Ozarks from the Forest City Basin or the Illinois Basin, differences by a factor of two or more are possible. In spite of inherent uncertainties, many cratonic unconformities unrelated to local tectonic irregularities suggest vertical uplift and concomitant erosion of a few hundred meters, which is within the range of sea-level rise and fall. Other unconformities of at least equal interregional scope are, however, difficult to resolve without consideration of a kilometer or more of uplift and erosion. Indeed, the magnitude of difference in the apparent amplitude of erosion from region to region at interregional unconformities is witness to the fact that cratons do not operate as monolithic rigid blocks subject only to transgression and withdrawal of eustatic seas.

Chronologic Variables

None of the preceding is likely to surprise many readers. When, however, we consider the temporal factors involved in the genesis and interpretation of unconformities the pace picks up because not many of us have emphasized the topic in sufficient depth to permit application to a number of theoretical and practical problems.

Lacuna, hiatus, and vacuity

Some decades back in the AAPG *Bulletin*, Harry E. Wheeler (1958) reviewed a number of trenchant observations by Eliot Blackwelder and some astute thoughts of James Barrell, A.W. Grabau, and Maurice Gignoux. Wheeler put these together with some very original thinking of his own to structure and codify the treatment of depositional and erosional episodes. Adapting and clarifying older language, Wheeler applied the term "lacuna" to the spatial extent and time value of chronostratigraphic units missing at an unconformity. Wheeler recognized, as had his predecessors, that the absence of a material sedimentary record of parts of geologic time at an unconformity is the product of both non-deposition and erosion; the non-depositional component was termed "hiatus" and the erosional component was called "erosional (later degradational) vacuity." I find lacuna, hiatus, and vacuity to be essential linguistic symbols in the discussion and comprehension of unconformities.

Wheeler diagrams

It is a natural extension of Wheeler's treatment of chronostratigraphy to display the area-time geometry of deposition and erosion at an unconformity by diagrams in which geography, as along a stratigraphic cross section, is plotted against geologic time. I call these "Wheeler diagrams" and I believe they are essential to a proper understanding of relationships, events, and conditions at unconformities. To illustrate this, I will examine Wheeler diagrams involving four interregional unconformities in the heart of the North American craton.

Sub-Sauk and sub-Tippecanoe unconformities

Figure 1 is a highly diagrammatic stratigraphic cross section of early and middle Paleozoic strata from the Anadarko Basin to the Wisconsin Arch. The dimensions are lengths with extreme vertical exaggeration; the succession is divided

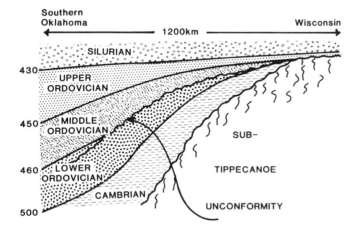

Figure 1: Grossly simplified stratigraphic diagram of early Paleozoic strata from the Anadarko Basin to the Wisconsin Arch. The chronostratigraphic boundaries shown are labelled in millions of years before present.

by lines of equal chronologic age (430 m.y., 450 m.y., etc.); all but the major sequence-bounding unconformities are ignored for purposes of simplification. Figure 2 is a Wheeler diagram of the same section with geologic time as the vertical dimension such that 10-m.y. intervals are shown by equally-spaced horizontal lines across the diagram. The age of the Precambrian basement over which Sauk strata were deposited ranges from 1,000 to 2,000 m.y. The base of the sub-Sauk lacuna lies several centimeters below the bottom of the diagram, and the diagram shows only the later part of the sub-Sauk hiatus, or non-depositional interval. As the diagram indicates, transgression leading to permanent accumulation of sediment began about 525 m.y. ago in the area of the Ozark Uplift, slightly earlier than in the present axis of the Anadarko Basin. From 525 m.y. ago to Early Ordovician (ca. 480 m.y. ago) Sauk onlap spread northward, with many interruptions not shown on this simplified sketch, to the flanks of the Wisconsin Arch. Accuracy of placement of the end of the sub-Sauk hiatus is limited only by chronostratigraphic resolving power of fossils found at, or close to, the

unconformity surface.

Figure 2 illustrates the entire sub-Tippecanoe lacuna. Again, progressive onlap, with many undisplayed interruptions, extends from the Anadarko Basin to the Wisconsin Arch and the placement of the end of the hiatus and initiation of Tippecanoe accumulation is as accurate as the chronostratigraphy of basal Tippecanoe units. Similarly, the bottom of the lacuna, the base of sub-Tippecanoe erosion, is defined by the chronostratigraphy of the youngest Sauk strata preserved. Definition of the beginning of the sub-Tippecanoe hiatus, the time at which this portion of the craton emerged above base level and ceased to accumulate sediment, is less obvious. In the present case, it appears that the sub-Tippecanoe unconformity virtually disappears in southern Oklahoma; here, the youngest beds of the Sauk Arbuckle Group are nearly conformable with the Joins Formation, the oldest strata of the Tippecanoe Simpson Group. Assuming that near conformity exists, then the minimum

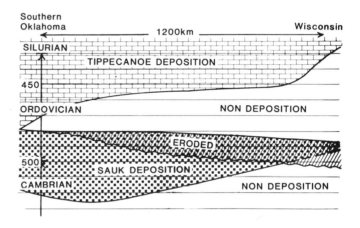

Figure 2: Wheeler diagram of the cross section shown on Figure 1. Hiatus (non-deposition) is shown by blank space, vacuity (erosion) by cross hatch.

age of the sub-Tippecanoe hiatus is established. The diagram assigns an age of approximately 480 m.y.; at all other points of observation of the sub-Tippecanoe unconformity the initiation of non-deposition (hiatus) must be older. Thus, in the absence of local complications, lines on Wheeler diagrams separating hiatus and vacuity always slope toward older values from cratonic margins to interiors and, in most cases, from basin centers to basin flanks. Note that the time value of the sub-Tippecanoe hiatus is about three times that of the associated vacuity. That is, for each million years during which the region was above base level, erosion stripped away the product of about one-third of a million years of deposition.

Sub-Kaskaskia unconformity

Figure 3 is a Wheeler diagram for the Ordovician through Pennsylvanian periods from the depths of the Illinois Basin in southern Illinois to the northern extremity of Pennsylvanian rocks preserved in northern Illinois. For the purposes of this study, ignore the pre-Pennsylvanian (sub-Absaroka) hiatus and vacuity and consider the Devono-Mississippian Kaskaskia Sequence and the underlying lacuna. Note that the sub-Kaskaskia lacuna is reduced to a time value of no more than a few million years as it approaches southern Illinois, permitting a reasonable estimation of the time of first passage of the craton above base level. As in the sub-Tippecanoe case, the base of the nondepositional hiatus must have a negative slope toward the Wisconsin Arch and Canadian Shield while the time value of stratigraphic units stripped (the vacuity) is small relative to the duration of the hiatus.

decrease in age from north to south. Note the degree to which pre-Pennsylvanian erosion has cut progressively into older chronostratigraphic levels from south to north across the state; also note the northwest lineation of erosional contacts along the west flank of the La Salle Anticline. The latter element is shown diagrammatically as a structure section along line A - A' on Figure 5. Although the vertical exaggeration is extreme the discordance across the unconformity is obvious, and this in a region spared widespread vertical dis-

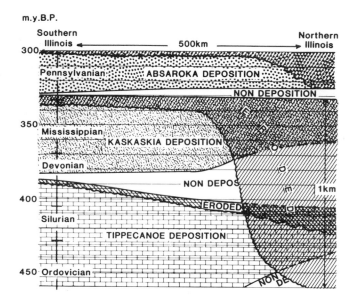

Figure 3: Wheeler diagram of Ordovician to Pennsylvanian rocks and time from southern to northern Illinois. Sub-Kaskaskia and sub-Absaroka lacunas and their contributing hiatuses and vacuities are shown.

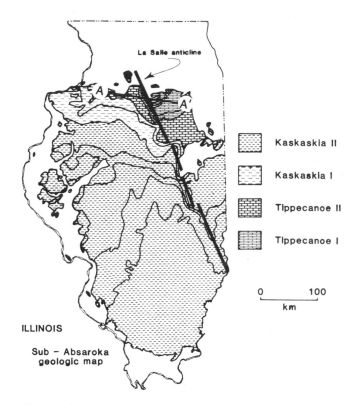

Figure 4: Subcrop map of the sub-Absaroka unconformity in Illinois. Modified after Willman et al (1967).

Sub-Absaroka unconformity

Stratigraphic patterns formed during late Paleozoic depositional and erosional episodes stand in sharp contrast to the relatively simple patterns of earlier Phanerozoic time. In Illinois, for example, although there was significant pre-Tippecanoe erosion, and channels approaching 100 m (328 ft) in depth were cut on the erosion surface, basal Tippecanoe beds rest almost everywhere within the state on upper Sauk units. Similarly, the sub-Kaskaskia pattern is marked by slightly greater erosion on the western, northern, and eastern margins of the Illinois Basin than in the basin interior; evidences of local pre-Kaskaskia uplift are rare except along the eastern flank of the Ozark Uplift. These stable conditions preceding Middle to Late Ordovician and Middle to Late Devonian onlap are typical of the remainder of the North American interior and of other cratons.

Figure 4 is a generalized sub-Absaroka subcrop map of Illinois (the subdivisions of Tippecanoe and Kaskaskia are those defined by Sloss, 1982, as adapted from Vail, Mitchum, and Thompson, 1977b); the lines crossing the area of subcropping Kaskaskia II are Mississippian units that

location for at least 700 m.y.

Figure 3 clarifies the chronostratigraphic relationships at the sub-Absaroka unconformity. The total interval of nondeposition (hiatus) cannot be accorded a time value of more than a few million years; yet the erosional vacuity created involves the products of over 100 m.y. of cratonic history and the stripping of about 1 km (.6 mi) of strata. The contrast with the two preceding major episodes of nondeposition is obvious.

Rates of offlap and onlap

Estimating the time required for the emergence of a previous area of deposition above base level is not easy. Erosion accompanying and following the transit across base level commonly removes the youngest deposits of the preceding episode; however, where the state of preservation is reasonably good, and where the environmental position of the preserved edge of the underlying strata with respect to a near or distant strandline can be determined, an approximation of the rate of regression and offlap can be made. In the case of the sub-Absaroka interruption discussed above (Figure 3) the time involved in putting all of Illinois above base level was minimal, perhaps less than 1 m.y., whereas the pre-Tippecanoe withdrawal of depositional environments from Wisconsin to Oklahoma cannot have exceeded a few million years. Lowering of a region or an entire craton back across the base-level surface to re-initiate deposition is well documented by the oldest onlapping strata that overstep an unconformity. The documentation abundantly available for the covering of the sub-Sauk, sub-Tippecanoe, and sub-Kaskaskia unconformities shows that tens-of-millions of years were involved in each transgression of depositional regimes across the cratonic interior. Such transgressions represent the algebraic sum of innumerable advances and retreats but the net effect is one of onlap at rates many times

domes and arches, and a lack of widespread brittle deformation manifested by faulting. The emergent episodes preceding each of these flexure-dominated times of deposition are the occasions for developing of the sequence-bounding unconformities and slow progressive transgression of the craton, as noted above.

A second type of depositional episode, called "oscillatory," is characterized by abrupt termination of a submergent episode through rapid cratonic uplift accompanied by high-angle faulting of basement crystalline rocks and, commonly, by faults that propagate from the basement to fracture and displace the overlying sedimentary cover. The Absaroka Sequence represents a typical oscillatory episode. The sub-Absaroka unconformity--with its evidence of a brief hiatus, deep erosion, and rapid transgression--appears

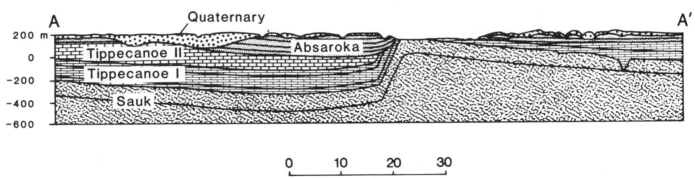

Figure 5: Cross section along line A - A' on Figure 4. After Willman et al (1967).

slower than those operating to produce net offlap (unconformities such as the sub-Absaroka are special cases with many quite distinct differences from the sub-Tippecanoe "standard," including differences in the rates of offlap and onlap).

Rapid regression and slow transgression patterns observed in connection with most cratonic unconformities are the same as those deduced by Vail, Mitchum, and Thompson (1977a) by analyzing seismic profiles, largely drawn from passive-margin examples. The rates and their variation during regression/transgression as seen on cratonic interiors also mimic, in part, the patterns for continental shelves predicted by the model developed by Pitman (1978). We consider the significance of rates of transit through baseline later in this paper.

Tectonic Variables

Some years ago Sloss and Speed (1974) attempted to show that the sequences of Sloss (1963) (the supersequences of Vail, Mitchum, and Thompson, 1977b) represent two kinds of depositional episodes that differ in their tectonic modes. One type, termed "submergent" and exemplified by the times of deposition of the Sauk, Tippecanoe, and Kaskaskia sequences, is dominated by flexure of the cratonic interior and margins. These sequences exhibit slow regional transgression and onlap as discussed above, gentle subsidence of interior basins separated by less subsident

to be the product of the same mode of vertical tectonics that pervades the complex history of Absaroka deposition.

GENESIS OF CRATONIC UNCONFORMITIES

Simply stated, a significant interregional cratonic unconformity requires only that a craton, or a major segment of a craton, be depressed relative to base level such that nondepositional or erosional regimes are replaced by the accumulation of sediment. Inasmuch as base level is closely tied to sea level and since global changes in sea level may result from continental glaciation and deglaciation and from variations in the displacement volume of thermally inflated mid-ocean ridges or of submarine volcanic edifices, it is convenient, even comforting, to consider the base-level transits represented by cratonic unconformities to be manifestations of eustatic sea-level rise and fall. However, it is nowhere stated that the evolution of cratons and their sedimentary covers need be convenient, comfortable, or capable of rationalization by simplistic hypotheses. Several difficulties stand in the path of accepting of eustatic controls. One such difficulty is found in the amplitude of the vertical movements of cratons with respect to base level; uplift can be determined by sediment thickness eroded at an unconformity and by the upward displacement of previously deeply-buried strata identified by vitrain reflectance studies and other measures of temperature and pressure. In all cases, the amplitudes involved greatly exceed the maxima attributable to any eustatic mechanism.

The tempo of vertical movement is also a problem; glaciation/deglaciation appears to operate at a scale of 10^3 to 10^5 years while thermally-driven ocean-floor displacing mechanisms require an order of magnitude longer time but less than 10^8 years. Cratonic transits of base level occur at many tempos, some clearly consonant with glacio-eustatics, others within the range of thermal inflation and deflation. Still others, however, have been active over time spans significantly longer than can be accounted for by known mechanisms for decrease and increase in the freeboard of continents.

Overriding all these negative factors leading to the rejection of eustatic control of cratonic unconformities is the clear relationship between cratonic episodes of interregional-unconformity development and concurrent episodes of tectonic change on cratons. The most dramatic of such concurrences are evidenced by the sub-Absaroka unconformity which documents a drastic change in the tectonic mode of cratons, including the emergence of previously inactive blocks, the subsidence of "new" basins, etc. Testimony of tectonic change during base level transit is not confined to the initiation of oscillatory episodes such as that represented by Absaroka deposition, however. Significant changes in the tectonic geography of cratons took place at each sequence boundary, (for example, between the times of Sauk and Tippecanoe accumulation; compare maps in Sloss, 1982, Figures 1 and 2). In sum, belief in a global tectonically-effective driving force is inescapable.

RELATIONSHIP OF CRATONIC AND OCEANIC UNCONFORMITIES

Before an abundance of stratigraphic data emerged from terrains exterior to cratons, particularly from the slopes and rises of passive margins, it was reasonable to predict a complementary relationship of erosional episodes on cratons and depositional episodes on continental margins. Granted that the legible history of continental margins is short when compared to the stratigraphic record of cratons, it appears, nevertheless, that certain erosional and depositional episodes have equal and synchronous representation on cratons and margins. For several of the prominent Cenozoic unconformities (e.g., sub-Upper Paleocene, sub-Upper Oligocene) interpretation from the seismic stratigraphy of passive margins are well developed at cratonic sites. The prominent "J" unconformity that separates Jurassic strata from "pre-rift" and "syn-rift" rocks of the North American Atlantic and Gulf margins is contemporaneous with the sub-Zuni unconformity of the cratonic interior. Other examples abound and most point to the fact that upward base-level transits on cratons tend strongly to place shelf/slope/rise areas above depositional base level at the same time such that the erosion products of cratons bypass continental mar-

gins and join with material displaced from slopes and rises to be dispersed in the deep ocean. Here, again, the pertubation of thermohaline and geostrophic current dynamics initiated by cratonic uplift may create oceanic unconformities in synchrony with those of cratons and margins.

Much remains to be learned about the mutual interactions of oceans and their margins and the stable interiors of continents. Thus far, there has been no broad community concern. If the bypassing and ultimate dispersal of craton-derived detrital sediments is clearly understood (is it really as simple as stated here?), what about the changes in ocean chemistry brought about by the solution of vast volumes of carbonates and evaporites during the erosional episodes associated with cratonic unconformities? Is there a record in oceanic sediments that documents shift in carbonate equilibria, salinity, the availability of nutrients, etc., that would result from alternate episodes of trapping and release on continents? These and related questions seek responses in greater communication between cratonic and oceanic stratigraphic tectonists.

REFERENCES CITED

Pitman, W.C., III, 1978, The relationship between eustacy and stratigraphic sequences of passive margins: Geological Society of America Bulletin, v. 89, p. 1389-1403.
Sloss, L.L., 1963, Sequences in the cratonic interior of North America: Geological Society of America Bulletin, v. 74, p. 93-113.
———, 1972, Synchrony of Phanerozoic sedimentary-tectonic events of the North American craton and the Russian Platform: 24th International Geological Congress, Section 6. p. 24-32.
———, 1982, The Midcontinent province: United States, in A.R. Palmer, ed., Perspectives in regional geological syntheses, DNAG Special Publication 1: Boulder, Colorado, Geological Society of America, p. 27-39.
———, and R.C. Speed, 1974, Relationships of cratonic and continental margin tectonic episodes, in W.R. Dickinson, ed., Tectonics and sedimentation: Society of Economic Paleontologists and Mineralogists Special Publication 22, p. 98-119.
Vail, P.R., R.M. Mitchum, Jr., and S. Thompson, III, 1977a, Relative changes of sea level from coastal onlap, in Seismic stratigraphy -- applications to hydrocarbon exploration: AAPG, Memoir 26, p. 63-82.
———, ———, and ———, 1977b, Global cycles of relative changes of sea level, in Seismic stratigraphy -- applications to hydrocarbon exploration, AAPG, Memoir 26, p. 83-98.
Wheeler, H.E., 1958, Time stratigraphy: AAPG Bulletin, v. 42, p.1047-1063.
Willman, H.B., et al, 1967, Geologic map of Illinois: Illinois State Geological Survey.

Relation of Unconformities, Tectonics, and Sea-Level Changes, Cretaceous of Western Interior, U.S.A.

Robert J. Weimer
Colorado School of Mines
Golden, Colorado

Intrabasin tectonics and sea-level changes influenced patterns of deposition and geographic distribution of major unconformities within the Cretaceous of the Western Interior. Nine major regional to near-regional unconformities have been identified. Previous workers have related five of these unconformities to sea-level changes and to well known regressive-transgressive cycles. The origin of the other four unconformities may be related either to tectonic movement or sea-level changes.

There is uncertainty in dating many of the unconformities. However, by use of the time scale of Obradovich and Cobban (1975), with subsequent minor revisions, the approximate dates for unconformities are estimated as follows (formations involved are in parenthesis; numbers are millions of years before the present, m.y.): 1) Late Neocomian to early Aptian, 112 (base lower Mannville, Lakota, Lytle); 2) late Aptian-early Albian 100 ± (upper Mannville, Fall River, Plainview); 3) Albian 97 ± (Viking, Muddy, Newcastle, or J Sandstone); 4) early Cenomanian, 95 ± (lower Frontier--Peay, and D); 5) Turonian, 90 ± (base upper Frontier or upper Carlile); 6) Coniacian, 89 ± (base Niobrara or equivalents); 7) early Santonian, 80 ± (Eagle, lower Pierre and upper Niobrara); 8) late Campanian, 73 ± (mid-Mesaverde, Ericson, base Teapot); 9) late Maestrichtian, 66 ± (top Lance or equivalents). Variations in the accuracy of the dating are probably within 1 million years because of problems in accurately defining the biostratigraphic level of the breaks and in the precision of radiometric dates.

The unconformities are grouped into three types: those completely within nonmarine strata such as at the base and top of the Cretaceous; those involving both marine and nonmarine strata; and, those within marine strata, as currently mapped.

Three examples are described as typical of the unconformities, all thought to be related primarily to drops in sea level, but with minor influence by tectonic movement. One is the 97 ± m.y. unconformity, with which the petroleum-producing J and Muddy Sandstone is related. A second is 90 ± m.y. unconformity which is recognized by relationships within the shelf, slope, and basin deposits of the Greenhorn, Carlile, and Frontier formations. The third is the 80 ± m.y. unconformity within the basin and shoreline regression associated with the upper Niobrara, lower Pierre, Eagle, and Shannon formations.

Several billion barrels of oil were found in sandstones associated with unconformities in the Cretaceous of the Rocky Mountain region. Future stratigraphic trap exploration is guided by a knowledge of tectonic influence on sedimentation during sea-level changes and how these factors control distribution of source rock, migration patterns, reservoir rock, and seal.

INTRODUCTION

Depositional systems on the margins of continents, both modern and ancient, have been related to sea-level changes. Two simplified diagrams by Vail, Mitchum, and Thompson (1977) illustrate highstand and lowstand conditions and associated unconformities (Figure 1). The highstand diagram represents a depositional system that might be observed in many coastal areas today or at times of highstand in the past. The four main components are the coastal plain, shelf, slope, and rise (or deep water basin). An unconformity related to coastal onlap during the rise of sea level to the highstand condition is shown. With a drop in sea level to the edge of the continental shelf (a lowstand condition), sediment bypasses the shelf and is deposited in deep water. The entire shelf is exposed to subaerial erosion and streams adjusting to the lower base level incise into the older shelf deposits. The depocenter shifts from the deltaic system under the highstand to marine subsea fans of the lowstand. Unconformities are present within the marine strata and also project into the coastal plain deposits. Several depositional models are possible in ancient strata between these two end members of sea-level conditions.

If concepts illustrated by the diagrams from continental margins are applied to ancient interior cratonic basins, then marine and associated deposits represent sea-level highstands and unconformities represent erosion during lowstands. Without considering tectonic influence on local

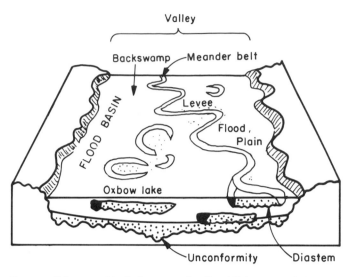

Figure 4: Schematic block diagram of valley-fill deposits of the Mississippi River (modified from Fisk, 1947). Breaks in the sequences are diastems at the base of meander-belt sands and the unconformity at the base of the valley fill. Braided channel deposits are widespread sands at base of valley fill; sand lenses in upper part are point-bar deposits with clay plugs of abandoned channel shown in black.

surface of erosion associated with a sea-level drop and drainage incisement (an unconformity; Figure 4).

Similar problems exist in defining and evaluating breaks in the record of marine sequences in ancient cratonic basins. The most easily recognized breaks commonly occur within shelf sequences. The breaks associated with sedimentation on modern shelves are documented for the Gulf of Mexico. The record of Holocene deposition on the shelf of the northwest Gulf of Mexico is summarized by Curray (1960, 1975). Most of the shelf was subaerially exposed during the Wisconsin glaciation lowstand. With the Holocene rise in sea level, mud, sand, or shell were deposited over shelf areas where a sediment supply was available. Pleistocene sediments are exposed over a large area of the shelf, or are covered by a thin veneer of relict or palimpsest sands, reworked by waves during the transgression. Thus, an erosional surface modified by marine processes occurs at the base of the Holocene.

Incisement of a drainage system into Pleistocene shelf deposits, with a subsequent Holocene valley fill, is well documented by Nelson and Bray (1970). An area of the Texas shelf encompassing 2,590 sq km (1,000 sq mi; Figure 5) was studied in detail by using bottom samples, marine sonoprobe profiles, and cores. A paleovalley from 9.6 to 12.8 km (6 to 8 mi) wide was mapped for a distance of approximately 80 km (50 mi; Figure 6). The valley was incised to an unknown depth by the combined flow of the Sabine and Calcasieu rivers (Figure 5) during the Wisconsin lowstand. The stratigraphic relations of three types of fill in the paleovalley are plotted on longitudinal and transverse sections (Figures 6 and 7). The lowest fill, older than 10,200 years, consists of fluvial-deltaic sands of unknown thickness because the base of the valley was not mapped. As the rising sea flooded the valley, estuarine and lagoonal clay and sandy mud were deposited in the valley. These deposits

vary in thickness from 6.1 to 15.2 m (20 to 50 ft) and range in age from 10,200 years to approximately 5,200 years. The final fill is composed of marine mud and sandy mud which varies in thickness from a wedge edge to 12.2 m (40 ft). During the past 5,000 years a marine sand bar complex known as the Sabine and Heald banks has formed approximately 32 km (20 mi) seaward from the present shoreline. The offshore bar is 9.6 to 12.8 km (6 to 8 mi) wide, 64 km (40 mi) long, and varies in thickness from a wedge edge to 6.1 m (20 ft; Figures 6 and 7). The sand bodies may have a transitional base with underlying Holocene marine mud or may rest on an erosional surface on the Pleistocene (Figure 7). Although the sand is dominantly detrital quartz, local concentration of shells make up 100 percent of the banks. The offshore sand bar has a northwest trend parallel with the present shoreline and a portion of the paleovalley; it overlaps the southern margin of the paleovalley (Figures 6 and 7).

This study by Nelson and Bray (1970) illustrates three types of Holocene sand bodies associated with sea-level changes that have potential as petroleum reservoirs. These are the fluvial-deltaic sands of the valley fill; the linear offshore marine bar and associated thin lag deposit; and, the thin sands of the shoreline zone. Important stratigraphic breaks are associated with these sand bodies. Similar sand bodies have been recognized in ancient sequences but not always have their origin been related to eustatic changes.

Criteria for recognition of sea-level changes in cratonic basins are listed in Table 1. When these criteria are used in conjunction with the factors listed above in evaluating unconformities, breaks which are the result of local tectonic influence can be separated from those caused by sea-level changes.

REGIONAL SETTING OF CRETACEOUS BASIN

The Western Interior Cretaceous basin of the North American continent is one of the largest cratonic (foreland or back-arc) basins in the world. Because of economic products, mainly petroleum and coal, strata deposited in this basin have been thoroughly studied, both in outcrop and subsurface occurrences. The original basin was 805 to 1,640 km (500 to 1,000 mi) wide and extended from the Arctic to the Gulf of Mexico (Figure 8). The relationship of the basin to other structural elements in the western portion of North America is outlined by Dickinson (1976; Figure 9). The foreland basin formed on a thick continental crust and was bordered on the west by a fold-thrust belt and on the east by the Canadian shield.

During the Early Cretaceous, sediments were derived from both sides of the basin, though the thickest strata are along the western margin. During the Late Cretaceous, the dominant source of sediment was along the western margin and lithofacies were controlled by changes in environments from coastal plain to shoreline to marine shelf and the deeper water of the basin. Intertonguing of nonmarine strata on the west with marine strata in the center of the basin is the dominant pattern of sedimentation (Figure 10). Thickness of the Cretaceous strata varies from 610 to 6,096 m (2,000 to 20,000 ft). The thinnest sections occur in the geographic center to the eastern margin of the basin because sedimentation rates were slower there. Total organic con-

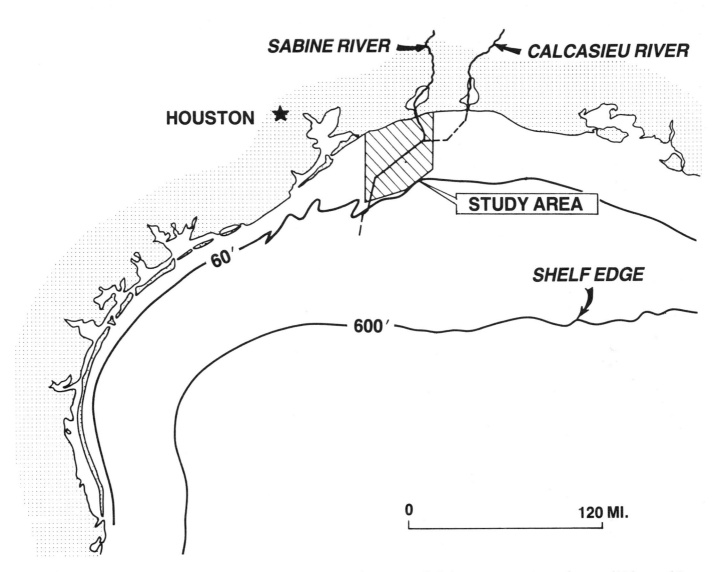

Figure 5: Index map with portion of shelf, Gulf of Mexico Holocene sedimentation. Ruled portion represents study area of Nelson and Bray (1970).

tent is higher in these strata because of the slow sedimentation rates and because of anoxic conditions in deeper water that favored preservation of organic matter.

The Cretaceous basin was deformed during the Laramide orogeny and segmented into the present-day intermontane basins of the Rocky Mountain region. Areas between these basins were uplifted and subsequent erosion has removed Cretaceous strata from the structural high areas. Hence, reconstruction of the entire original depositional basin requires correlation among the intermontane Laramide structural basins. Because of widespread faunas and floras and closely spaced subsurface well control, accurate correlations are possible within much of the stratigraphic section. Therefore, time-stratigraphic units can be mapped, breaks accurately evaluated and facies changes and depositional models reconstructed.

Stratigraphic concepts derived from studies of this Cretaceous depositional basin have widespread application to understanding detrital sequences in all ancient basins which had a structural setting on, or marginal to, cratonic regions of continental plates.

UNCONFORMITIES WITHIN THE CRETACEOUS BASIN

Unconformities within the Cretaceous basin have been recognized by many investigators. The best published synthesis of the Cretaceous system for the Rocky Mountain region, U.S.A., is the Geologic Atlas published by the Rocky Mountain Association of Geologists. The Cretaceous chapter, compiled by McGookey (1972, p. 195), identifies many unconformities within the system, but only 8 or 9 within the different structural basins can be placed in a regional framework. A restored section (Figure 10) is plotted from the western margin of the Cretaceous basin (generally western Wyoming and western Montana) to the geographic center of the basin (eastern Colorado, Black Hills area, and eastern Alberta). Strata are dominantly nonmarine in the western portion of the basin and dominantly marine in the geographic center of the basin.

Unconformities are in three positions: those completely within nonmarine strata such as at the base and top of the Cretaceous; those involving both marine and nonmarine strata; and, those totally within the marine strata, as currently mapped. Wavy lines extending completely across the diagram represent times when the entire basin was subjected to subaerial or submarine erosion. In general, the amount of erosion of underlying strata associated with each break is less than one hundred meters. Where regional beveling occurs, the angularity is too small to recognize on a local basis.

On Figure 10 strata are plotted relative to age, and thickness of section is not considered. Uncertainty exists in the dating of many of the unconformities. The time scale and faunal zones of Obradovich and Cobban (1975) and modified by Fouch et al (1983), were used to date the major unconformities. Their approximate dates (± 1 million years), together with associated formations (in parenthesis), are estimated as follows (numbers are millions of years, m.y., before the present): 1) late Neocomian to early Aptian, 112 (base lower Mannville, Lakota, or Lytle); 2) late Aptian--early Albian, 100 (upper Mannville, Fall River, Plainview); 3) Albian, 97 (Viking, Muddy, Newcastle, or J); 4) early Cenomanian, 95 (lower Frontier--Peay and D); 5) late Turonian, 90 (base upper Frontier, upper Carlile or Juana Lopez); 6) Coniacian, 89 (base Niobrara or Fort

Table 1

1. Regression of shoreline with incised drainage followed by overlying marine shale.
2. Valley-fill deposits (of incised drainage system) overlying marine shale:
   A. Root zones at or near base of valley-fill sequence.
   B. Paleosoil on scour surface.
3. Unconformities within basin:
   A. Missing faunal zone.
   B. Missing facies in a normal regressive sequence (e.g., shoreface or delta front sandstones).
   C. Paleokarst with regolith or paleosoil.
   D. Concentration of one or more of the following on a scour surface: phosphate nodules, glauconite, recrystallized shell debris to form thin lenticular limestone layers.
4. Thin widespread coal layer overlying marine regressive delta front sandstone deposits = rising sea level.
5. Correlation with the record of seal-level changes from other continents.

Table 1: Criteria for recognition of sea-level changes in cratonic basins.

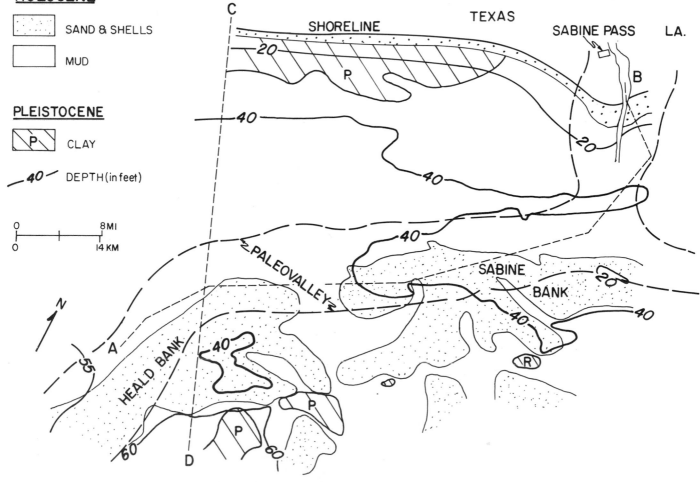

Figure 6: Map of offshore Texas (area shown on Figure 5) with distribution of Holocene and Pleistocene sediments on sea floor. Approximate locations of paleovalleys of Sabine and Calcasieu rivers are shown (after Nelson and Bray, 1970, p. 55).

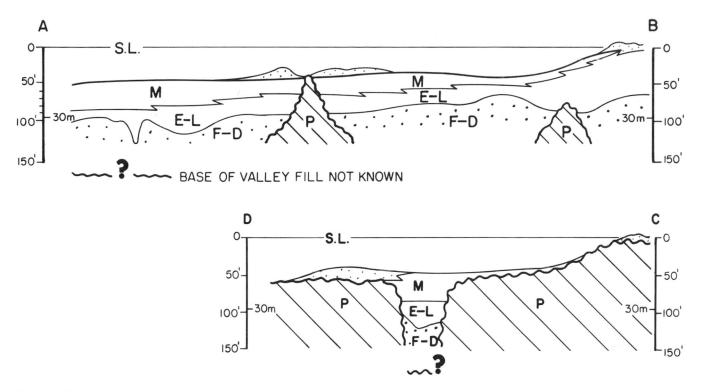

Figure 7: Cross sections A-B within the paleovalley of the Sabine River and C-D across the valley (locations on Figure 6) (after Nelson and Bray, 1970, p. 59, 63). S.L. = sea level; M = marine; E-L = estuary-lagoon; F-D = fluvial-deltaic; P = Pleistocene.

Hayes); 7) early Santonian, 80 (Eagle, lower Pierre--upper Niobrara); 8) late Campanian--early Maestrichtian, 73 (mid-Mesaverde, base Ericson, base Teapot); and, 9) late Maestrichtian, 66 (top Lance or equivalents). The ages of stage boundaries are based on work by Obradovich and Cobban (1975) for the Western Interior Cretaceous, and modified by Lanphere and Jones (1978) and Fouch et al (1983). The positions of some of the stage boundaries relative to this radiometric time scale are not in agreement with those published by other workers, for example Van Hinte (1976) or Kauffman (1977). These variations in the age of stage boundaries are plotted on Figure 11.

Many of the above unconformities can be related to sea-level changes and to well known regressive and transgressive cycles. However, one major problem is determining if some of the breaks are associated with regional or local tectonics instead of sea-level changes. The difficulties in relating transgressions and regressions to sea-level changes or tectonics in the Canadian portion of the Cretaceous basin were discussed by Jeletzky (1978).

INFLUENCE OF SEA-LEVEL CHANGES ON DEPOSITIONAL SYSTEMS

Changes in sea level have a direct influence on base levels of erosion and deposition, which control sedimentation. The influence varies among major environments of deposition but the most noticeable effect is in nonmarine and shallow-marine environments. Overall, drainages adjust to a lower base level (lowstand) by incisement (erosion) into underlying strata; when base level rises, streams aggrade

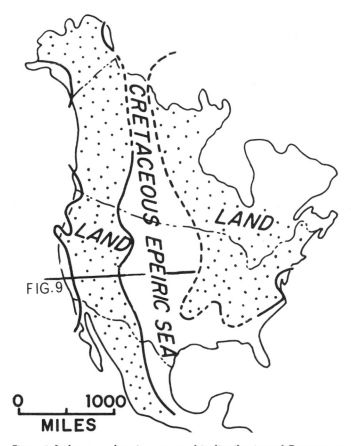

Figure 8: Index map showing geographic distribution of Cretaceous seaway in interior of continent. Location of cross section (Figure 9) with representative tectonic elements is indicated.

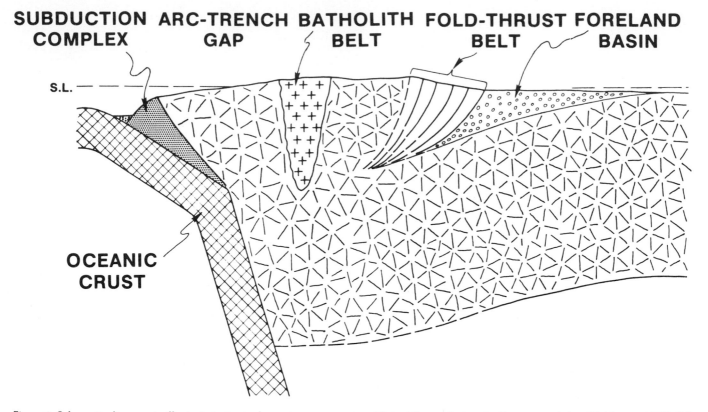

Figure 9: Schematic diagram to illustrate tectonic elements across western United States during the Cretaceous (after Dickinson, 1976). S.L. = sea level.

and deposition resumes in valleys and marginal areas. Shorelines normally regress during lowstands and transgress during highstands. Depending on the magnitude of change, marine-shelf sedimentation may be influenced only slightly, or widespread erosion may occur either in a subaerial or submarine setting. Generally, the deposits that are studied in cratonic basins are associated with highstands, whereas lowstands are represented by breaks which may or may not have been identified, especially in nonmarine strata.

Many of the unconformities in the Cretaceous basin are associated with overall regressive cycles of shoreline movement. The widespread regressive shoreline and shallow-marine sandstones may be related to a stillstand or a slow rise or lowering of sea level where a high rate of sediment supply prevails. The Holocene Mississippi River delta complex is an example of a regressive event during a Holocene still stand, or slightly rising sea level, because of a high sediment input to the basin. The shoreline has prograded seaward approximately 160 km (100 mi) during the last 5,000 years. Thus, a shoreline regression in an ancient sequence need not be related to a sea-level drop. The best single indicator of an ancient falling sea level is incised drainage with root zones or paleosoils on the surface of erosion at the base of valley-fill deposits, as summarized in Table 1. Other criteria that are useful in establishing eustatic changes are also listed.

Sea level curves for the Cretaceous have been published by Hancock (1975), Kauffman (1977), and Hancock and Kauffman (1979). These authors relate major transgressive and regressive cycles for the Western United States with those of north Europe and relate these recorded events to sea-level changes. Based on criteria discussed in this paper, a modified sea-level curve for the Western Interior has been prepared (Figure 11) and compared with Hancock's curve for north Europe. The comparison of events between continents allows for discrimination of those shoreline movements caused by sea-level changes. However, when the unconformities are added, a more accurate dating can be determined for the lowstands.

Transgressions in the Cretaceous are represented by widespread marine shale strata between sandstone units of regressive events, some of which are capped by unconformities (Figure 10). From oldest to youngest these shale or chalk formations are as follows: 1) Clearwater of Canada, an event represented by nonmarine strata in the United States (the Lakota or Lytle); 2) Skull Creek; 3) Mowry--Huntsman; 4) Greenhorn; 5) upper Carlile; 6) Niobrara; 7) Claggett; and 8) Bearpaw. Because of their wide distribution, these formations have been related to highstand conditions. However, underlying valley-fill deposits, where present, are also related to rising sea level and transgressive events. Because of the unconformities, in many areas regional sedimentation patterns cannot be easily related to symmetrical cycles of transgression and regression, although locally this has been done (Kauffman, 1977). Because the breaks are generally on top of the regressive phases, cycles, if present, are asymmetric in favor of the regression event.

TECTONIC INFLUENCE ON SEDIMENTATION

Depositional models proposed for the Cretaceous do not generally consider whether or not syndepositional tectonic

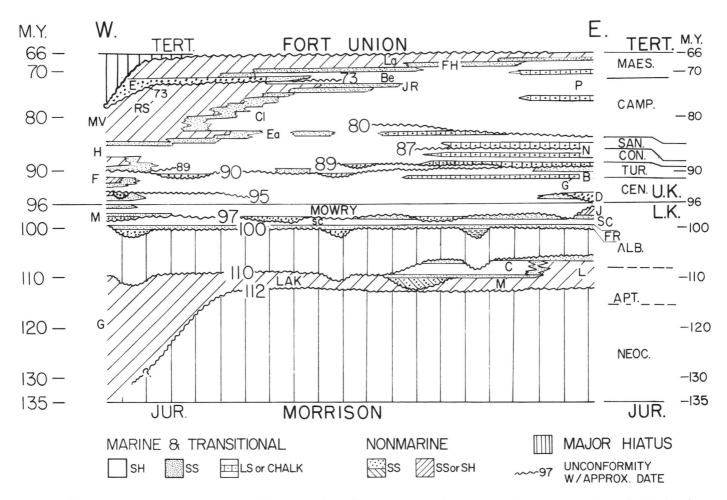

Figure 10: Diagrammatic east-west section across Cretaceous basin showing stratigraphic position and approximate dates of major intraba-sin unconformities (modified after McGookey, 1972). Formations or groups to the west are: G = Gannett; SC = Skull Creek; M = Mowry; F = Frontier; H = Hilliard; MV = Mesaverde; RS = Rock Springs; E = Ericson; Ea = Eagle; Cl = Claggett; JR = Judith River; Be = Bearpaw; FH = Fox Hills; La = Lance. To the east, formations are: L = Lytle; LAK = Lakota; FR = Fall River; SC = Skull Creek; J and D = Sandstones of Denver basin; G = Greenhorn; B = Benton; N = Niobrara; P = Pierre; M and C are the McMurray and Clearwater of Canada. The vertical ruled lines represent unconformities where a major hiatus is recognized. When the gap is a million years or less, vertical ruling is omitted.

movement occurred. Detailed studies in eastern Colorado have clearly established that structural elements had peri-odic movement on the Cretaceous sea floor (Weimer, 1978, 1980). Major northeast-trending basement fault blocks in the northern Denver basin are mapped as extensions of well-documented Precambrian shear zones observed in Front Range outcrops (Figure 12). Recurrent movement occurred on several of these paleostructures during the Pennsylva-nian, Permian, and Cretaceous (Sonnenberg and Weimer, 1981).

The shear zones, Precambrian in age, are "weak rock" and bound rigid fault blocks. At times during the Phanero-zoic when the crust was highly stressed, the stress was relieved primarily by movement along these pre-existing lines of weakness. Cretaceous strata clearly show that fault movement was sporadic, affecting some layers but not oth-ers. The strata which record movement are referred to as "tectonically sensitive intervals." They are the keys to reconstructing the size and distribution of paleotectonic ele-ments. Recurrent movements on the basement fault blocks are normally in the same direction, but important reversals

in movement (structural inversions) along faults have been recorded.

Fault-block boundaries may be recognized in the sedimen-tary sequence overlying the basement by direct offset of sed-imentary layers, abrupt change in strike and dip related to drape folding, and closely spaced fractures in competent layers. Strata overlying major basement blocks generally have uniform dip over the extent of the block. Different sized blocks are referred to as first, second, or third order features.

One of the most important premises in establishing a tec-tonic control on sedimentation is that fault-block movement controls topography and bathymetry (referred to as struc-tural topography). We can make the following generalities concerning sedimentation. In continental deposits, rivers flow on topographic lows whereas interchannel areas gener-ally occur over higher structural areas. However, because of the leveeing process associated with channels, interchannel deposits may be deposited in areas topographically lower than the channel.

In shoreline deposits with a high sediment influx, deltas

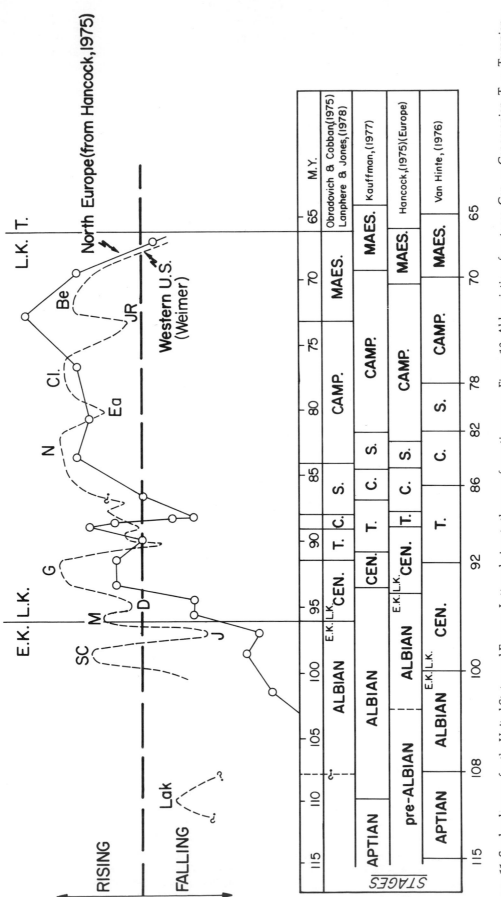

Figure 11: Sea-level curves for the United States and Europe. Letters designate the same formations as on Figure 10. Abbreviations for stages: Cen. = Cenomanian; T. = Turonian; C. = Coniacian; S. = Santonian; Camp. = Campanian; Maes. = Maestrichtian; E.K. = Early Cretaceous; L.K. = Late Cretaceous; T. = Tertiary.

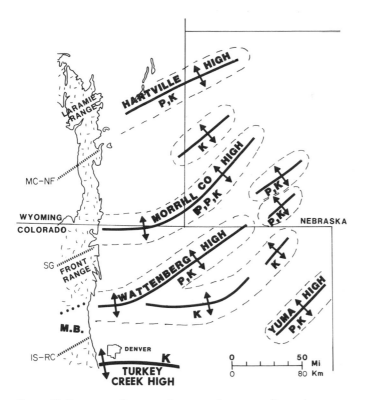

Figure 12: Summary diagram of east-northeast trending paleostructures (basement fault blocks) in northern Denver basin. Times of dominant movement are: IP = Pennsylvanian; P = Permian; K = Cretaceous. Outcrop of Precambrian with major shear zones along left side of diagram. MC-NF = Mullen Creek-Nash Fork; SG = Skin Gulch; IS-RC = Idaho Springs-Ralston Creek; M.B. = Colorado mineral belt.

develop in structural and topographic low areas and interdeltaic deposits occur over and around the more positive structural blocks. When sediment influx is low the structural and topographic low areas (deltas) may become estuaries. In marine deposits, topographically high blocks may be shoal areas and thus control the distribution of sand bodies or reefs. Moreover, in deeper water deposits, sand turbidites (both calci-clastic and silici-clastic) are deposited in bathymetrically low areas, which may coincide with downthrown blocks. In summary, thin successions are associated with paleohighs and thick successions are associated with paleolows, but sand deposits may occur in either setting depending upon the depositional environment.

Depositional topography may have developed within the Cretaceous basin because of thickness variations in units related to rates of sedimentation (Asquith, 1970). Thick and thin sediment accumulations, associated with depositional topography, can be confused with thickness patterns related to tectonics, and a careful analysis of depositional environments, processes, and subtle breaks is needed to determine what controlled thickness variation.

Tectonic movements can influence the accuracy of time correlations. Older units can be elevated to the same or a higher stratigraphic level than younger strata. An example of this relation is demonstrated in the discussion of the J Sandstone.

By applying the above concepts, the cause of an uncon-

formity can be evaluated--tectonic, sea-level change, or a combination of both. Depositional models constructed for all ancient depositional basins should consider these factors to reconstruct accurately the recorded geologic events.

EXAMPLES OF UNCONFORMITIES ASSOCIATED WITH CRETACEOUS SEA-LEVEL CHANGES

Although field, paleontologic, and subsurface data are available to support all of the unconformities in the Cretaceous shown on Figure 10, only three breaks, regarded as typical, are described in the following discussion. The 97 m.y. and 90 m.y. unconformities can be related to events on the continental margins and are, therefore, interregional in nature. The 80 m.y. break is now best defined within the marine basin deposits but probably extends to the margin of the basin and elsewhere on the continent.

BASIN-WIDE INCISEMENT OF DRAINAGE DURING SEA-LEVEL CHANGE
(97 m.y. ago)

An important transgressive-regressive-transgressive sequence is recorded in the Albian strata of the Western Interior Cretaceous basin. A widespread marine shale, mapped throughout the basin, is known in different areas as the Skull Creek, Kiowa, Thermopolis, and Joli Fou (Canada) shales (Figures 10 and 13). Equivalent strata in western Wyoming in the basin-margin area are generally included in the Dakota Group or Bear River formation. The shale deposits, which accumulated during a high stand of the Albian Sea, are correlated over large areas, either by contained faunas or by stratal continuity. The shales, generally 30.4 to 60.8 m (100 to 200 ft) thick, represent the first widespread transgression of the Cretaceous sea into the United States portion of the Western Interior basin. Overlying regressive sandstone units names the Muddy, J, or Viking (Canada) sandstones are widespread and productive of petroleum in stratigraphic or structural traps. Generally less than 30.4 m (100 ft) thick, these sandstones were deposited in a range of environments from freshwater to marine. They are generally regarded as deposits related to a lowering of sea level. The following transgression is recorded by the widespread marine Mowry Formation and other high stand deposits. When the history of these strata is related to radiometric dates from associated bentonite beds, the sequence spans the time interval of approximately 96 to 98 m.y. ago. The major event correlates with the worldwide sea-level drop 97 m.y. ago, reported by Vail, Mitchum, and Thompson (1977) and Hancock, (1975).

The events described above are known largely from detailed stratigraphic studies in the Powder River basin, Wyoming, and the Denver basin, Colorado, Wyoming, and Nebraska. Only the Denver basin work is summarized in this paper but literature describing the Powder River basin is extensive (Weimer, 1982).

Denver Basin

Petroleum was discovered in upper sandstones (D and J) of the Dakota Group in the Denver basin in 1923. Since the

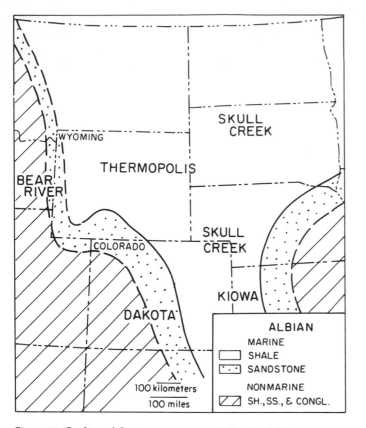

Figure 13: Outline of Cretaceous seaway during the Albian.

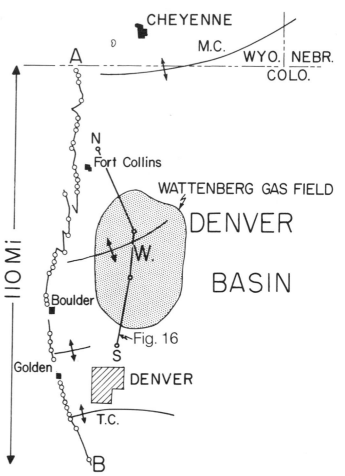

Figure 14: Map of Denver basin showing outcrop sections of Dakota Group and paleotectonic elements: M.C. = Morrill Co. High; W. = Wattenberg; T.C. = Turkey Creek. Locations of measured surface sections indicated by O.

discovery of the Wattenberg gas field in 1970, more than 1,000 wells have been drilled to the J Sandstone in an area covering 600,000 acres across the deeper portion of the Denver basin (Figure 14). Production is about 24 km (15 mi) east of outcrop sections which are described by MacKenzie (1971), Clark (1978), and Suryanto (1979). Details of the Wattenberg field have been published by Matuszczak (1976) and Peterson and Janes (1978). Because of new data from the outcrop and subsurface, a reinterpretation of the depositional model for the J Sandstone is now possible. An important new element is the recognition of the Wattenberg area as a paleostructure which influenced J Sandstone deposition (Weimer, 1980, and Weimer and Sonnenberg, 1982).

The J Sandstone (Muddy) is present in outcrop sections from the Wyoming state line to the South Platte River southwest of Denver (Figure 14). Two types of sandstone bodies comprising the J Sandstone in the Denver basin were described by MacKenzie (1965) from outcrops in the Fort Collins area (Figure 15). One type, named the Fort Collins Member by MacKenzie, is a very fine-grained to fine-grained sandstone containing numerous marine trace fossils and is interpreted to be delta front, shoreline, and marine bar sandstone which was deposited during rapid regression of the shoreline of the Skull Creek sea. A second type is fine- to medium-grained, well-sorted, cross-stratified sandstone (channels) containing carbonized wood fragments and associated shales and siltstones. These lithologies were named the Horsetooth Member by MacKenzie. Valleys of an extensive drainage system were incised into the Fort Collins Member or the underlying Skull Creek Shale and contain a

fill of sandstone, siltstone, and claystone varying in origin from alluvial plain to shoreline deposits (Horsetooth Member). The J varies in thickness from 6.1 to 45.8 m (20 to 150 ft) (Figures 15 and 16).

The thick dominantly fresh to brackish water J Sandstone facies in the Golden area was interpreted by Waage (1953), Haun (1963), MacKenzie (1971), and Matuszczak (1976) to be laterally equivalent to the thin marine sandstone facies north of Boulder. This interpretation led the concept of a northwest-trending marine basin in the area between Boulder and Fort Collins.

The interpretation shown on Figure 15 correlates the Golden area sections with the Horsetooth Member. The Kassler Sandstone, the lower unit of the J, where present, rests on an erosional surface cut into the Skull Creek Shale. Sandstone of the Fort Collins Member is interpreted as having been removed by erosion prior to deposition of the Kassler. Root zones are found in the Kassler only a few feet above the base. In addition, conglomeratic sandstones with chert pebbles up to 1 cm (.4 in) in diameter are sporadically present in the Kassler (Poleschook, 1978). Thus, lower J Sandstone is interpreted as a valley-fill complex of a major drainage system. Cross-strata in the Kassler Sandstone

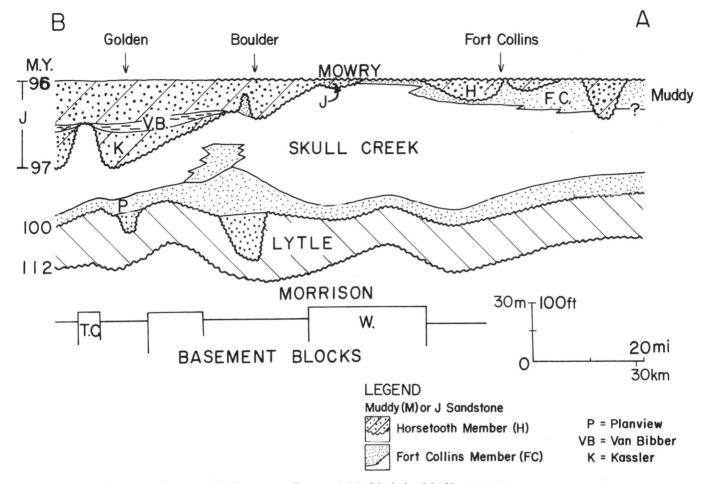

Figure 15: Stratigraphic restored section A-B (location on Figure 14). Modified after MacKenzie, 1971.

(lower J) indicate a dominantly southeast transport direction (Poleschook, 1978, Lindstrom, 1979) (Figure 17). The drainage patterns are interpreted as tributary rather than distributary as previously described.

A north-south electric log section east of the outcrop across the Wattenberg field shows a similar interpretation of the J Sandstone (Figure 16). The widespread gas-bearing sandstone at Wattenberg is the Fort Collins Member of the J, which is transitional with the underlying Skull Creek. Major channel sandstones of the Horsetooth Member are present to the north and south of Wattenberg (Figures 16 and 17). In the Third Creek field (section 19, T. 2 S., R. 67 W.; Figure 16), root zones are preserved below the J channel sandstone and above the erosional surface on top of the marine Fort Collins Member. Based on core interpretation of facies, at least 10 m (30 ft) of the Fort Collins Member was eroded prior to deposition of the fluvial channel sandstone that is oil-productive. Previous interpretations have shown the channel sandstone to be the lateral equivalent of the Fort Collins Member at Wattenberg. Since the channel sandstone can be demonstrated to be younger than the erosional surface cut on the Fort Collins Member, such a facies interpretation is in error.

Tectonics and Sedimentation Model
for J Sandstone, Denver Basin

The following model for tectonic influence on J Sandstone sedimentation is proposed for the Denver basin. At the end of Skull Creek deposition (T_1, Figure 18), a regressive event began which deposited shoreline and shallow-marine sandstones with a transitional contact with underlying Skull Creek Shale. Depositional patterns over basement fault blocks, where slight fault-block movement influenced sedimentation, are illustrated. Rivers and associated deltas positioned themselves in structural and topographically low areas, grabens, whereas delta-margin or interdeltaic sedimentation occurred along an embayed coast over structural horst blocks. Delta-front and shoreface sands extended seaward from the shoreline to a distance controlled by effective wave base. The shoreline prograded seaward to position T_2 and a sheet-like sand body was deposited over a large area (Wattenberg pay sandstone or Fort Collins Member).

A drop in sea level occurred (T_3) during which all, or a large portion, of the depositional basin (Skull Creek seaway) was drained (Figure 19). River drainages were incised into marine shales and/or the regressive shoreline sandstones in topographic lows which generally correspond to

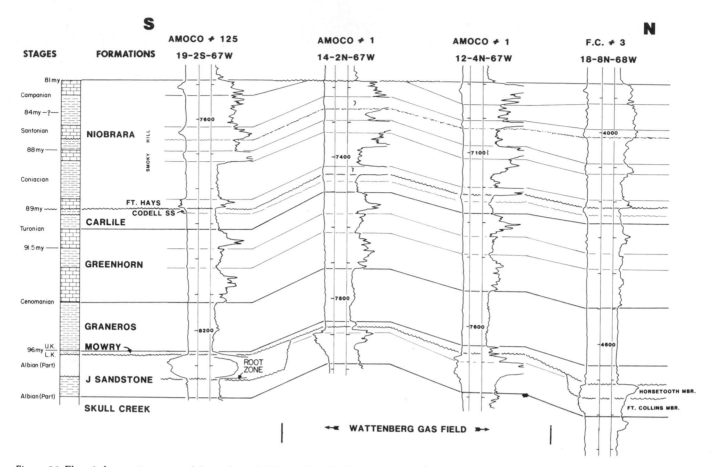

Figure 16: Electric log section across Wattenberg field from Fort Collins to Denver (location on Figure 14).

the graben fault-block area. Over much of the Denver basin the base of the incisement is on the Fort Collins Member (T_1 or T_2 sand complex). Only locally did the erosional surface cut into the Skull Creek Shale.

A rise in sea level occurred during which the incised valleys were probably modified and filled with fluvial and estuarine sandstone, siltstone, and shale. Vertically, the valley fill has a wide variety of fluvial environments in the lowermost part and estuarine or deltaic environments in the upper part (Figure 20). With a rising sea level the earliest fluvial deposits were deposited in narrow valleys as upper meander belt sandstones. As sea level increased, the lower meander belt environments shifted landward and channel meandering widened the valley by scour. These sandstones may overlie either the upper meander belt sandstone or the Fort Collins Member. The final deposits of the J are transitional estuarine, deltaic, or shoreline deposits. After the valleys filled, deposition covered the interstream divide areas. With a continued rise in sea level (T_4, Figure 21) and minor renewed fault block movement, strata were eroded from the top of the horst blocks and an extensive thin transgressive-lag deposit of conglomeratic or coarser-grained sandstone formed over the horst blocks on a surface of erosion. Following T_4 the entire region received marine siltstone and shale (Mowry or Graneros shales).

In the above model, an important unconformity separates T_1 and T_2 deposits from T_4 deposits. This basin-wide unconformity (T_3, Figure 19) may be within sandstone deposits

(i.e., valley-fill sandstones rest on older regressive sandstones), or between sandstone and marine shale deposits (i.e., valley-fill sandstones rest on Skull Creek Shale). In portions of the Wattenberg field, the unconformity is at the base of the Mowry Shale or top of the regressive sandstone.

Previous correlations which show the J Sandstone to be deposited across the basin during one major regressive event need to be modified. Sandstones above the unconformity (Horsetooth Member) are younger than the regressive sandstone at the top of the Skull Creek (Fort Collins Member), although because of tectonic movement the older sandstones are now at a stratigraphic high position (Figures 15 and 16). This model has important implications for future petroleum exploration in the Denver basin.

A relation exists between major northeast-trending Precambrian shear zones mapped in the Front Range and paleostructure in the northern Denver basin (Figure 12, Sonnenberg and Weimer, 1981). Recurrent movement on these old fault zones has controlled thickness variations and depositional facies in Paleozoic and Mesozoic strata. Five major east-to-northeast trending paleostructures occur in the northern Denver basin which had recurrent movement during the Cretaceous and some have documented movement during the Permian. The major paleostructures are the Wattenberg high, the Morrill County high, the Hartville high, the Turkey Creek high, and the Yuma high (Figure 12). These paleohighs vary in width from 32 to 40 km (20 to 40 mi) and in length from 80 to 288 km (50 to 180 mi). Sev-

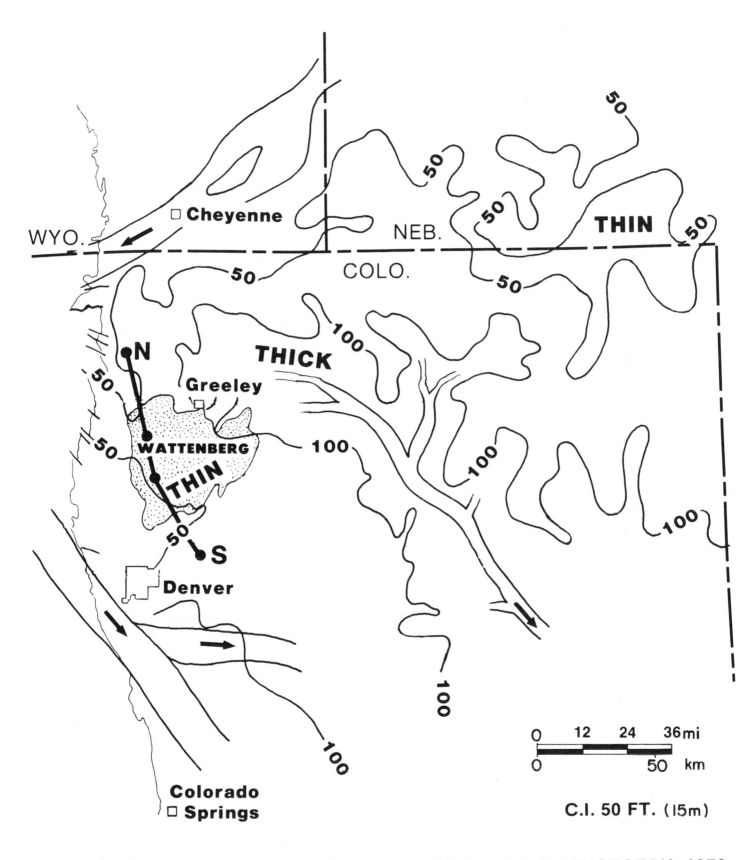

Modified from HAUN, 1963; MATUSZCZAK, 1976.

Figure 17: Isopach map of J Sandstone (includes Fort Collins and Horsetooth Members) with location of major incised valleys as indicated by lower J (equivalent of Kassler Sandstone of outcrop--Figure 15). Stippled pattern is area of gas production from J Sandstone.

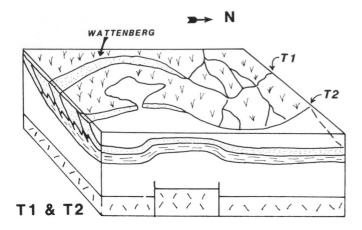

Figure 18: Depositional and tectonic model for Fort Collins Member of J Sandstone showing highstand regression over basement fault blocks (Wattenberg high) with penecontemporaneous movement. T_1 = time 1; T_2 = time 2. Rate of sediment supply exceeds rate of subsidence or submergence.

eral important northwest- and north-trending paleostructures are omitted from Figure 12 (Sonnenberg and Weimer, 1981). Each structural paleohigh should be investigated to determine if recurrent movement influenced sedimentation as documented in the Wattenberg area.

Recurrent movement on paleostructural elements affected the Cretaceous seaway in a broader sense. The five paleostructural elements (Figure 12) collectively have been grouped together as a broad structural arch referred to as the Transcontinental arch (Weimer, 1978). Structural movement on this broad arch during the time of J or Muddy Sandstone deposition created a topographic high which

divided the drainage in the Cretaceous basin during the low sea-level stand (T_3; Figure 19). A general south-flowing drainage developed in southwest Wyoming and eastern Colorado (Figure 22), whereas a north-flowing system developed in northern Wyoming (Powder River basin).

EROSION OF SHELF STRATA DURING SEA-LEVEL CHANGE
(90 m.y. ago)

Unconformities within marine strata have been described in the Cretaceous by Reeside (1944, 1957), McGookey (1972), and Merewether and Cobban (1972, 1973), Merewether, Cobban, and Spencer (1976), and Merewether, Cobban, and Cavanaugh (1979). The stratigraphic positions of the documented unconformities are shown on Figure 10. Uncertainty exists as to the cause of the marine unconformities: Are they the result of sea-level change with subaerial or submarine erosion. Do they result from submarine erosion associated with tectonic movement, nondeposition; or, a combination of these processes? To what extent is the geographic distribution of the unconformities influenced by depositional topography, which in turn is controlled by the interplay of rates of sedimentation and oceanic processes?

Model for Depositional Topography

Depositional topography commonly developed in ancient basins which had a high input of terrigenous sediment. Asquith (1970) described depositional topography in the Western Interior Cretaceous basin for Campanian and Maestrichtian strata in Wyoming. A modified model from

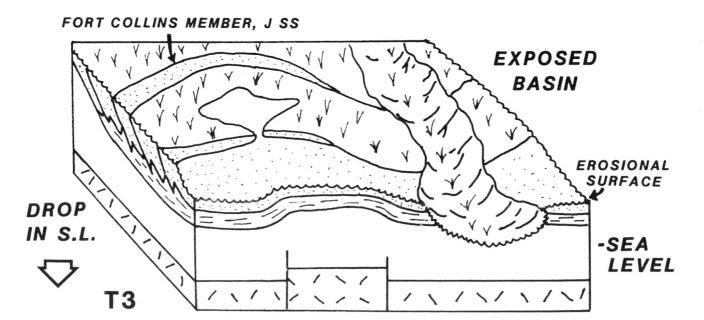

Figure 19: Lowstand sea level (time 3, T_3) recorded as basin-wide erosional surface resulting from subaerial exposure. Root zones form on exposed marine shales and sandstones.

Asquith for high-stand sea-level depositional topography includes environments of deposition related to coastal plain, shoreline, shallow marine/shelf, slope, and basin (Figure 23A). Because of high rate of sedimentation and lateral accretion (progradation), higher than normal depositional dip formed in two areas: the delta front-prodelta, and the slope. The primary dips (clinoforms) in these areas of dominantly clay and silt deposition are generally less than 1°, as determined from closely spaced well log correlations (Figure 23B). Because of their low dip, the clinoforms are not generally recognized on seismic sections or on outcrop.

The slowest rates of sedimentation existed in deeper water areas (basin, Figure 23A) where sedimentation of organic rich chalk and clay, related to pelagic sedimentation, was dominant. Water depths are difficult to reconstruct but, in general, shelf depths are estimated to vary from 30.5 to 91.5 m (100 to 300 ft). Depths for chalk sedimentation in the basin are controversial but estimates range from 61 to 487.5 m (200 to 1600 ft) (Kent, 1967; Eicher, 1969; Hattin, 1975a, and Kauffman, 1977). Based on the physical evidence for the shelf, slope, and basin model (Figure 25), I favor basin water depths in the range of 183 to 305 m (600 to 1000 ft) during sea-level high stands.

Water depths greatly influenced sedimentation or erosion during sea-level changes. During low stand events erosion by wave energy or by subaerial processes may have removed shelf and slope deposits, depending on the magnitude of the sea-level drop (Figure 23C). Sand normally con-

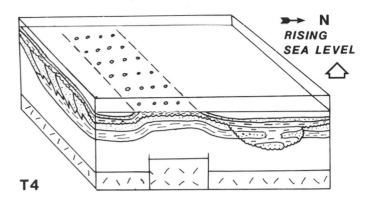

Figure 21: Rising sea level during Time 4 (T_4) with fill of incised valley and deposition of marine shale and sandstones. A thin transgressive lag (generally less than .3 m; 1 ft thick) of conglomeratic sandstone (indicated by circles) occurs in association with basement-controlled horst block.

fined to the coastal plain and shoreline areas was transported to the shelf area. During a subsequent sea-level rise, the sand may have been reworked into marine shelf complexes, either as narrow linear marine bars, as thin transgressive sheet sands, or as a final fill of incised drainages. The shelf deposits of the Gulf of Mexico illustrate these

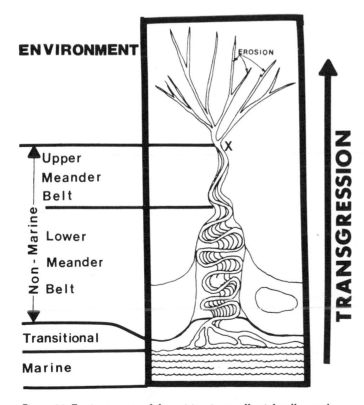

Figure 20: Environments of deposition in an alluvial valley and shoreline setting. X marks reference point for vertical changes in lithology in valley-fill deposits as sea level rises and environments shift landward during transgression.

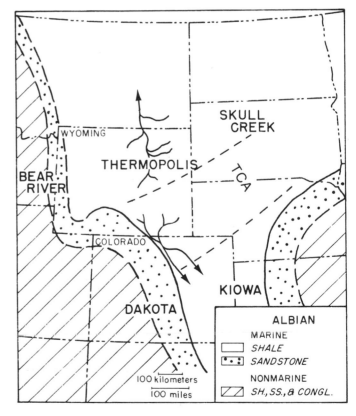

Figure 22: Map showing shoreline trends during high stand of sea level during Skull Creek deposition with direction of flow of incised drainage during subsequent sea-level drop (lines within marine area). A drainage divide developed across the basin in the area of the Transcontinental arch (TCA).

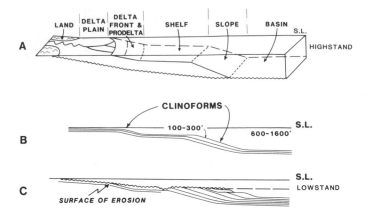

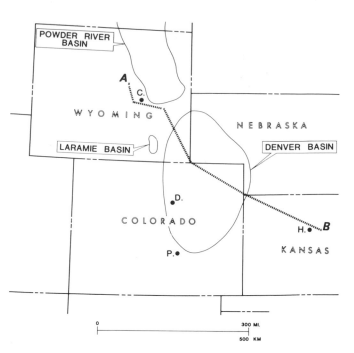

Figure 23: Cross section showing depositional topography in Cretaceous basin during highstand of sea level (A). Areas of depositional dip (clinoforms) develop by lateral accretion during deposition (B). Erosional surface develops across shelf during lowstand of sea. S.L. = sea level; numbers give water depth in feet.

Figure 24: Index map for stratigraphic section from central Wyoming to central Kansas (Figure 25). C = Casper; D = Denver; P = Pueblo; H = Hays.

types of sand bodies and were discussed previously (Figures 5, 6, 7).

Although the above model applies generally to interpreting all strata in the Cretaceous basin, an example illustrating the components in the marine phase of the model has been developed for Cenomanian and Turonian strata in the Denver basin and adjacent areas. A stratigraphic section from near Hays, Kansas, to central Wyoming incorporates surface and subsurface data (Figures 24 and 25). Ages of formations at each end of the section are based on ammonite correlations. Elsewhere subsurface correlations are by stratal continuity of bentonite layers in shale units, and thin limestones (chalks) identified in thousands of well logs in the Denver basin. Cobban and associates have developed faunal zones (Table 2) for the Cenomanian, Turonian, and Coniacian stages (Merewether, Cobban and Cavanaugh, 1979). These faunal zones have been related to lithologic markers and traced over a large area as time-stratigraphic units (Weimer, 1978; Sonnenberg and Weimer, 1981; Weimer, 1983). Shelf sedimentation for the Cenomanian and Turonian is well illustrated by the interbedded sandstones and shales of the Frontier Formation (Figure 25) in central Wyoming (Merewether, Cobban, and Cavanaugh, 1979, p. 94). They estimate water depths on the shelf of less than 130 m. Synchronous basin sedimentation is represented by chalks of the Greenhorn and Fairport of Kansas (Figure 25). A major unconformity associated with shelf sedimentation (90 million year lowstand) is present within the Carlile and Frontier formations.

Carlile Formation

The Carlile Formation in the Great Plains province consists of four widespread members which, in ascending order, are: Fairport chalk; Blue Hill Shale; Codell Sandstone; and the Juana Lopez. A fifth member, the Sage Breaks Shale, is present as the upper Carlile in the northern Denver basin and Powder River basin area. The Juana Lopez member is a lenticular limestone unit which is locally conspicuous in outcrop along the west flank of the Denver basin but is gener-

ally too thin (less than 1 m; 3 ft) to map in the subsurface. The unit is ubiquitous in outcrop along the southern margin of the Denver basin and adjacent areas to the south. The unit is absent in outcrop (Figure 25) over most of central Kansas (Hattin, 1975b, p. 198).

The Carlile contains one or more unconformities in the upper portion which play a significant role in the thickness and depositional patterns. The thickness of the Carlile varies from less than 15.3 m (50 ft) to more than 61 m (200 ft) (Figures 25 and 26). The Fairport and Blue Hill members in Kansas are progressively cut out westward by an unconformity at the base of the Codell Sandstone, or in its absence the unconformity at the base of the Fort Hays Limestone (Figure 25). A similar relationship was observed in north-south correlations from the more complete Pueblo, Colorado, section (similar to central Kansas) with the strata in the northern Denver basin (Weimer, 1978).

If a model for a shelf, slope, and basin is used for deposition of the Carlile (Figure 23), the associated regional unconformities can be easily explained by a combination of eustatic changes and tectonic movement. Several important stratigraphic relations in the area from central Wyoming to Kansas support this model.

The Fairport chalk of Kansas (basin deposit) thickens westward and changes across the Denver basin to siltstone and shale with the well-developed clinoforms of slope deposits (Figure 25; especially in southeastern Wyoming and western Nebraska).

The Codell Sandstone in portions of central Kansas and in the Pueblo, Colorado, area has a transitional contact with the underlying Blue Hill Shale (Hattin, 1962; Merriam, 1957; Pinel, 1977; Merriam, 1963). However, over much of the Denver basin the Codell is sporadically developed, and where present, an unconformity is observed at the base

Table 2

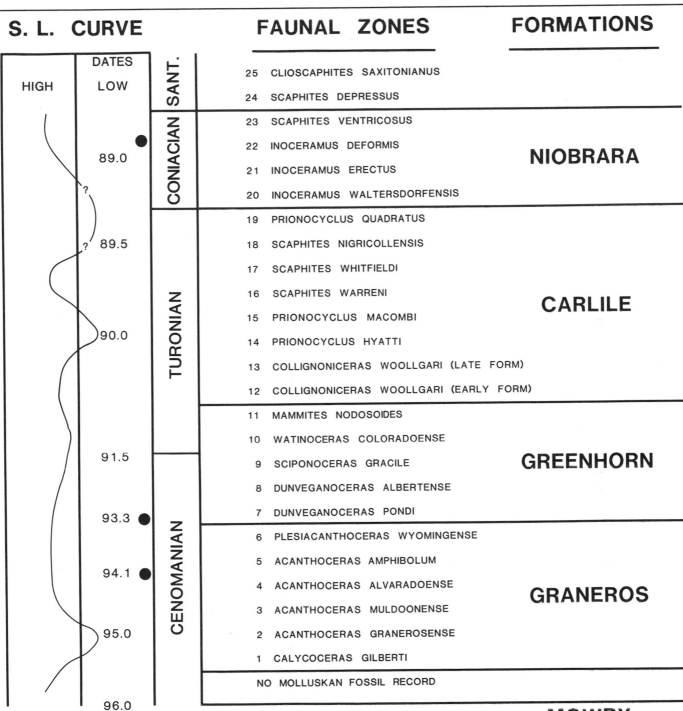

S. L. CURVE				FAUNAL ZONES	FORMATIONS
HIGH	DATES		SANT.	25 CLIOSCAPHITES SAXITONIANUS	
	HIGH / LOW			24 SCAPHITES DEPRESSUS	
	89.0 ●		CONIACIAN	23 SCAPHITES VENTRICOSUS	NIOBRARA
				22 INOCERAMUS DEFORMIS	
	?			21 INOCERAMUS ERECTUS	
				20 INOCERAMUS WALTERSDORFENSIS	
	? 89.5		TURONIAN	19 PRIONOCYCLUS QUADRATUS	
				18 SCAPHITES NIGRICOLLENSIS	
				17 SCAPHITES WHITFIELDI	
				16 SCAPHITES WARRENI	CARLILE
	90.0			15 PRIONOCYCLUS MACOMBI	
				14 PRIONOCYCLUS HYATTI	
				13 COLLIGNONICERAS WOOLLGARI (LATE FORM)	
				12 COLLIGNONICERAS WOOLLGARI (EARLY FORM)	
				11 MAMMITES NODOSOIDES	
				10 WATINOCERAS COLORADOENSE	
	91.5		CENOMANIAN	9 SCIPONOCERAS GRACILE	GREENHORN
				8 DUNVEGANOCERAS ALBERTENSE	
				7 DUNVEGANOCERAS PONDI	
	93.3 ●			6 PLESIACANTHOCERAS WYOMINGENSE	
				5 ACANTHOCERAS AMPHIBOLUM	
	94.1 ●			4 ACANTHOCERAS ALVARADOENSE	
				3 ACANTHOCERAS MULDOONENSE	GRANEROS
	95.0			2 ACANTHOCERAS GRANEROSENSE	
				1 CALYCOCERAS GILBERTI	
				NO MOLLUSKAN FOSSIL RECORD	
	96.0				MOWRY

Table 2: Faunal zones in lower part of Upper Cretaceous (after Merewether, Cobban, and Cavanaugh, 1979, p. 70). Dots are faunal zones with radiometric dates (after Obradovich and Cobban, 1975; modified by Fouch et al, 1983).

which has a hiatus increasing in magnitude to the west (Figure 25). A subcrop map of formations beneath the surface of erosion illustrates these relationships (Figure 26). In the northern Front Range area and the Laramie basin, Codell Sandstone rests on strata equivalent to the middle Greenhorn (faunal zone 8 or 9).

The regional subcrop pattern of the Fairport and Greenhorn beneath the unconformity shows a broad eastward bulge in southern Wyoming and northern Colorado (Figure 27). This feature is related to a structural doming with the beveling of faunal zones 9 through 13 from central Wyoming to Kansas. The regional distribution of regressive

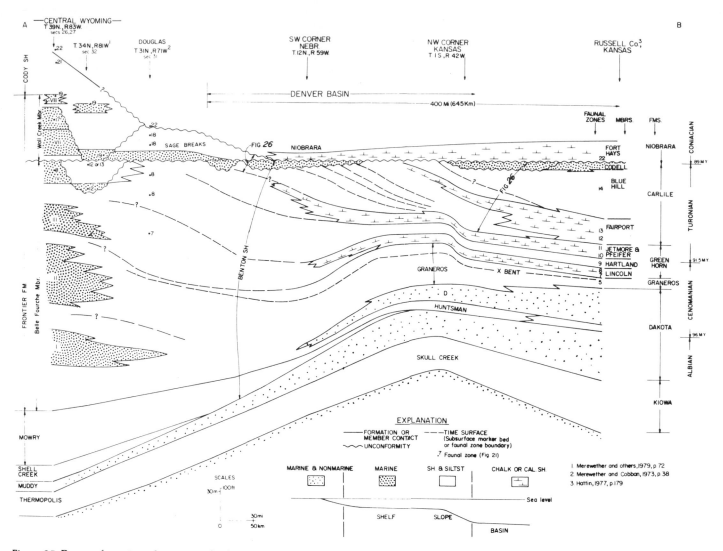

Figure 25: Restored stratigraphic section for lower part of Cretaceous from central Wyoming to central Kansas. Unconformities associated with J and D sandstones are not shown.

sandstone during faunal zone 14 (*Prionocyclus hyatti*) suggests a major sea-level drop of short time duration (Table 2). The scour and fill pattern in central Wyoming (Figure 25), showing remnants of faunal zones 12 and 13 (Fairport equivalents) on faunal zone 8 (lower Greenhorn equivalents), suggests movement of local tectonic elements superimposed on the broad doming. Structural movement, with erosion, started during or after the deposition of faunal zones 12 and 13 (the unnamed member of the Frontier Formation of Merewether, Cobban, and Cavanaugh, 1979, p. 82). Units 12 and 13 were either deposited over the entire area and subsequently removed by erosion, or they were deposited only in topographic (structural) low areas. Regional correlations suggest that sea level remained high during faunal zones 9 through 13 (Table 2) so the erosion is thought to be related to structural movement and submarine scour. The clinoform pattern of slope and basin deposits in the upper Greenhorn and Fairport formations (Figure 25) may be the result of rapid deposition because of sediment recycling by erosion on the top and deposition on the margin of the broad structural element.

Erosion by subaerial or submarine processes on the shelf during a sea-level drop (faunal zone 14) removed strata and erosional depressions (valleys?) were cut into shelf deposits as base level was lowered. Chert pebbles and fine- to coarse-grained sand were transported across the shelf by streams and currents. These scours are observed mainly in the northernmost Denver basin.

With the subsequent rise in sea level (the time of either late faunal zone 14 or zone 15) three types of sandstone were deposited above the surface of erosion: 1) thick sand deposits of fresh or brackish water origin accumulated in the scour depressions; 2) thin widespread fine-grained bioturbated shelf sand; 3) coarsening upward fine- to medium-grained sand reworked into marine bar complexes. These sands are best preserved in the northern Denver basin. These types of sand occurrences, deposited during a changing sea level, are believed similar to those observed on the modern shelf of the Gulf of Mexico (Figures 5, 6, and 7) as described by Nelson and Bray (1970).

In the southern Denver basin and Kansas, where sedimentation was continuous from the underlying Blue Hill

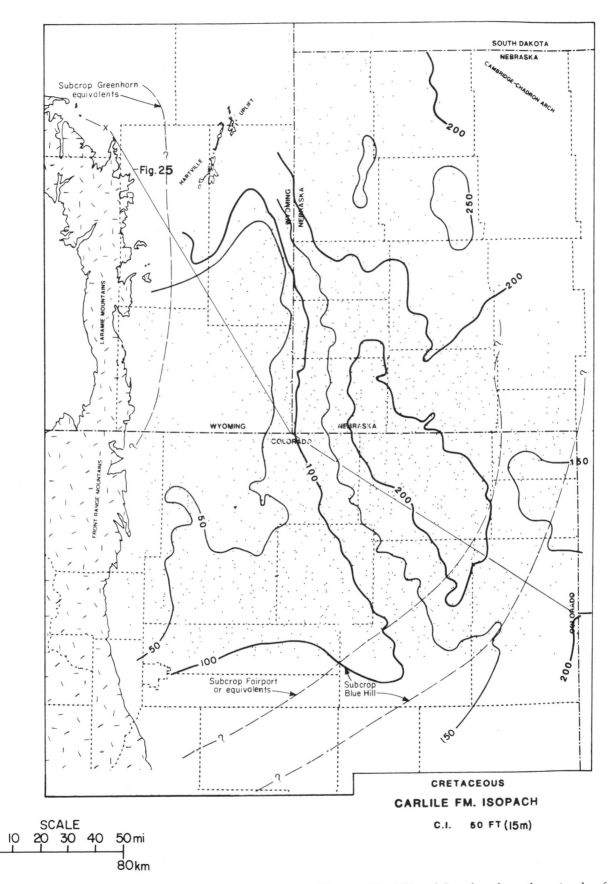

CRETACEOUS

CARLILE FM. ISOPACH

C.I. 50 FT (15m)

SCALE

0 10 20 30 40 50 mi

0   80 km

Figure 26: Isopach map of Carlile Formation with areas of subcrop of Fairport, Blue Hill, and Greenhorn beneath erosional surface at base of Codell Sandstone or Fort Hays Limestone (see Figure 25).

through the Codell, the sand was deposited either as a shallow-marine bar sequence or along a regressing shoreline during the low stand of sea level. These sands are slightly older than the sands above the surface of erosion in the central and northern Denver basin (Figure 25).

The Juana Lopez in Colorado, a thin relict or palimpsest deposit, rests on an erosional surface with chert and phosphate pebbles and coarse sand occurring as a lag in the lower portion. Bioclastic recrystallized limestone with shark teeth occurs in the upper portion. This unit, where well developed (for example, southern Colorado and northern New Mexico), contains faunal zones 15, 16, and 17 (Kaufman, 1977, p. 82; Hook and Cobban, 1979, p. 41). The Frontier units 6 and 7 (Wall Creek Member, Figure 25) record this high stand in central Wyoming.

Erosion in central Wyoming described by Merewether, Cobban, and Cavanaugh (1979, p. 72) developed during or near the end of faunal zone 18 and removed sandstone and shale containing zones 16 and 17. The scour was filled by marine deposits containing faunal zone 19 (Figure 25). These relations are interpreted to result from tectonic movement and submarine scour similar to those previously described in the same area for faunal zones 8 through 13. However, an alternative interpretation is for erosion to have occurred on the shelf during minor sea-level drops at the times of faunal zones 10 or 11 and 18 or 19. Some regional evidence supports a possible drop during 18 or 19.

Over much of the central Denver basin and eastward into Kansas (Figure 25) a widespread surface of erosion is also present above the Codell Sandstone with the upper part of the Fort Hays Member of the Niobrara Formation (either faunal zone 21 or 22) resting on the surface of erosion. This unconformity has been related by Weimer (1978) to result from erosion and then marine onlap on a broad northeast-trending structural element called the Transcontinental arch. Sparce faunal evidence suggests that the hiatus represents the time of faunal zones 15 through 21 or 22. The erosion may have been associated with the possible sea-level drops during faunal zones 18 or 19.

The approximate positions of the western shoreline of the basin for highstands during faunal zones 5, 9, 12-13, 17, and 21 (Figure 27) have been mapped by Hook and Cobban (1977, 1979, 1981a) and Cobban and Hook (1979, 1981). Because of the lack of detailed faunal data on a regional basis, the pattern of deposition and erosion is preliminary in nature and no attempt has been made to change shoreline trends. Some of these shorelines must be significantly modified because of the erosion of faunal zones from the paleo-high (Figure 25), especially for faunal zones 9 through 14.

EXAMPLE OF TECTONIC MOVEMENT AND SEA-LEVEL CHANGES AFFECTING BASIN DEPOSITS
(80 to 81 m.y. ago)

Niobrara Formation

Following the late Turonian (Carlile and equivalents of previous section) depositional patterns changed significantly. During the Coniacian, the Niobrara Formation (composed of chalk and organic rich shale deposition, simi-lar in lithology to the Greenhorn) was deposited over eastern Colorado, Kansas, and adjacent areas (Figures 10, 28). This lithologic change from Carlile to Niobrara was the result of a significant increase in water depth as the shoreline zone shifted to western Wyoming and Utah (highstand 21, Figure 27). The shelf-slope break shifted westward in a corresponding manner to central Wyoming (Figure 29). Thus, the northern Denver basin, where shelf and slope deposits were dominant in the Carlile, became an area of chalk and shale deposits of the basin setting.

Whereas thickness changes associated with the Greenhorn chalks follow a regional pattern (Sonnenberg and Weimer, 1981), the Niobrara shows many thickness anomalies of a more local distribution (Figure 28). The Niobrara Formation in the northern Denver basin varies in thickness from 76.3 m (250 ft) to more than 152.5 m (500 ft). The formation or equivalent strata are thicker to the south in the Pueblo area and to the north in the Powder River basin (Figure 28).

Within the Niobrara, four limestone intervals and three intervening shale intervals occur regionally and are easily recognized on geophysical logs (Figure 16). The lower limestone is named the Fort Hays and the overlying units are grouped together as the Smoky Hill Member (Scott and Cobban, 1964). The lower boundary is an erosional surface where the limestone is in contact with either Carlile Shale or, in some places, strata as old as the Greenhorn Formation. The upper boundary of the Niobrara is placed at the contact with the noncalcareous black shales of the Pierre Shale. Locally, the contact is an erosional surface; elsewhere the contact is transitional.

Regionally, the northern Denver basin has a thin Niobrara section that is related to structural movement on the Transcontinental arch (Weimer, 1978) and depositional thinning because of slow rates of deposition. Superimposed on the broad pattern are the major northeast-southwest axes of thinning (Figure 28), which are related to second-order paleotectonic elements (Sonnenberg and Weimer, 1981). Thinning over the paleohighs results from erosion of the upper chalk and underlying shale (Figure 16), whereas these strata are preserved in structural and topographic lows between the structural highs (Weimer, 1980).

The time of major movement on the paleohighs was near the end of Niobrara deposition or early in the deposition of the Pierre Shale. The maximum thinning occurs over the Wattenberg paleostructure where up to 30.5 m (100 ft) of Upper Niobrara was removed by sea-floor erosion prior to deposition of the overlying Pierre (Figure 28).

Minor thinning also occurs at the base of each of the four limestones of the Niobrara in local areas. Regional unconformities can be mapped at the base of the Fort Hays and the base of the second chalk from the top of the Niobrara (Figure 16). Because these breaks occur within the deep water sediments generally during highstand conditions, the breaks may be associated with marine onlap processes as diagrammed by Vail, Mitchum, and Thompson (1977).

A major question is what changed the pattern of structural movement from a broad regional uplift centered in Wyoming during Carlile deposition (Figures 26 and 27) to the movement on northeast structural trends centered in Colorado during late Niobrara and early Pierre deposition (Figure 28). The answer may be related to the direction and

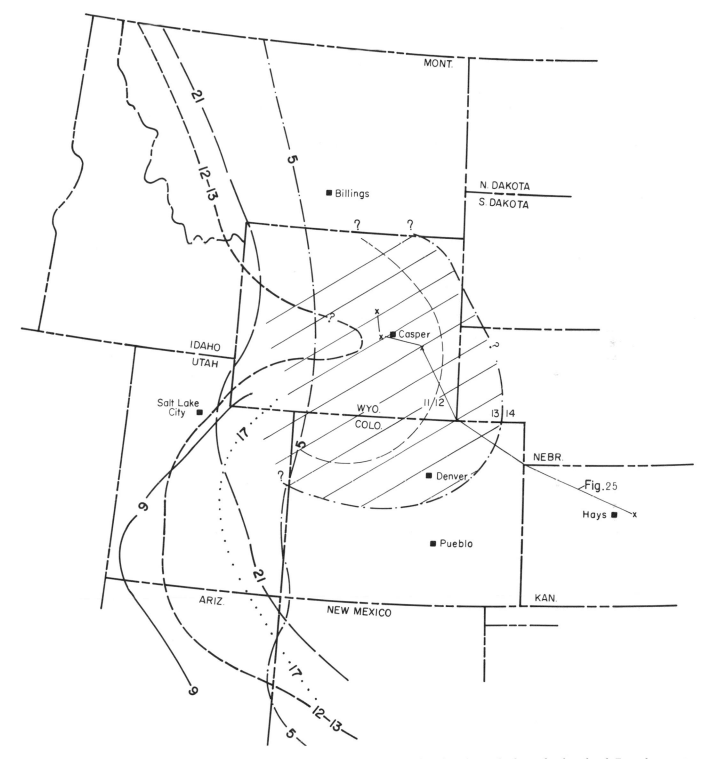

Figure 27: Regional distribution of faunal zones indicating geographic position of shoreline during highstands of sea level. Faunal zones 5 and 12-13 after Cobban and Hook (1979, 1980); faunal zones 9, 17, and 21, after Hook and Cobban (1977, 1979, 1980). Diagonal ruling indicates area of erosion because of structural movement and a low stand of sea level (faunal zone 14).

rate of movement of the American plate. Normally the plate had western movement to establish the tectonic framework shown on Figures 9 and 29. Either a more rapid rate of westward movement or a more northwesterly movement may have caused vertical movement on northeast-trending basement faults with submarine erosion recurring over the higher standing blocks.

At approximately the same time as fault blocks were reactivated in the northern Denver basin, volcanic activity on the sea floor, during deposition of the Austin chalk,

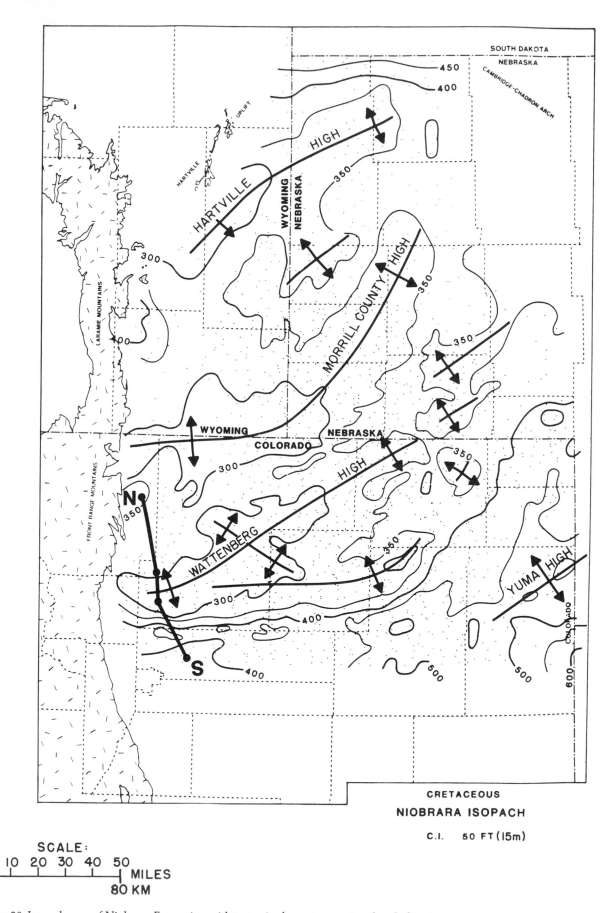

CRETACEOUS

NIOBRARA ISOPACH

C.I.   50 FT (15m)

SCALE:

0   10   20   30   40   50
├────┼────┼────┼────┼────┤ MILES
0   80 KM

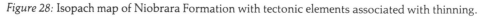

Figure 28: Isopach map of Niobrara Formation with tectonic elements associated with thinning.

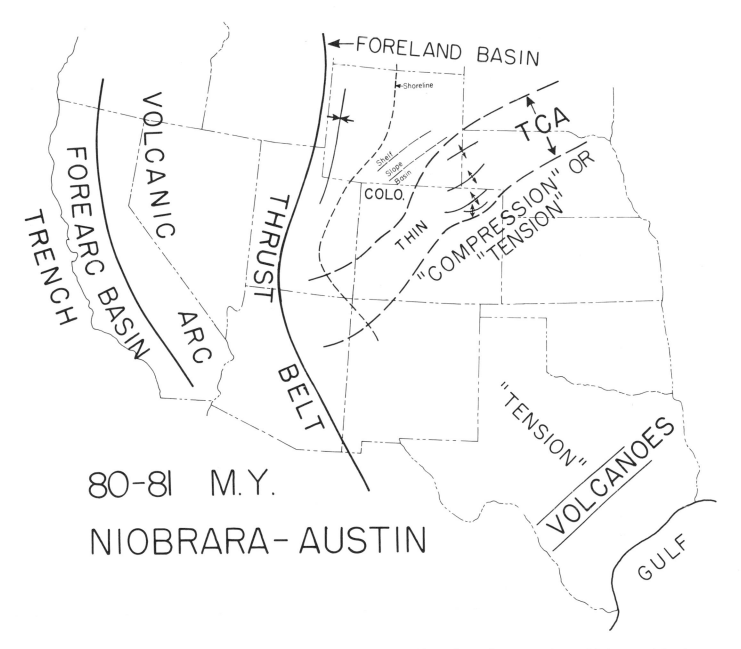

Figure 29: Map of western United States showing events in Cretaceous foreland basin during deposition of upper Niobrara and Austin strata.

occurred in Texas (Figure 29; Simmons, 1967). Ancient submarine volcanoes occur along a northeast trend across central Texas. The lithology of the igneous material is ultrabasic to basic in composition and the features are referred to as "serpentine plugs" altered from igneous rocks rich in olivine. Carbonate reefs are found fringing or capping the volcanoes. Depth of the sea during deposition of the upper Austin is estimated to have been between 61 to 244 m (200 to 800 ft) (Simmons, 1967, p. 126).

The northeast trend of the volcanoes was related by Simmons (1967) to extrusion above basement fault systems. Tension along the Balcones fault zone opened the crust to allow mantle-derived material to be intruded in the strata or to be erupted on the sea floor as volcanoes.

The Austin chalk of Texas is correlated with the Niobrara

Formation of eastern Colorado and Kansas. Thus, at the time of the volcanic events in Texas, basement fault block movement occurred on second order structures along the ancestral Transcontinental arch (Figure 29). Moreover, major thrusting was in progress along the Meade-Crawford thrust sheets in western Wyoming and Utah (Royse, Warner, and Reese, 1975; Jordan, 1981). The timing of these events is approximately 80 to 81 m.y. ago.

Uplift of the fault blocks on the sea floor in the northern Denver basin could have resulted from compression with the blocks forming topographic and structural highs; or, a broad arch related to compression may have been modified by tension with some blocks subsiding to lower structural positions than adjacent more positive blocks.

The structural movement was synchronous with a possible sea-level drop. The upper Niobrara chalk is a basin facies equivalent of the Eagle and Shannon sandstones of Montana and Wyoming. The wide distribution of these shelf sandstones may have been related to events similar to the processes described previously for the Carlile Formation. A lower sea level would have affected storm wave base and associated scour. By a lowering of storm wave base, strata were removed by erosion over the topographic (structural) highs but not in the intervening structural and topographic lows. Thus, the Niobrara isopach pattern and associated unconformities is interpreted to result from a northwest direction of plate motion, the development of compression to uplift portions of the sea floor, and submarine scour during a contemporaneous sea-level drop. Following these events, a normal pattern of tectonics and sedimentation for the Western Interior Cretaceous was resumed for the time of lower Pierre sedimentation.

PETROLEUM OCCURRENCE IN GEOLOGIC MODELS

Modern petroleum exploration integrates geologic factors which relate to origin, migration, and accumulation of oil and gas. The factors are source beds, generation area, migration paths, reservoir, seal, time of formation, and preservation of trap.

The principal source beds in the Denver basin have been identified as the Benton Group (Graneros, Greenhorn, and Carlile) by Clayton and Swetland (1980). They believe that oil was generated in the deeper portions of the Denver basin and that migration occurred from these organically rich layers into the J (Muddy) Sandstone. In the proposed depositional model (Figures 23 and 25), the organic-rich source beds originated as basin and slope deposits. These are deeper water deposits where anoxic conditions favored preservation of organic matter.

Reservoir rocks in the J (Muddy) Sandstone are of two main types: One is the widespread marine--delta front sandstone (Fort Collins Member) illustrated by the Wattenberg field. The second is the fluvial channel sandstone facies of the valley-fill deposits (Horsetooth Member). Because most of the individual sand bars within the channel complex are small in geographic distribution, the oil fields are small (generally less than 2 million barrels of reserves per field).

The trap for petroleum may be either facies changes within the valley fill from reservoir to nonreservoir rock, or the unconformity surface. In the latter case, wedge out of sandstone against an impermeable valley wall occurred, or erosion of porous sandstone with impermeable shales placed in the scour forming a trap.

Reservoir rocks in the Frontier Formation in central Wyoming are lenticular marine shelf sandstones which occur both above and below the major surface of unconformity (Figure 25). Petroleum occurs where offshore marine sandstones are in favorable structural condition for entrapment. Over 400,000,000 bbls of oil have been produced form the Second Wall Creek Sandstone at the Salt Creek field, Wyoming (Barlow and Haun, 1966). This sandstone is shown as Unit 3 of the Frontier Formation on Figure 25 (Merewether, Cobban, and Cavanaugh, 1979). Many other fields in central Wyoming also produce from the Frontier Formation.

CONCLUSIONS

Stratigraphic evaluation of ancient sequences has been directed principally toward the construction of a depositional model to explain the origin and distribution of formations, facies, and time-stratigraphic units. For some sequences, little attention has been given to the possible influence of intrabasin deformation on sedimentation and the development of unconformities, especially during times of major eustatic changes.

This paper presents data and concepts which illustrate unconformities associated with the 97, 90, and 80 to 81 m.y. sea-level changes. During these times, complex relationships existed among the following: sea-level fluctuations and related changes in base level of erosion and deposition; tectonic movements and/or climatic changes in the source area which influenced the rate of sediment supply to the basin; unconformities within the basin; distribution of sandstone reservoirs related to environments of the shelf, shoreline, and alluvial valleys; and, the influence of intrabasin recurrent movement of basement fault blocks. The stratigraphic record from studies of inland cratonic basins should be coordinated with studies of plate margin areas and of other continents.

By evaluating the above factors, a more complete geologic model can be constructed and used as a powerful predictive tool in stratigraphic trap exploration for petroleum. Improved modeling will aid significantly in the interpretation and use of seismic data both in the older mature areas of exploration and in frontier areas.

ACKNOWLEDGMENTS

I appreciate S.A. Sonnenberg's assistance in compiling and interpreting subsurface data, and am grateful to the Getty Oil Company for making the research possible by generous financial support. Barbara Brockman typed the manuscript and Craig Corbin drafted many of the illustrations. I thank M. Reynolds, T.D. Fouch, and P.C. Weimer for helpful suggestions in improving the manuscript.

REFERENCES CITED

Asquith, D.O., 1970, Depositional topography and major marine environments, Late Cretaceous, Wyoming: AAPG Bulletin, v. 54, n. 7, p. 1184-1224.

Barlow, J.A., Jr., and J.D. Haun, 1966, Regional stratigraphy of Frontier Formation and relation to Salt Creek field, Wyoming: AAPG Bulletin, v. 50, n. 10, p. 2185-2196.

Blackwelder, E., 1909, The valuation of unconformities: Journal of Geology, v. 17, p. 289-300.

Clark, B.A., 1978, Stratigraphy of the Lower Cretaceous J Sandstone, Boulder County, Colorado - a deltaic model, in J.D. Pruit and P.E. Coffin, eds., Proceedings, Symposium on mineral resources of the Denver basin: Rocky Mountain Association of Geologists, p. 237-246.

Clayton, J.L., and P.J. Swetland, 1980, Petroleum generation and migration in Denver basin: AAPG Bulletin, v. 64, p. 1613-1634.

Cobban, W.A., and S.C. Hook, 1979, *Collignoniceras woollgari woollgari* (Mantell) ammonite fauna from

Upper Cretaceous of western interior, United States: New Mexico State Bureau of Mines and Mineral Resources, Memoir 37, p. 5-51.

———, 1980, Occurrence of *Ostrea beloiti* Logan in Cenomanian rocks of Trans-Pecos Texas: New Mexico Geological Society Guidebook, 31st Field Conference, p. 169-172.

———, ———, 1981, New turrilitid ammonite from mid-Cretaceous (Cenomanian) of southwest New Mexico: New Mexico State Bureau of Mines and Mineral Resources Circular 180, p. 22-35.

Curray, J.R., 1960, Sediments and history of Holocene transgression, continental shelf, northwest Gulf of Mexico, *in* F.P. Shepard, F.B. Phleger, and T.J. van Andel, eds., Recent sediments, northwest Gulf of Mexico: AAPG Special Publication, p. 221-266.

———, 1975, Marine sediments, geosynclines and orogeny *in* A.G. Fischer, and S. Judson, eds., Petroleum and global tectonics: Princeton, New Jersey, Princeton University Press, p. 157-217.

Curry, W.H., III, 1979, Type section of the Wall Creek Sandstone Member of the Frontier Formation: Wyoming Geological Association, Earth Science Bulletin, v. 12, n. 2, p. 49-57.

Dickinson, W.R., 1976, Plate tectonic evolution: AAPG Continuing Education Course Note Series 1, p. 46.

Eicher, D.L., 1962, Biostratigraphy of the Thermopolis, Muddy, and Shell Creek Formations, *in* R.L. Enyert, and W.H. Curry, III, eds., Symposium on Early Cretaceous rocks of Wyoming and adjacent areas: Wyoming Geological Association 17th Annual Field Conference Guidebook, p. 72-93.

———, 1965, Foraminifers and biostratigraphy of the Graneros Shale: Journal of Paleontology, v. 39, n. 5, p. 875-909.

———, 1969, Paleobathymetry of Cretaceous Greenhorn sea in eastern Colorado: AAPG Bulletin, v. 53, n. 5, p. 1075-1090.

——— and P. Worstell, 1970, Cenomanian and Turonian foraminifera from the Great Plains, United States: Micropaleontology, v. 16, n. 3, p. 269-324.

Fisk, H.N., 1947, Fine-grained alluvial deposits and their effects on Mississippi River activity: Vicksburg, Mississippi, U.S. Corps of Engineers, Waterway Experiment Station, 98 p.

Fouch, T.D., 1983, Patterns of synorogenic sedimentation in Upper Cretaceous rocks of central and northeastern Utah, *in* M. Reynolds and E. Dolly, eds., Mesozoic paleogeography of west-central United States: Denver, Colorado, Society of Economic Paleontologists and Mineralogists, Rocky Mountain Section Special Publication, p. 305-336.

Hancock, J.M., 1975, The sequence of facies in the Upper Cretaceous of northern Europe compared with that in the Western Interior, *in* W.G.C. Caldwell, ed., The Cretaceous System in the Western Interior of North America: Geological Association of Canada Special Paper No. 13, p. 82-118.

——— and E.G. Kauffman, 1979, The great transgressions of the Late Cretaceous: Geological Society of London Journal, v. 136, p. 175-186.

Hattin, D.E., 1962, Stratigraphy of the Carlile Shale (Upper Cretaceous) in Kansas: Kansas Geological Survey Bulletin, n. 156, 155 p.

———, 1965, Stratigraphy of the Graneros Shale (Upper Cretaceous) in central Kansas: Kansas Geological Survey Bulletin, n. 178, 83 p.

———, 1971, Widespread, synchronously deposited, burrow-mottled limestone beds in Greenhorn Limestone (Upper Cretaceous) of Kansas and southeastern Colorado: AAPG Bulletin, v. 55, n. 3, p. 412-431.

———, 1975a, Stratigraphy and depositional environment of Greenhorn Limestone (Upper Cretaceous) of Kansas: Kansas Geological Survey Bulletin, n. 209, 128 p.

———, 1975b, Stratigraphic study of the Carlile-Niobrara (Upper Cretaceous) unconformity in Kansas and northeastern Nebraska: Geological Association of Canada Special Paper No. 13, p. 195-210.

———, 1977, Upper Cretaceous stratigraphy, paleontology, and paleocology of western Kansas: The Mountain Geologist, v. 14, n. 3 and 4, p. 175-218.

Haun, J.D., 1963, Stratigraphy of Dakota Group and relationship to petroleum occurrence, northern Denver basin, *in* P.J. Katich, and D.W. Bolyard, eds., Geology of the northern Denver basin and adjacent uplifts: Rocky Mountain Association of Geologists Guidebook, p. 119-134.

Hook, S.C., and W.A. Cobban, 1977, *Pycnodonte newberryi* (Stanton)--common guide fossil in Upper Cretaceous of New Mexico: New Mexico State Bureau of Mines and Mineral Resources Annual Report, p. 48-54.

———, ———, 1979, *Prionocyclus novimexicanus* (Marcou)--common Upper Cretaceous guide fossil in New Mexico: New Mexico State Bureau of Mines and Mineral Resources Annual Report, p. 35-42.

———, ———, 1980, Some guide fossils in Upper Cretaceous Juana Lopez member of Mancos and Carlile shales, New Mexico: New Mexico State Bureau of Mines and Mineral Resources Annual Report, p. 38-49.

———, ———, 1981a, *Lopha sannionis* (White)--common Upper Cretaceous guide fossil in New Mexico: New Mexico State Bureau of Mines and Mineral Resources Annual Report, p. 52-56.

———, ———, 1981b, Late Greenhorn (mid-Cretaceous) discontinuity surfaces, southwest New Mexico, *in* S.C. Hook, compiler, Contributions to mid-Cretaceous paleontology and stratigraphy of New Mexico: New Mexico State Bureau of Mines and Mineral Resources, Circular 180, p. 5-36.

Jeletzky, J.A., 1978, Causes of Cretaceous oscillations of sea level in western and arctic Canada and some general geotectonic implications: Geological Survey of Canada, Paper 77-18, 38 p.

Jordan, T.E., 1981, Thrust loads and foreland basin evolution, Cretaceous, western United States: AAPG Bulletin, v. 65, p. 2506-2520.

Kauffman, E.G., 1977, Upper Cretaceous cyclothems, biotas, and environments, Rock Canyon anticline, Pueblo, Colorado: The Mountain Geologist, v. 14, n. 3 and 4, p. 129-152.

Kent, H.C., 1967, Microfossils from the Niobrara Formation (Cretaceous) and equivalent strata in northern and western Colorado: Journal of Paleontology, v. 41, n. 6, p. 1433-1456.

Lanphere, M.A., and D.L. Jones, 1978, Cretaceous time scale from North America, in G.V. Cohee, and M.F. Glaessner, eds., The geologic time scale: AAPG Studies in Geology, n. 6, p. 259-268.

Lindstrom, L.J., 1979, Stratigraphy of the South Platte Formation (Lower Cretaceous), Eldorado Springs to Golden, Colorado, and channel sandstone distribution of J Member: Master's thesis, Golden, Colorado, Colorado School of Mines, 142 p.

MacKenzie, D.B., 1965, Depositional environments of Muddy Sandstone, Western Denver basin, Colorado: AAPG Bulletin, v. 49, p. 186-206.

———, 1971, Post-Lytle Dakota Group on west flank of Denver basin, Colorado: The Mountain Geologist, v. 8, n. 3, p. 91-131.

MacMillan, L.T., 1974, Stratigraphy of the South Platte Formation (Lower Cretaceous), Morrison-Weaver Gulch area, Jefferson County, Colorado: Golden, Colorado, Master's thesis, Colorado School of Mines, 131 p.

Matuszczak, R.A., 1976, Wattenberg Field: a review, in R.C. Epis and R.J. Weimer, eds., Studies in Colorado field geology: Golden, Colorado, Colorado School of Mines Professional Contribution 8, p. 275-279.

McGookey, D.P., 1972, Cretaceous system, in W.W. Mallory, ed., Geologic atlas Rocky Mountain region: Rocky Mountain Association of Geologists Special Publication, p. 190-228.

Merewether, E.A., and W.A. Cobban, 1972, Unconformities within the Frontier Formation, northwestern Carbon County, Wyoming: U.S. Geological Survey Professional Paper 800-D, p. D57-D66.

———, ———, 1973, Stratigraphic sections of the Upper Cretaceous Frontier Formation near Casper and Douglas, Wyoming: Wyoming Geological Association Earth Science Bulletin, v. 6, n. 4, p. 38-39.

———, ——— and C.W. Spencer, 1976, The Upper Cretaceous Frontier Formation in the Kaycee-Tisdale mountain area, Johnson County, Wyoming: Wyoming Geological Association 28th Annual Field Conference Guidebook, p. 33-44.

———, ——— and E.T. Cavanaugh, 1979, Frontier Formation and equivalent rocks in eastern Wyoming: The Mountain Geologist, v. 16, n. 3, p. 67-101.

Merriam, D.F., 1957, Subsurface correlation and stratigraphic relation of rocks of Mesozoic age in Kansas: Kansas Geological Survey, Oil and Gas Investigations n. 14, 25 p.

———, 1963, The geologic history of Kansas: Kansas Geological Survey Bulletin 162, 309 p.

Nelson, H.F., and E.E. Bray, 1970, Stratigraphy and history of the Holocene sediment in the Sabine-High Island area, Gulf of Mexico, in J.P. Morgan and R.H. Shaver, eds., Deltaic sedimentation modern and ancient: Society of Economic Paleontologists and Mineralogists Special Publication, n. 15, p. 48-77.

Obradovich, J.D., and W.A. Cobban, 1975, A time scale for the Late Cretaceous of the Western Interior of North America, in W.G.A. Caldwell, ed., The Cretaceous System in the Western Interior of North America: Geological

Association of Canada Special Paper, n. 13, p. 31-54.

Peterson, W.L., and S.D. Janes, 1978, A refined interpretation of depositional environments of Wattenberg Field, Colorado, in J.D. Pruit and P.E. Coffin, eds., Rocky Mountain Association of Geologists Symposium, Energy Resources of the Denver Basin: Denver, Colorado, Rocky Mountain Association of Geologists, p. 141-147.

Pinel, M.J., 1977, Stratigraphy of the upper Carlile and lower Niobrara formations (Upper Cretaceous), Fremont and Pueblo counties, Colorado: Golden, Colorado, Master's thesis, Colorado School of Mines, 111 p.

Poleschook, D., Jr., 1978, Stratigraphy and channel discrimination, J Sandstone, Lower Cretaceous group, south and west of Denver, Colorado: Golden, Colorado, Master's thesis, Colorado School of Mines, 226 p.

Reeside, J.B., Jr., 1944, Map showing thickness and general character of the Cretaceous deposits in the Western Interior of the United States: U.S. Geological Survey, Oil and Gas Investigations Map OM-10.

———, 1957, Paleoecology of the Cretaceous seas of the Western Interior of the United States: Geological Society of America Memoir 67, p. 505-542.

Royse, F., Jr., M.A. Warner, and D.L. Reese, 1975, Thrust belt structural geometry and related stratigraphic problems, Wyoming-Idaho-northern Utah, in D.W. Bolyard, eds., Deep drilling frontiers of the central Rocky Mountains: Denver, Colorado, Rocky Mountain Association of Geologists, p. 41-55.

Scott, G.R., and W.A. Cobban, 1964, Stratigraphy of the Niobrara Formation at Pueblo, Colorado: U.S. Geological Survey Professional Paper 454-L, p. L1-L30.

Simmons, K.A., 1967, A primer on "serpentine plugs" in south Texas, in W.G. Ellis, ed., Contributions to the geology of south Texas: San Antonio, Texas, South Texas Geological Society, v. 7, n. 2, p. 125-132.

Sonnenberg, S.A., and R.J. Weimer, 1981, Tectonics, sedimentation, and petroleum potential, northern Denver basin, Colorado, Wyoming, and Nebraska: Colorado School of Mines Quarterly, v. 76, n. 2, 45 p.

Suryanto, Untung, 1979, Stratigraphy and petroleum geology of the J Sandstone in portions of Boulder, Larimer, and Weld Counties, Colorado: Golden, Colorado, Master's thesis, Colorado School of Mines, 173 p.

Vail, P.R., R.M. Mitchum, Jr., and S. Thompson, III, 1977, Seismic stratigraphy and global sea-level changes, part 3: AAPG Memoir 26, p. 63-82.

Van Hinte, J.E., 1976, A Cretaceous time scale: AAPG Bulletin, v. 60, n. 4, p. 498-516.

Waage, K.M., 1953, Dakota group in northern Front Range foothills, Colorado: U.S. Geological Survey Professional Paper 274-B, p. 15-51.

Weimer, R.J., 1962, Late Jurassic and Early Cretaceous correlations, south-central Wyoming and northwestern Colorado, in R.L. Enyert, and W.H. Curry, III, eds., Symposium on Early Cretaceous rocks in Wyoming and adjacent areas: Wyoming Geological Association 17th Annual Field Conference Guidebook, p. 124-130.

———, 1978, Influence of transcontinental arch on Cretaceous marine sedimentation: a preliminary report, in J.D. Pruit, and P.E. Coffin, eds., Energy resources of the Denver Basin: Denver, Colorado, Rocky Mountain Association of Geologists Symposium, p. 211-222.

———, 1980, Recurrent movement on basement faults, a tectonic style for Colorado and adjacent areas, *in* H.C. Kent, and K.W. Porter, eds., Colorado geology: Denver, Colorado, Rocky Mountain Association of Geologists Symposium, p. 23-35.

———, 1981, Relation of unconformities, tectonics, and sea-level changes, Cretaceous of Western Interior, United States and Canada (abs): AAPG Bulletin, v. 65, n. 5, p. 1006.

———, 1982, Tectonic influence on sedimentation, Early Cretaceous, east flank, Powder River basin, Wyoming and South Dakota: Colorado School of Mines Quarterly, v. 73, n. 4.

———, 1983, Relation of unconformities, tectonics, and sea-level changes, Cretaceous of Denver basin and adjacent area, *in* M. Reynolds and E. Dolly, eds., Mesozoic paleogeography of west-central United States: Denver, Colorado, Society of Economic Paleontologists and Mineralogists, Rocky Mountain Section Special Publication, p. 359-376.

———, and C.B. Land, Jr., 1974, Field guide to Dakota Group (Cretaceous) stratigraphy Golden-Morrison area, Colorado: The Mountain Geologist, v. 9, n. 2 and 3, p. 241-267.

——— and S.A. Sonnenberg, 1982, Wattenberg field, paleostructure-stratigraphic trap, Denver basin, Colorado: Oil and Gas Journal, v. 80, n. 12, p. 204-210.

——— et al, 1982, Tectonic influence on sedimentation, Early Cretaceous, east flank, Powder River basin, Wyoming and South Dakota: Colorado School of Mines Quarterly, v. 73, n. 4.

Outcrop Features and Origin of Basin Margin Unconformities in the Lower Chesapeake Group (Miocene), Atlantic Coastal Plain

Susan M. Kidwell
University of Arizona
Tucson, Arizona

This paper describes unconformable stratigraphic relations within basin-margin deposits from an outcrop rather than seismic perspective, emphasizing the physical characteristics, stratigraphic context, age, origin, and usefulness of unconformities in stratigraphic subdivision and correlation. Stratigraphic relations within the outcropping Miocene of the Salisbury Embayment, middle Atlantic Coastal Plain, are more complex than previously described. The Calvert (Plum Point Member) and Choptank formations are subdivided by a series of erosional surfaces into six depositional sequences traceable over a 9,000-sq-km (3,475-sq-mi) study area in Maryland and Virginia. The disconformities take the form of burrowed firmgrounds in outcrop, but exhibit up to 14 m (46 ft) of topographic relief locally and represent transgressive and regressive ravinement surfaces. One of the surfaces records sediment starvation in a distal marine environment. Internal facies relations within the depositional sequences are complex but basically cyclic, consisting usually of a basal condensed shell or bone deposit formed under prolonged conditions of reduced net sedimentation and grading upward into less fossiliferous and siltier facies in regressive sequence. Each sequence consists of two or more of the original lithologic zones of Shattuck (1904).

Diatom biostratigraphic data indicate that the disconformities are not measurably diachronous within the study area and represent less than one biozone except where erosion has enlarged the vacuity; they thus provide a basis for fine-scale chronostratigraphic subdivision and correlation within the Miocene outcrop belt. Diatom data published elsewhere identify a major, 2.5 m.y.-long hiatus within the Fairhaven Member of the Calvert Formation that encompasses the Burdigalian-Langhian stage boundary, where Vail and Hardenbol (1979) have identified a minor interregional unconformity in seismic profiles (16.5 m.y. event). The disconformity at the base of the Plum Point Member of the Calvert Formation corresponds in age to another minor reflector at the Langhian-Serravallian boundary (15.5 m.y. event), and a disconformity between the Choptank and St. Marys formations may correspond to Vail and Hardenbol's (1979) mid-Serravallian reflector (13.0 m.y. event). Other disconformities within the studied section, including the disconformable contact of the Calvert and Choptank formations, are expressions of stratigraphic complexity relating to transgressive-regressive migrations of marginal marine and open marine strandlines within the basin margin record.

INTRODUCTION

Although seismic stratigraphic interpretation has evolved into a major exploration tool for industry and has become widely appreciated since publication of the AAPG Memoir 26 (Payton, 1977), several aspects of the method and its application remain controversial. The most fundamental of these involves the accuracy of age assignments and the chronostratigraphic significance of unconformities capable of generating seismic reflectors; also the applicability of the approach to the interpretation of outcrop scale features. This paper describes the outcrop features and origins of a series of disconformities in the Miocene Calvert and Choptank formations, Atlantic Coastal Plain, as a means of evaluating the precepts of seismic stratigraphic interpretation under conditions of high resolution. Biostratigraphic data permit the comparison of these basin-margin events of erosion and non-deposition with seismic reflectors of proposed interregional extent.

The Calvert and Choptank formations, comprising the lowermost strata of the Neogene Chesapeake Group, are advantageous units for such an investigation. These relatively undeformed and unlithified marine terrigenous silts and sands crop out in a series of spectacular cliffs and lesser bluffs in an arcuate belt from the Delaware border to Richmond, Virginia (Figure 1). The very gentle regional dip of less than 1° and the downdip alignment of shoreline exposures permit direct lateral tracing of beds over great distances. These exposures include the historic Calvert Cliffs, forming the Chesapeake Bay shoreline of Calvert County, Maryland, and the Westmoreland and Nomini Cliffs along the Virginia shore of the Potomac River. Miocene strata rest

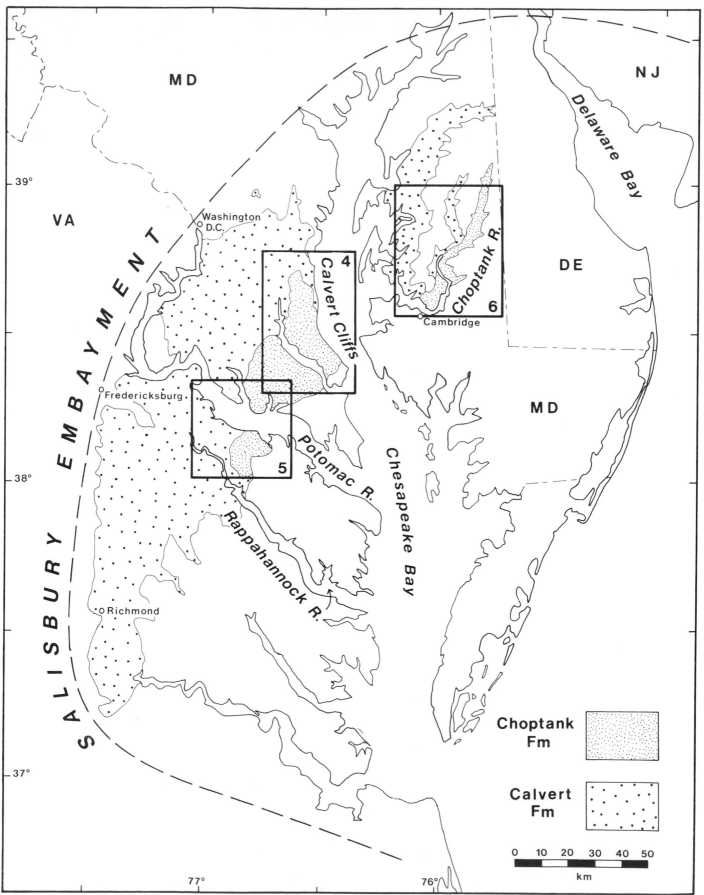

upon and locally onlap early Tertiary units along the western edge of the Salisbury Embayment, a landward extension of the Baltimore Canyon Trough bounded on the north by the South Jersey High and on the south by the Norfolk High (Poag, 1979)(Figure 1).

The Maryland Miocene is famous for its diverse, well-preserved, and abundant fauna of molluscs, echinoids, bryozoans, cirripeds, brachiopods, corals, reptiles, fish, and marine mammals (Glenn, 1904; Martin, 1904; Ulrich and Bassler, 1904; Gazin and Collins, 1950, Kellogg, 1965-1969, 1969; Gernant, 1970, 1971). The distribution of benthic macroinvertebrates is strongly controlled by lithofacies, but several micropaleontologic zonations permit correlation within the basin and with standard sections (Abbott, 1978; Andrews, 1978; Gibson, 1962; Malkin, 1953). Correlations indicate that the two formations span the Burdigalian, Langhian, and Serravallian stages, and thus have potential to record the 16.5, 15.5, and 13.0 m.y. eustatic events inferred by Vail and others (Vail and Hardenbol, 1979; Vail et al., 1980) from offshore seismic records.

Because the study was designed to analyse the stratigraphic context and origin of molluscan shell beds, the study area was restricted to the northern part of the outcrop belt where fossil preservation and exposures are best. The study therefore concentrated on a 9,000 sq km (3,475 sq mi) area from the Maryland-Delaware border to the Rappanhannock River, Virginia (Figure 1). The fossil-poor Fairhaven Member of the Calvert Formation was excluded throughout its outcrop belt. Most of the 194 stratigraphic sections measured for the study are located in Figures 4-6 (locality register available upon request). Informal field units were distinguished on the basis of bioclastic fabric (orientation, close-packing, size distribution and extent of fragmentation) and faunal dominants as well as sedimentary textures, physical and biogenic structures, bedding, and color. These units ranged in thickness from tens of centimeters to a few meters. Field descriptions were supplemented by wet-sieving of approximately 200 sediment samples at 0.5 phi (ϕ) intervals to determine modal sand size and percent admixed mud (less than $1/16$ mm diameter) for the terrigenous fraction. Ninety large (8 to 10 kg, or 18 to 22 lb) samples collected from the most fossiliferous beds yielded quantitative taphonomic and paleoecologic data on macroinvertebrates, and about 100 sediment samples were analyzed for biostratigraphically significant diatom species by W.H. Abbott (South Carolina Geological Survey). All of these data are tabulated in Kidwell (1982a).

UNCONFORMITIES AND DEPOSITIONAL SEQUENCES

Previous Investigations

Despite a fairly stable nomenclature, stratigraphic relations within the Calvert and Choptank formations have been the subject of continued controversy (Figure 2). Shattuck (1904) interpreted the section as internally conformable with the exception of the Calvert-Choptank formational boundary, which he recognized as unconformable owing to downdip beveling of the Calvert Formation and onlapping relations of Choptank beds. Subsequent workers have employed Shattuck's (1904) subdivision but many have found the formational contact difficult to identify and Shattuck's evidence for an unconformity equivocal (Dryden, 1930; Gibson, 1962; Gernant, 1970, Blackwelder and Ward, 1976). Dryden (1936), Schoonover (1941), and Blackwelder (1981) considered the contact to be conformable. Unconformities have also been recognized between the Fairhaven and Plum Point members of the Calvert Formation (Dryden, 1930, 1936; Gernant, 1971; Blackwelder and Ward, 1976; Andrews, 1978; Blackwelder, 1981), and between the Choptank and St. Marys formations (Gibson, 1971; Gernant, 1970, 1971; Blackwelder and Ward, 1976; Blackwelder, 1981). Newell and Rader (1982) reinterpreted all of these breaks within the lower part of the Chesapeake Group as minor, diastemic features and thus describe the Calvert-Choptank-St. Marys sequences as entirely conformable.

Stratigraphic Methods and Terminology

My initial hypothesis regarding stratigraphic relations was based on a belief that Shattuck's (1904) zones were not lithologically homogeneous layers as traditionally conceived, each recording a unique time interval, but instead were facies arranged within a largely conformable sequence. However, this was largely rejected when field examination revealed a series of erosional surfaces subdividing the section into groups of his zones. These disconformity-bound units are referred to here as depositional sequences (Figure 2). Facies relations as evidenced by lateral gradation, intertonguing, and marker-bed tie-ins could be demonstrated only within single zones and among those zones grouped within a single depositional sequence.

Qualitatively, the Calvert Formation appears to be a siltier and less fossiliferous unit than the Choptank Formation, and quantitative analysis bears this out (Kidwell, 1982a, p. 136). However, each formation contains a similar range of lithologies from silty clays to well-sorted fine sands with 0 to 70% shell carbonate by volume as a distinct coarse mode (Figure 3). The formations are thus distinguished primarily on the basis of proportional representation of sediment types.

Both the Plum Point Member of the Calvert Formation and the Choptank Formation exhibit strong cyclic trends in sediment types, with clean shell-rich sands alternating with shell-poor silty sands, sandy silts, and silty clays (Figure 3). For informal subdivision, an effective solution is to denote four laterally persistent and visually dramatic shell beds and a fifth bone bed as key beds (Figures 2 and 3). The five key beds are assigned geographic names to minimize confusion with existing nomenclatural schemes using numbers and

Figure 1: Outcrop belt of the Calvert and Choptank formations in the Salisbury Embayment, Atlantic Coastal Plain. The study was restricted to the Plum Point Member of the Calvert Formation and Choptank Formation north of the Rappahannock River. Boxes mark location of Figures 4, 5, and 6. Geology based on Cleaves et al (1968) and Calver and Hobbs (1963).

Lithologic Zones Shattuck 1904	Harris 1893	Shattuck 1904	Dryden 1930	Dryden 1936	Spangler & Peterson 1950	Gibson 1971	Gernant 1970, 1971	Blackwelder & Ward 1976 / Blackwelder 1981	Newell & Rader 1982	Kidwell, This Report (informal units)
24		St.Marys Fm	(not studied)	St.Marys Fm	St. Marys Fm	St. Marys Fm	St. Marys	St.Marys Fm	St. Marys Fm	(not studied)
23	St. Marys Fauna (g)									
22								Little Cove Point unit		SM-0
21										
20		Choptank Fm	Choptank Fm		Choptank Fm	Choptank Fm	Camp Conoy Mbr	Choptank Fm	Choptank Fm	unnamed / CT-1
19	Jones Wharf Fauna (f)						Boston Cliffs Mbr			Boston Cliffs
18							St.Leonard Mbr			Mytilus
17	(e)						Drumcliff Mbr			Drumcliff / CT-0
16							Calvert Beach Mbr			Turritella-Pandora
15	(d)	Calvert Fm / Plum Point Mbr	Calvert Fm / Plum Point Mbr	Choptank Fm	Calvert Fm / Plum Point Mbr	Calvert Fm / Plum Point Mbr	Plum Point Mbr	Calvert Fm / Plum Point Mbr	Calvert Fm / Plum Point Mbr	Kenwood Beach
14										PP-3
13	(c)									Glossus-Chione
12				Choptank Fm						Parker Creek / PP-2
11	Plum Point Fauna (b)									barren
10										Camp Roosevelt / PP-1
9							Plum Point Mbr			
8										
7										Ostrea-Corbula
6										
5										
4	(a)			Calvert Fm						PP-0
3		Fairhaven Mbr	Fairhaven Mbr		Fairhaven Mbr	Fairhaven Mbr	Fairhaven Mbr	Fairhaven Mbr	Fairhaven Mbr	(not studied)
2										
1										

Figure 2: Nomenclature and stratigraphic relations of the Maryland Miocene. Column at far right summarizes informal lithostratigraphic nomenclature used in this paper, consisting of highly fossiliferous key beds with geographic names and sparsely fossiliferous silty intervals with taxonomic (generic) names. Disconformities in the section are numbered sequentially within each formal unit (PP = Plum Point Member of Calvert Formation; CT = Choptank Formation; SM = St. Marys Formation). The informal alpha-numeric name of each disconformity also denotes the immediately overlying depositional sequence, all but one of which consist of a key bed and its succeeding interval. Lithologic column not to thickness scale.

letters (Harris, 1893; Shattuck, 1904; Dryden, 1930, 1936; Abbott, 1978; Andrews, 1978). The geographic names of the two key beds in the Choptank Formation, coinciding approximately with Zones 17 and 19 of Shattuck (1904), are the same as those used by Gernant (1970) when he elevated these units to formal member status; these are the Drumcliff and Boston Cliffs shell beds. Key beds in the Plum Point Member of the Calvert Formation, corresponding closely to Zones 10, 12, and 14 of Shattuck (1904), are assigned new geographic names; these are the Camp Roosevelt shell bed, Parker Creek bone bed, and Kenwood Beach shell bed. These units are not proposed as members or submembers; such formalization would unnecessarily fragment the firmly established nomenclature of the Plum Point Member, whose lithologic heterogeneity is already accepted by regional workers.

Shell-poor strata lying between key beds are assigned to informal intervals. These are named for abundant and distinctive macroinvertebrate genera in Calvert Cliffs exposures in order to avoid redundant lithologic terms among sequences. Geographic names were rejected in order to avoid confusion with key-bed nomenclature. In ascending order, the Plum Point and Choptank intervals are designated: *Ostrea-Corbula*, barren, *Glossus-Chione*, *Turritella-Pandora*, and *Mytilus* intervals (Figures 2 and 3). The interval whose base is defined by the Boston Cliffs shell bed remains unnamed pending further investigation of the Choptank-St. Marys transition.

Shattuck's (1904) scheme of informal zones, utilized by most previous workers, is not used in this report for several reasons. Most importantly, because the units recognized in this field study do not correspond exactly to Shattuck's zones in all situations. In addition, previous workers have had difficulty identifying Shattuck's zones in localities away from his 12 reference sections as well as in some reference sections (Dryden, 1930; Schoonover, 1941; Blackwelder and Ward, 1976); and finally, the zonal scheme carries the connotation of a layer-cake series of laterally invariant strata inconsistent with the findings of this study.

Facies are designated within intervals exhibiting notewor-

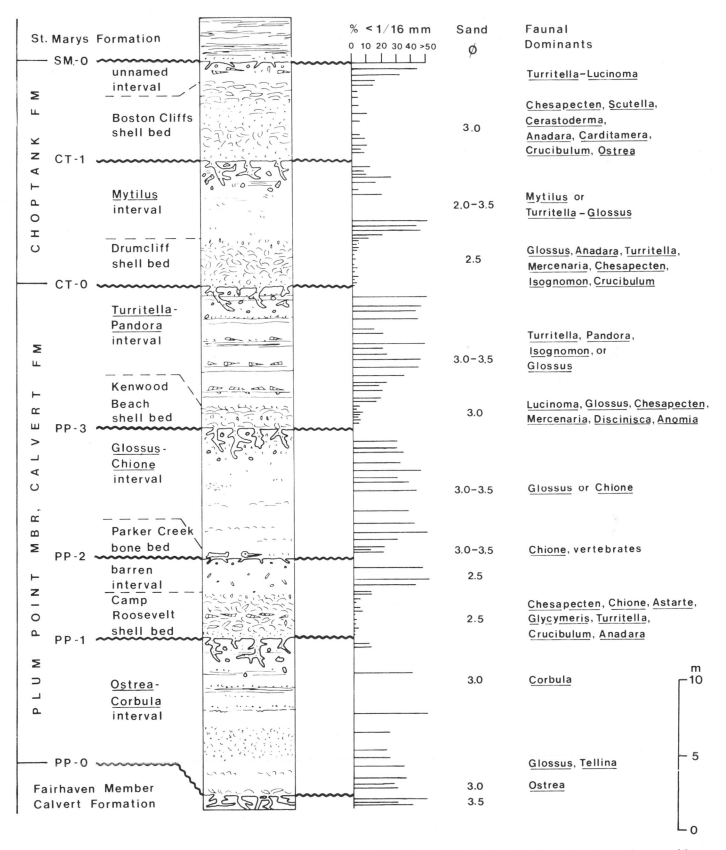

Figure 3: Composite stratigraphic column of the Plum Point Member and the Choptank Formation. The disconformities are burrowed firm-grounds in individual exposures and, with the exception of the PP-0 surface, are overlain by a richly fossiliferous clean sand that grades into much less fossiliferous and usually silty sediments. Only the most abundant taxa in faunal assemblages are noted; key shell beds usually contain 40 or more species. Sediment and faunal data are tabulated in Kidwell (1982a).

thy lateral or vertical variation in sedimentary textures, structures, or fossil content (abundance and generic composition). To avoid unwieldy lithologic descriptors and to circumvent the nomenclature problem of recurrent lithologic types, facies are named for abundant invertebrate genera. For example, the *Turritella-Pandora* interval contains a *Turritella*-dominated facies and a *Pandora*-dominated facies in the Calvert Cliffs, an *Isognomon* facies in the Patuxent River area, and a *Glossus* facies along the Potomac River. This nomenclatural solution transforms neither the intervals nor their facies into biostratigraphic units or biofacies: lithologic as well as paleontologic features are used to define the units, and there is no chronostratigraphic intent.

The erosional surfaces that group key beds and intervals into depositional sequences are numbered sequentially within the Plum Point Member (PP-0, PP-1, PP-2, PP-3) and the Choptank Formation (CT-0, CT-1) (Figures 2 and 3). The seventh surface truncates the Choptank Formation and is provisionally labeled as the basal disconformity of the St. Marys Formation (SM-0).

Depositional sequences are denoted by their basal disconformity and, with the exception of the PP-0 sequence, each consists of a basal key bed overlain by a less fossiliferous interval. The sequences are thus conspicuously cyclic on a scale of approximately 10 m (33 ft), both fining and losing shell carbonate upward. A few coarsen upward in the uppermost meter or so (Figure 3). Despite this basic cyclic pattern, stratigraphic relations within depositional sequences are usually complex in detail owing to pinchout, intertonguing, and internal truncation of facies and the erosional relief of the bounding disconformities.

In the following descriptions of depositional sequences, paleoenvironmental interpretations are based on sedimentary structures and paleoecology. Structures alone are used to estimate water depths in terms of wave base and tidal exposure. The sublittoral (subtidal) zone is divided into very shallow (above fair-weather wave base), shallow (between fair-weather and storm-wave bases), and intermediate-to-deep conditions (below storm-wave base). Absolute water depths of wave base depend on geomorphology and oceanographic facing, but by analogy with modern western Atlantic environments, fair-weather wave base can be taken as approximately 20 m (66 ft) in shelf environments and as shallow as 5 m (16 ft) in marginal marine environments (Howard and Reineck, 1981). Invertebrate paleoecology is used to infer mass properties of the substratum and water salinity. Paleoenvironmental results are presented in detail in Kidwell (1982a; see also Gernant, 1970, 1971 for comparison).

PP-0 Disconformity and Depositional Sequence

The PP-0 disconformity is a burrowed contact of clean brownish sand resting on very silty dark gray sand. It marks the boundary between the Fairhaven and Plum Point members of the Calvert Formation (Figure 3). It can be traced for 5 km (16 ft) in the northern Calvert Cliffs and also along the Patuxent River (Figures 7 and 8). Evidence for the disconformity includes truncation of Fairhaven beds, lithologic change and presence of a firmground along the contact, topographic irregularity of the surface, distribution of overlying facies with respect to the surface, and coincidence with

biostratigraphic zone boundaries (see next section).

Beneath the disconformity, the Fairhaven Member is a tightly consolidated, dark gray, very silty, very fine sand containing a sparse and low-diversity molluscan fauna (Figure 3). Physical sedimentary structures are absent; sharply defined spiral burrows (*Gyrolithes*; Dryden, 1933; Gernant, 1972) are notable in the otherwise mottled-to-homogeneously bioturbated sediment[1]. The sedimentary texture of the Fairhaven Member changes slightly toward the south in the Calvert Cliffs suggesting bedding (Dryden, 1930; Kidwell, 1982a). Large diameter (3 to 5 cm, or 1.2 to 2 in) branching *Thalassinoides* burrow systems extend from the PP-0 surface 1 m (3.3 ft) or more into the Fairhaven Member and indicate firmground conditions on the PP-0 sea floor (*Glossifungites* trace assemblage of Frey and Seilacher, 1980). These burrows are filled with the basal sand of the Plum Point Member (Zone 4 of Shattuck, 1904), a brownish, clean to slightly-silty fine sand containing abundant oysters (*Ostrea percrassa*). The sand is pervasively mottled and marked by spreiten; small-scale cross sets, usually truncated or disrupted by burrows, are rare.

The PP-0 firmground exhibits 2 m (6.6 ft) of topographic relief within the northern Calvert Cliffs, rising in the section from near sea level at Chesapeake Beach (section 22, Figure 7) to maximum elevation at Locust Grove Beach (section 13), a distance of 3 km (1.9 mi). South of Locust Grove Beach, the PP-0 surface exhibits a normal, southerly dip of 1 m/km (5 ft/mi) and disappears below beach level near Willows Beach (section 16). At Fairhaven, 8 km (5 mi) north of the Calvert Cliffs (Figure 4), the PP-0 surface lies at least 10 m (33 ft) above mean sea level, indicating considerable topographic relief on the surface.

Stratigraphic relations of the three facies of the PP-0 sequence reflect paleotopographic relief on the PP-0 surface in the Calvert Cliffs. In the northern Cliffs where the PP-0 surface exhibits a reversed dip, the basal sand of the PP-0 sequence (*Ostrea* facies; Zone 4 of Shattuck, 1904) also has a reversed dip, and eventually pinches out against the rising PP-0 surface (Figure 7). A muddier fine-sand facies (species-rich facies, Figures 3 and 7; Zone 5 of Shattuck, 1904) that rests on the *Ostrea* facies also pinches out to the south against the PP-0 surface. The species-rich facies is additionally thinned toward the south by beveling, evidenced by the truncation of shell stringers within the facies. South of Locust Grove Beach (section 13) where the PP-0 surface resumes a normal southward dip, the PP-0 sequence consists only of the *Corbula* facies, a thinly to thickly bedded sand and muddy sand characterized by the small bivalve *Corbula elevata* (Figures 3 and 7; Zones 6 through 9 of Shattuck, 1904). This facies dips regularly to the south from Chesapeake Beach to just north of Plum Point. Over this distance of about 8 km (5 mi), however, it is reduced from 8 to 4 m (26 to 13 ft) by internal thinning of sand beds and overall

---

[1]The term burrow denotes fully three-dimensional biogenic sedimentary structures with distinctive outlines. All other penetrative biogenic structures, including mottled and homogeneous fabrics, are described by the general term bioturbation. Burrowed contacts are thus well-defined but microtopographically complex surfaces; bioturbated contacts will be gradational.

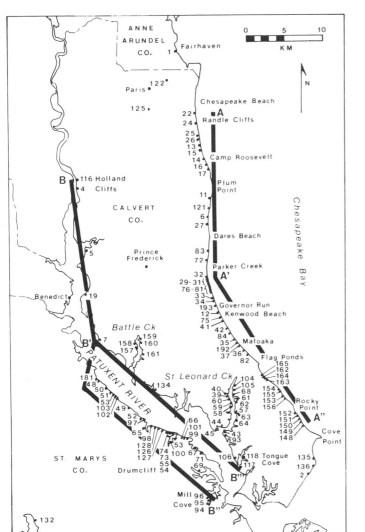

Figure 4: Index map of measured sections in the Calvert Cliffs and along the Patuxent River, Maryland. Cross section along line A - A'' is in Figure 7; cross sections along lines B - B'' and B' - B''' are in Figure 8.

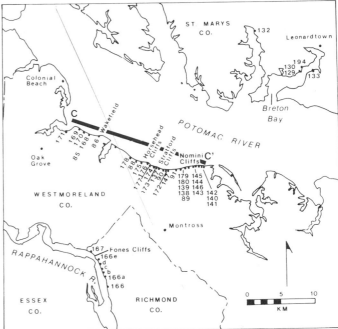

Figure 5: Index map of measured sections along the Potomac and Rappahannock rivers, Virginia, and Breton Bay, Maryland. Cross section along C - C' is in Figure 9.

nance of typically polyhaline genera such as Ostrea and its associates (Martesia, Mytilus), Corbula, and the trace fossil Gyrolithes found in the species-rich facies (see references cited in Kidwell, 1982a; Gernant, 1972). Sand-mud couplets and overall coarsening up within the Corbula facies might indicate a delta-influenced environment.

PP-1 Disconformity and Depositional Sequence

The PP-1 disconformity is a burrowed contact of massive clean shelly sand on bedded silty sand, and lies along the base of the Camp Roosevelt shell bed (Zone 10 of Shattuck, 1904), the lowest key bed of the Plum Point Member (Figure 3). The contact is exposed in the northern Calvert Cliffs and in the Holland Cliffs along the Patuxent River (Figures 7 and 8).

Like the PP-0 disconformity, large diameter (3 to 5 cm, or 1.2 to 2 in) the Thalassinoides burrows characterize the PP-1 surface. These penetrate the muddy sands of the underlying PP-0 sequence (Corbula facies) to a depth of 0.7 m (2.3 ft) at Chesapeake Beach (sections 22 and 24) and a depth of 2 m (6.6 ft) or more at Plum Point (section 17; Figures 4 and 7). The burrows are filled with the very shelly (to 40 percent by weight), clean fine sand of the Camp Roosevelt shell bed; shell material in the Camp Roosevelt shell bed is always more abundant and far more diverse in species than that of the PP-0 sequence, making the contact easy to recognize. Where the uppermost PP-0 sequence is very sandy (for example, at Holland Cliffs and at Chesapeake Beach), pods of diverse shells in the upper 2 m (6.6 ft) of the sequence impart a gradational appearance to the PP-1 contact. However, on closer inspection these pods are clearly part of Thalassinoides burrows originating at the PP-1 surface.

truncation by the PP-1 surface. The dotted contact in the Corbula facies just below the PP-1 surface in Figure 7 marks a shell stringer (Zone 9 of Shattuck, 1904) that demonstrates this downdip truncation of the PP-0 sequence.

The PP-0 disconformity is tentatively identified at Holland Cliffs (section 4, Figures 5 and 8) on the Patuxent River. There, it also has the form of a burrowed, sand on muddy sand contact and is inferred to dip relatively steeply to the south, comparable to the PP-0 surface from Fairhaven south to the Calvert Cliffs. The basal Ostrea facies of the PP-0 sequence at Holland Cliffs is a very clean, lithified fine sand and supports an oyster bioherm rather than isolated oyster specimens. The Corbula facies of the PP-0 sequence crops out further south near Benedict Bridge (section 19).

The PP-0 sequence records a variety of polyhaline environments from very shallow (Ostrea facies and upper part of the Corbula facies) to shallow (lower Corbula facies) and intermediate sublittoral water depths (species-rich facies). A freshwater influence is indicated by the numerical domi-

In the Calvert Cliffs, the PP-1 contact dips regularly to the south at about 1.6 m/km (8 ft/mi) for a distance of 8 km (5 mi). Although the contact does not exhibit any reversals in dip, the downdip termination of a thin *Corbula* shell horizon against the PP-1 surface (dotted line near top of *Corbula* facies in Figure 7) indicates beveling of at least the uppermost 2 m (6.6 ft) of the PP-0 sequence during formation of the PP-1 surface. At Holland Cliffs on the Patuxent River (section 4, Figures 4 and 8), the disconformity is marked by an undulatory scour surface developed on the indurated *Ostrea* facies (PP-0 sequence), and dips southward at 1.0 m/km (5 ft/mi). The PP-1 surface is inferred to steepen in dip to at least 3 m/km (15 ft/mi) in the Benedict Bridge (section 19) to Sandy Point area (section 7; Figures 4 and 8).

In all of its exposures, the PP-1 disconformity is overlain directly by the Camp Roosevelt shell bed, a clean fine sand containing a densely packed and diverse assemblage of whole and fragmental shells (Figure 3). The 3 m-thick (10 ft-thick) shell bed contains a microstratigraphic sequence of fossil assemblages characterized by both soft-bottom and shell-gravel invertebrate species (Kidwell, 1982a; Kidwell and Jablonski, 1983). In the Plum Point area of the Calvert Cliffs, the shell bed grades up into an interbedded facies of clean fine sand and muddy sand with thin shell layers and clay bands (transitional sand facies, Figure 7). Further north in the Camp Roosevelt and Chesapeake Beach area, this transitional sand facies is absent and the Camp Roosevelt shell bed is overlain immediately by a barren clay interval (Zone 11 of Shattuck, 1904) (Figure 3). Unlike other intervals of the Calvert and Choptank formations, the barren interval cannot be subdivided into facies but is a massive, blue-gray sandy clay throughout its exposed extent in the study area (to the Virginia shore of the Potomac River). With the exception of rare, poorly-preserved shell fragments and echinoid spines, its fauna consists only of traces: color and textural mottling; small-diameter (1 to 2 cm, or .4 to .8 in) unbranched burrows with clay fill; and 5 cm (2 in) diameter pods of clay packed with dark fecal pellets.

The PP-1 sequence records fully marine depositional conditions in shallow sublittoral (Camp Roosevelt shell gravel) and deep sublittoral water depths (barren interval). An open-shelf setting is indicated both by the fauna and the sheet-like geometry of the units.

PP-2 Unconformity and Depositional Sequence

The PP-2 surface is a burrowed contact along which the brownish, slightly muddy to clean fine sand of the Parker Creek bone bed (Zone 12 of Shattuck, 1904) is juxtaposed against the blue-gray sandy clay of the barren interval (Figure 3). The disconformity and overlying depositional sequence can be traced throughout the study area with the exception of the eastern shore of Maryland (Figures 7, 8, and 9).

Burrows in the PP-2 surface tend to be short (15 cm, or 6 in, maximum length), less than 2 cm (.8 in) in diameter, unlined simple tubes oriented either vertically or at a steep angle to bedding. The burrows are distinguished by the brownish, loosely-packed sand of the Parker Creek bone bed that fills them. This sand contains sparsely disseminated bone and teeth and local concentrations of poorly preserved

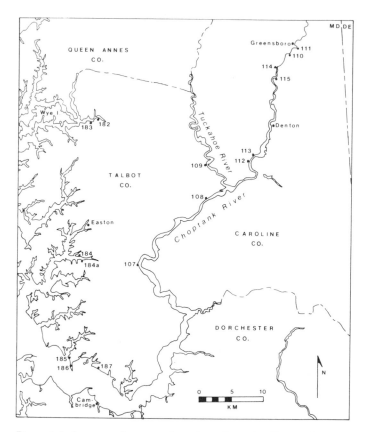

Figure 6: Index map of measured sections along the Choptank River and other tributaries, eastern shore of Maryland.

shell dominated by the bivalve *Chione parkeria*. Densely interpenetrating *Thalassionoides* burrows are developed in the PP-2 surface along the Potomac River (sections 169 and 171; Figures 5 and 9), and penetrate more than 2 m (6.6 ft) into the underlying barren interval. The PP-2 contact is gradational in extreme northern Calvert Cliffs outcrops (sections 25 and 26) owing to bioturbation.

The PP-2 surface dips at a uniform rate of 1 m/km (5 ft/mi) in the Calvert Cliffs, where it is most continuously exposed, and is probably best described as an omission or non-depositional surface inasmuch as it lacks evidence of sea-floor erosion. Channel-shaped features and sharp changes in dip are lacking, and, because the underlying barren interval is massive, erosional beveling cannot be demonstrated. The only suggestion of erosional truncation is thinning of the PP-1 barren interval in the northern part of the Calvert Cliffs. Evidence for stratigraphic omission, either by sediment starvation or dynamic bypassing, includes the concentration of fossil vertebrate bones and teeth and the presence of authigenic glauconite in the winnowed quartz sand of the Parker Creek bone bed. In his detailed taphonomic analysis of the bone bed in the Potomac River exposures, Myrick (1979) concluded that the marine mammal assemblage of the bed is a condensed record of at least two ecologically distinct faunas, and resulted from reduced sedimentation over a period of thousands to tens-of-thousands of years.

The PP-2 depositional sequence consists of the thin (0.6 m, or 2 ft) Parker Creek bone bed and the lithologically het-

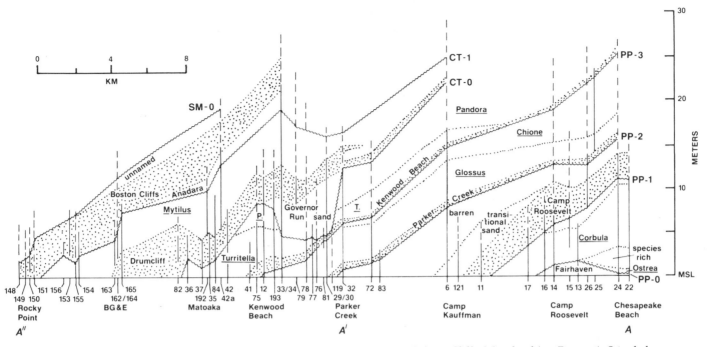

Figure 7: Cross section of the Plum Point Member and the Choptank Formation in Calvert Cliffs, Maryland (see Figure 4). Stippled pattern indicates key beds. White areas are silty, less fossiliferous intervals; taxonomic terms indicate facies names; vertical lines mark measured sections and are dashed where elevation data are estimated. Vertical exaggeration is 250:1. True dips range from 0.4 to 5 m/km (2 to 25 ft/mi); all are less than 1 degree.

erogeneous *Glossus-Chione* interval (Figure 3). The *Glossus-Chione* interval consists of a thickly- to very thickly-bedded series of muddy sands and sandy muds with sparse, relatively low-diversity invertebrate assemblages and can be traced throughout the study area (Figures 7, 8, and 9). It is equivalent to zone 13 and, in northern outcrops of the Calvert Cliffs, to both zone 13 and the lower part of zone 14 of Shattuck (1904). Two facies are recognized within the interval in the Calvert Cliffs. The *Glossus* facies consists of massive sandy mud and muddy sand with disarticulated specimens of the bivalve *Glossus fraterna* arranged in widely-spaced shell stringers. The stratigraphically higher and overlapping *Chione* facies consists of thickly- to very thickly-bedded muddy sand characterized by the small bivalve *Chione parkeria*. *Chione* specimens tend to be clumped within the sandier layers rather than uniformly disseminated, and common associates include the gastropod *Turritella plebeia*, *Glossus fraterna*, and the scallop *Chesapecten nefrens*. This stratigraphic pattern of a basal clayey facies and an upper sandier and shellier facies recurs in the few exposures of this interval along the Patuxent River, although shell preservation there is poor.

A third, sandier facies represents the *Glossus-Chione* interval in the Potomac region. This facies consists of non-cyclic alternations of bioturbated thickly- to very thickly-bedded sand, sandy clay, and clayey sand, and contains relatively abundant wood and vertebrate material. The facies takes its name from burrows attributable to ghost crabs (*Ocypode*; see Radwanski, 1977) found along some bedding planes in the uppermost part of the interval. Clay-rich beds often contain abundant fecal pellets within small burrows; surficial gypsum crystals and yellow sulfurous

crusts, and fetid odor of fresh cuttings indicate finely-disseminated pyrite.

The deep sublittoral water depths responsible for the barren interval (PP-1 sequence) persisted at least until the initial phases of Parker Creek bone bed accumulation. In addition to taphonomic and paleoecologic evidence within the bed itself, the context of the bone bed between the assuredly deep-water barren interval and the intermediate sublittoral *Glossus* facies also indicates a deep depositional environment more distal from source areas than any other Plum Point or Choptank unit. The shallow sublittoral *Chione* facies succeeds the *Glossus* facies everywhere but in the Potomac region, where the very shallow sublittoral to littoral *Ocypode* facies constitutes the upper PP-2 sequence.

PP-3 Disconformity and Depositional Sequence

The PP-3 surface is an erosional surface in the Calvert Cliffs, but is a simple bedding plane elsewhere in the study area. The shelly sand on silty sand contact is burrowed and marks the base of the Kenwood Beach shell bed, the stratigraphically highest key bed of the Calvert Formation (Zone 14 of Shattuck, 1904; Figures 3, 7, 8, and 9).

The surface dips southward with a uniform dip of 1.2 m/km (6 ft/mi) in the Calvert Cliffs. North of Parker Creek (section 32, Figures 4 and 7), burrows are unbranched, small diameter (less than 3 cm, or 1.2 in), and sharply defined vertical tubes extending less than 20 cm (8 in) into underlying strata. However, south of Parker Creek as far as Kenwood Beach where the PP-3 surface drops below beach level, burrows are larger in diameter, branched, and penetrate 1 m (3.3. ft) or more into the *Glossus-Chione* interval. These

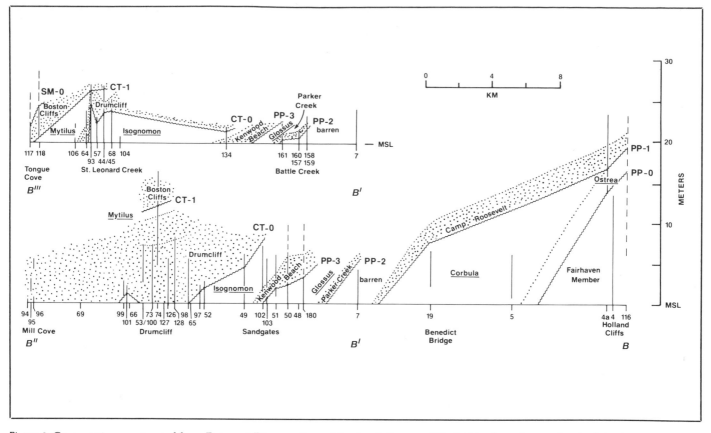

Figure 8: Cross section constructed from Patuxent River sections, Maryland. Segment B' - B'' is along the west bank of the river; segment B' - B''' describes the east bank (Figure 4). Conventions as in Figure 7.

Thalassinoides burrows are accompanied by *Gyrolithes* burrows, especially in the Kenwood Beach area. An erosional origin for the PP-3 surface is indicated by: (1) truncation of bedding and downdip beveling of the *Glossus-Chione* interval (Figure 7); and, (2) by the appearance in the base of the Kenwood Beach shell bed of specimens of *Chione parkeria* reworked from the PP-2 sequence, especially in the Parker Creek to Kenwood Beach area. Elsewhere in the study area, the PP-2 and PP-3 depositional sequences are entirely conformable and it is difficult to identify a single bedding surface as the basal PP-3 contact.

The PP-3 depositional sequence consists of the Kenwood Beach shell bed and the several facies of the *Turritella-Pandora* interval (Figure 3). In the northern Calvert Cliffs (Chesapeake Beach to Parker Creek; Figure 7), the Kenwood Beach shell bed is a thin (0.5 m, or 1.6 ft) densely-packed shell accumulation having a clean fine sand matrix. It is equivalent to only the upper 0.5 m (1.6 ft) of strata assigned by Shattuck (1904) to Zone 14. South of Parker Creek, the shell bed is thicker (to 2 m, or 6.6 ft) and contains interbeds of less fossiliferous, muddy fine sand. Burrowed discontinuities with firmground trace assemblages separate as many as four discrete shell layers within the bed. The Kenwood Beach shell bed is thicker (to 7 m, or 23 ft) in the Patuxent region (Figure 8) where it includes a greater number of shell layers interbedded with muddy sand. Identification of the bed is only tentative in the Potomac region because of poor shell preservation and great lithologic varia-

bility throughout the entire upper part of the Calvert Formation. There, the shell bed consists of a 7-m-thick (23-ft-thick) unit distinguished from adjacent units by its more diverse fossil assemblage and by the less silty matrix found in its thin shell layers (Figure 9).

Throughout the study area, the Kenwood Beach shell bed grades up into the *Turritella-Pandora* interval, a coarsening-up sequence of thin interbedded sands, muddy sands, and clays with a sparse and low diversity fauna that is typically concentrated into thin shell bands within sandy layers (Figure 3). The interval includes Zones 15 and 16 of Shattuck (1904), except in updip outcrops of the Calvert Cliffs where it includes only the lower portion of the great thickness of strata assigned by Shattuck to Zone 15 (1904; his reference sections 5, 7, and 8).

Four laterally disposed facies are present in the *Turritella-Pandora* interval. In the southern Calvert Cliffs, including the type section for these strata (Gernant, 1970; sections 35, 36, and 37; Figures 4 and 7), the interval consists of thickly-bedded and bioturbated muddy, very fine sand with 8- to 10-cm-thick (3.2- to 3.9-in-thick) bands of homogeneous clay. The top surfaces of the clay bands are bored, and are usually overlain by a thin stringer of the gastropod *Turritella plebeia* (Figure 3). This *Turritella* facies grades into, and in the northern Calvert Cliffs outcrops is entirely replaced by, a thin bedded muddy sand with sand-laminated clay bands (*Pandora* facies, Figure 7). The small inequivalve bivalve *Pandora crassidens* occurs as fine fragmental shell

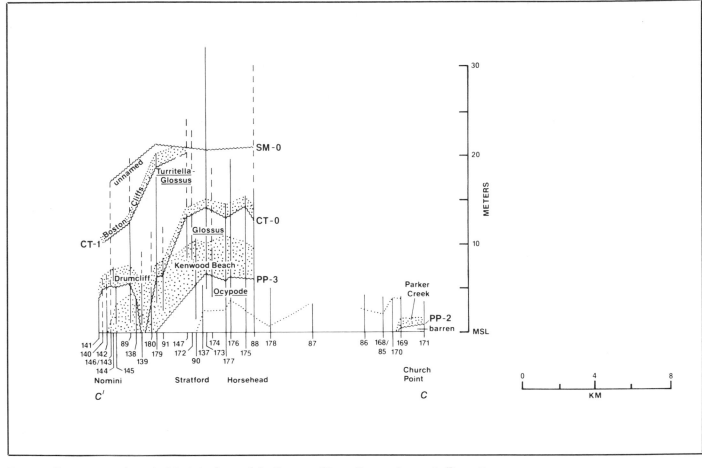

Figure 9: Cross section along the Virginia shore of the Potomac River. Conventions as in Figure 7.

material in the sand layers of that facies along with other small-sized species.

The *Turritella-Pandora* interval is indicated in Patuxent River exposures by interbedded silty- to very-silty very-fine sand with widely-spaced bored-clay bands. The large bivalve *Isognomon maxillata* dominates the macrofauna and occurs primarily in shell pavements. In the St. Leonard Creek region along the Patuxent (Figure 8), *I. maxillata* is densely packed into a clean fine sand (mistakenly assigned to the CT-0 sequence in Kidwell, 1982a). Finally, in the Potomac area, the *Turritella-Pandora* interval consists of massive clay and sandy mud that coarsen upward by the inclusion of muddy sand layers (*Glossus* facies, Figure 9). Shell material is very sparse except for thin, laterally discontinuous shell stringers dominated by *Glossus fraterna.*

Shallow sublittoral, polyhaline conditions are indicated for the Kenwood Beach shell bed by its trace- and body-fossil assemblage of venerid, lucinid, and corbulid bivalves, and *Gyrolithes.* The *Turritella-Pandora* facies tract includes: a littoral-to-sheltered, very shallow sublittoral *Pandora* facies; a soft-bottom, shallow sublittoral but fully marine *Turritella* facies; a firm bottom, shallow sublittoral polyhaline(?) *Isognomon* facies (both sandy shoal and adjacent muddy sea floor); and a soft-bottom, intermediate sublittoral marine *Glossus* facies.

CT-0 Disconformity and Depositional Sequence

The CT-0 disconformity is a burrowed firm ground, developed on the dark-gray silty fine sands of the *Turritella-Pandora* interval (PP-3 sequence) and overlain by the very shelly, well-sorted fine sand of the Drumcliff shell bed (Zone 17 of Shattuck, 1904; Figure 3). The burrowed surface coincides with an unconformity recognized by Shattuck (1904) between his Zones 15 and 17 in the Parker Creek area of the Calvert Cliffs, and used by him to define the Calvert-Choptank formational contact. The CT-0 disconformity is exposed over a distance of 15 km (9.3 mi) in the Calvert Cliffs, and along the Patuxent, Potomac, Rappahannock, and Choptank rivers (Figures 7, 8, and 9; Kidwell, 1982a).

Throughout much of its geographic extent, the CT-0 disconformity supports large-diameter (3 to 5 cm, or 1.2 to 2 in), interbranching *Thalassinoides* burrows that penetrate 0.7 to 2 m (2.3 to 6.6 ft) into underlying PP-3 strata (Figure 11). Exceptionally, the disconformity is an unburrowed scour surface in a 4-km (2.5-mi) stretch of the Calvert Cliffs between Parker Creek and Governor Run (sections 32 and 193 respectively; Figure 7). In that area, the disconformity describes a broad channel that cuts as much as 7 m (23 ft) into the underlying *Turritella-Pandora* interval and, at its lowest point, truncates the upper part of the Kenwood Beach shell bed. The erosional origin of this feature is evi-

Figure 10: Bedding plane exposure of the burrowed PP-0 disconformity at Chesapeake Beach (section 22, Figure 4). Waves have preferentially removed the basal, oyster-bearing sand of the Plum Point Member from *Thalassinoides* burrows developed in the top of the tightly consolidated Fairhaven Member. Shells of *Ostrea percrassa* in photograph are up to 20 cm (7.9 in) in length.

denced by the truncation of beds in the PP-3 sequence, lithofacies variation in the lower CT-0 sequence, and a thin (10 to 30 cm, or 4 to 12 in) basal lag of reworked calcitic shell debris and bone. The channel is filled with a body of interbedded clean fine sand and sandy clay informally named the Governor Run sand—clay interbeds characterized by the mussel *Mytilus incurvus* and the barnacle *Balanus concavus* thicken and become more closely spaced almost to the exclusion of sand beds along the flanks of the channel.

The basal lag of fragmental shells is immediately overlain by as much as 1 m (3.3 ft) of muddy sediments characterized either by algal laminations (?) or by the irregular urchin *Echinocardium cordatum* and its feeding traces. As a result, the CT-0 disconformity is marked by a notch of easily eroded sand between two muddy units in this Parker Creek to Governor Run stretch of the Calvert Cliffs, where workers most frequently choose to study and sample the formations. Truncation of bedding, development of a basal lag, and lithologic and paleontologic dissimilarity indicate that the relief on the CT-0 surface and the Governor Run sand resulted from primary erosion and deposition on the Miocene sea floor (Dryden, 1930; Kidwell, 1982a), not from post-depositional sagging of the section (Dryden, 1936) or

diagenetic alteration of the PP-3 sequence (Gernant, 1970). Shattuck (1904) and subsequent workers have interpreted the Governor Run sand as a sandy updip facies of the *Turritella-Pandora* interval of the PP-3 sequence (that is, Zone 16).

South of Governor Run in the Calvert Cliffs, the CT-0 surface has its usual burrowed appearance and is immediately overlain by the Drumcliff shell bed (Figure 11). North of Parker Creek, the Drumcliff shell bed interfingers with an updip facies of the *Mytilus* interval of the CT-0 sequence (Figure 7). The CT-0 disconformity in these northern Calvert Cliffs exposures is overlain by a 0.7 m-thick (2.3 ft-thick) fine sand containing only scattered shell material.

The CT-0 depositional sequence consists of the very shelly, well-sorted sand of the Drumcliff shell bed and the several sparsely fossiliferous facies of the *Mytilus* interval (Zone 18 of Shattuck, 1904; Figure 3). In the Calvert Cliffs, the lower part of the *Mytilus* interval is a massive, mottled, very muddy fine sand to sandy clay, and contains typically deposit-feeding bivalve species. It grades into a muddy to slightly muddy, thickly-bedded sand facies containing sparsely disseminated shell and shell lenses dominated by the bivalves *Mytilus incurvus* and *Anadara staminea*. In

Figure 11: Weathered exposure of *Thalassinoides* burrow systems originating from the CT-0 disconformity at Matoaka, Calvert Cliffs of Maryland (section 35). The large diameter burrows are filled with loosely-consolidated, well-sorted shelly sand from the Drumcliff shell bed. Chisel at lower right is 45 cm (17.7 in) long.

extreme downdip exposures in the Calvert Cliffs (Figure 7; sections 165 and 154), the exposed interval consists of a clean, very bluish and unfossiliferous sand. Clean sand containing abundant calcitic shell hash characterizes the *Mytilus* interval along the Patuxent River and infills 2 m-deep (6.6 ft-deep) channels cut into the Drumcliff shell bed. The fossil assemblage, consisting of both whole and broken specimens of *Balanus concavus, Chesapecten nefrens, Ostrea* sp., *Abertella aberti* (sand dollar), and bryozoans, closely resembles the calcitic portion of the upper sand facies of the *Mytilus* interval in Calvert Cliffs. The cliffs along the Potomac River expose a fourth facies consisting of bioturbated, muddy very fine sand and sandy mud dominated by *Turritella plebeia*, which occurs in thin lenses as much as 0.5 m (1.6 ft) long.

The characteristically littoral (*Mytilus-Balanus*; algal mats) to very shallow sublittoral (*Echinocardium*) assemblages and sedimentary structures of the Governor Run sand, as well as reworked *Mytilus* along the base of the Drumcliff shell bed, indicate extremely shallow conditions during initial accumulation of the CT-0 sequence. The Drumcliff shell bed itself contains a stratigraphically condensed series of polyhaline and fully-marine, shallow- to

very shallow-sublittoral, soft-bottom, and shell-gravel molluscan assemblages (Kidwell, 1979; 1982a). The overlying *Mytilus* interval consists of a progradational sequence of intermediate sublittoral (deposit-feeding fauna) to very shallow sublittoral to littoral facies (*Anadara* and *Mytilus* sands), with the exception of the *Turritella-Glossus* facies in the Potomac region, which arose in open-marine, shallow- to intermediate-sublittoral depths.

CT-1 Disconformity and Depositional Sequence
The CT-1 disconformity is a burrowed firmground along which the very shelly sands of the Boston Cliffs shell bed (Zone 19 of Shattuck, 1904) are superposed upon the lithologically variable sediments of the *Mytilus* interval of the CT-0 sequence (Figure 3). The surface can be traced in the Calvert Cliffs and along the Patuxent and Potomac rivers (Figures 7, 8, and 9).

In the southern Calvert Cliffs and along the Patuxent River, *Thalassinoides* burrows characterize the CT-1 disconformity. Smaller diameter but similarly deep (2 m, or 6.6 ft) burrows are present in the northern Calvert Cliffs, and in Potomac River exposures, non-branching small-diameter

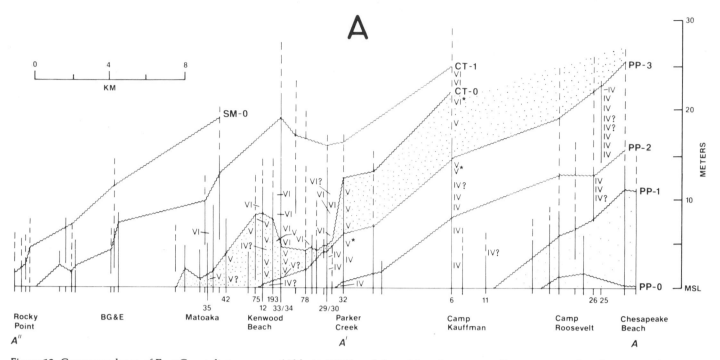

Figure 12: Correspondence of East Coast diatom zones (Abbott, 1978) and depositional sequences. Roman numerals indicate zone designations of individual samples. Asterisked samples are probably contaminated by burrow piping of younger material; question marks indicate doubt in zone assignment for some other reason. Zones II and V are stippled. Location and conventions of cross sections a, b, and c are identical to those in Figures 7, 8, and 9.

tubes penetrate less than 1 m (3.3 ft) into the *Mytilus* interval. The disconformity steepens locally in dip to 2 to 2.5 m/km (10 to12 ft/mi), often mimicking the topography of the underlying CT-0 disconformity (Figures 7, 8, and 9). Evidence for the erosional origin of the CT-1 surface includes the truncation of beds and facies of the underlying CT-0 sequence, and channel-like irregularities in the CT-1 surface. In addition, the Boston Cliffs shell bed varies in thickness and shell-packing density with topographic irregularities in the CT-1 surface. This variation is less well established than that of the Drumcliff shell bed because the CT-1 disconformity is exposed over a smaller outcrop area than the CT-0 surface, and secondly, the shell bed is often inaccessible in high cliffs.

The Boston Cliffs shell bed usually rests directly on the CT-1 disconformity (Figures 7, 8, and 9). From Matoaka (section 35) north to at least Governor Run (section 193) in the Calvert Cliffs, however, a lensoidal sand body (*Anadara* sand; Figure 7) intervenes. Unlike the Governor Run sand, which lies between the CT-0 surface and the Drumcliff shell bed, the *Anadara* sand does not appear to be a channel fill but simply rests on a stretch of the burrowed CT-1 surface, and grades upward into the Boston Cliffs shell bed.

The Boston Cliffs shell bed grades upward into a less fossiliferous, very muddy unnamed interval (Figure 3). The interval has a *Turritella*-dominated facies in the Calvert Cliffs (Figure 7) and a *Glossus*-dominated facies in the Potomac region (Figure 9). These massive sediments are a portion of the strata assigned by Shattuck (1904) and Gernant (1970) to Zone 20 of the Choptank Formation. Although relatively thick in Potomac outcrops (3 to 5 m, or 10 to 16 ft; Figure 9), this part of the Choptank Formation is largely or

entirely truncated elsewhere by a disconformity (SM-0) provisionally identified here as the lower contact of the St. Marys Formation (*sensu* Shattuck, 1904).

At Paris, Maryland (section 122), located northwest of the Calvert Cliffs (Figure 4), the CT-1 depositional sequence rests on or very near strata assigned with certainty to the *Glossus-Chione* interval (PP-2 sequence, Calvert Formation). The *Chione* facies is exposed in the extreme headwaters of a branch of Fishing Creek; nearby exposures of the Boston Cliffs shell bed are indicated by float specimens of indurated sandstone containing a Boston Cliffs fauna. These float specimens could not be derived from strata more than 3 m (10 ft) above the highest *Chione* exposure because of topographic constraints. The absence, or minimal thickness, of PP-3 and CT-0 strata in this area indicates that these strata were truncated during CT-1 erosion or were never deposited in this northern part of Calvert County. Shattuck (1904) and Gernant (1969, 1970) made similar observations in this area, interpreting the stratigraphic relations as evidence of unconformable Choptank onlap of Calvert strata.

The Boston Cliffs shell bed accumulated in shallow to very shallow sublittoral environments; the abundance of *Ostrea, Martesia, Isognomon,* and *Gyrolithes* suggests at least intermittent polyhaline conditions. The shell bed grades up into the fully marine shallow to intermediate sublittoral *Turritella* facies and intermediate sublittoral *Glossus* facies of the unnamed interval.

Rappahannock and Choptank River Exposures

In Fones Cliffs along the northern bank of the Rappahannock River (Figure 5), the exposed Calvert Formation con-

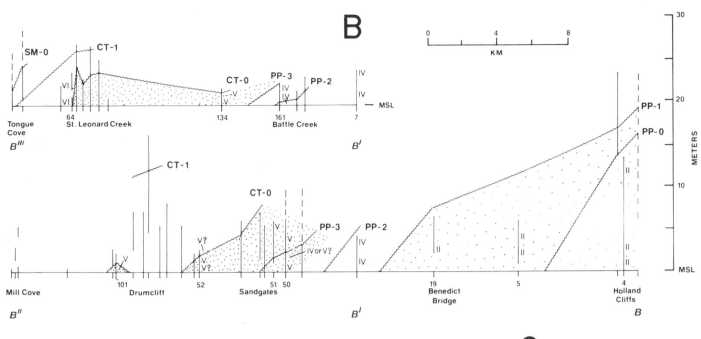

sists of a 3- to 4-m-thick (10- to 13-ft-thick) section of massive diatomaceous clay to sandy clay that most closely resembles the Fairhaven Member or the barren interval (PP-1 sequence) of the Calvert Formation in southern Maryland. Its upper contact is heavily burrowed by interpenetrating, sub-horizontal *Thalassinoides* filled with clean, very fine sand. This contact and infilling sand are nearly identical to the CT-0 surface and lower Drumcliff shell bed in the extreme eastern end of Nomini Cliffs, Potomac River (sections 140 and 141; Figure 9). The "Drumcliff" sand that rests on this burrowed surface at Fones Cliffs is leached of skeletal carbonate, varies from 2.7 to 2.9 m (8.9 to 9.5 ft) in thickness along the 5-km-long (3.1-mi-long) exposure and grades rapidly into 14 m (46 ft) of thickly- to very-thickly-bedded muddy fine sand with thin shell bands. Tentatively identified as undifferentiated Choptank Formation, this thick interval is truncated by a sharp burrowed surface and overlain by clayey sediments of the Eastover Formation (stratigraphically above the St. Marys Formation; see Ward and Blackwelder, 1980; Newell and Rader, 1982). The contact is marked by a lag of bone, wood, and phosphatic debris.

In the Rappahannock River area, it thus appears that the PP-0 (or possibly PP-2) through CT-0 surfaces are combined into a single unconformity that represents a large part of the Plum Point Member as developed in southern Maryland. Determining the precise nature of stratigraphic relations within the Choptank-St. Marys-Eastover interval here will require additional fieldwork. Preliminarily, the simplest interpretation is that the lag-marked lower contact of the Eastover Formation incorporates the CT-1, SM-0, and any intraformational discontinuities that might be present within the St. Marys Formation in the Potomac region. The

convergence of the CT-1 and SM-0 disconformities in Nomini and Stratford Cliffs along the Potomac River (Figure 9) points to such a relationship.

The scarcity and small size of outcrops on Maryland's eastern shore hinder stratigraphic subdivision and analysis of Calvert-Choptank strata. Of the six depositional sequences, the CT-1 sequence can be identified with greatest confidence based on exposures of the Boston Cliffs shell bed along the Choptank River (section 107; Figure 6). The CT-1 surface itself is not exposed. Further north along the Choptank River are a series of small outcrops of homogeneous clay which may be a facies of the *Mytilus* interval of the CT-0 sequence (sections 108, 112, and 113; Figure 6). Maximum exposed thickness of the clay is 5 m (16.4 ft); neither the upper nor the lower contact is exposed; fossils consist only of the bivalve molds not identifiable to species level. Further upstream on the Choptank River near Greensboro (sections 110 and 111; Figure 6), a clean, very fine sand with abundant shell molds crops out at river level and is identified tentatively as the Drumcliff shell bed. The burrowed basal contact of this shell bed (provisionally, the CT-0 surface) is coated with ferricrete, and the shell bed itself is heavily stained by iron oxides and indurated in patches. Exposed below the Drumcliff shell bed are 0.4 m (1.3 ft) of interlaminated and thinly-interbedded silts and micaceous sands that Gernant (1970) assigned to Zone 16 of Shattuck (1904; equivalent to a part of the *Turritella-Pandora* interval, PP-3 sequence). They are here identified only as part of the undifferentiated Calvert Formation. All exposures of the Calvert Formation examined on the eastern shore were only a few meters thick; stratigraphic subdivision comparable to that throughout the rest of the study area is thus impossible (see also Spangler and Peterson, 1950).

CHRONOSTRATIGRAPHIC SIGNIFICANCE AND ORIGIN

Biostratigraphic Evidence

Microfossil samples were evaluated independent of lithostratigraphic information by W.H. Abbott using his East Coast Diatom Zone (ECDZ) scheme (Abbott, 1978). This scheme uses an assortment of range, partial range, and concurrent range zones whose boundaries are defined both by first and last appearance data. Some of the samples collected from burrowed intervals immediately below the disconformities must be rejected (asterisked samples in Figures 12a, 12b, and 12c). In these burrowed intervals, guide species from superjacent zones have clearly been piped into underlying strata; elsewhere in the study area, these guide species are found only above the disconformities. Since previous workers have not recognized the burrowed nature of these contacts, many of their samples collected less than 2 m (6.6 ft) below the disconformities may also be contaminated with younger species.

ECDZ I is restricted to the lowest part of the Fairhaven Member of the Calvert Formation (Abbott, 1978; 1982) and was not encountered. ECDZ II diatoms were found in the upper Fairhaven Member and in the PP-0 sequence along the Patuxent River (Figure 12b) and in the Calvert Cliffs (Abbott, 1978). ECDZ III diatoms were not recovered in the course of this study, possibly because of the scarcity of any diatoms within the Camp Roosevelt shell bed. However, Abbott (1982) identified them in the top of the PP-0 sequence at Chesapeake Beach (section 22, Figure 12a; burrow piping?), and in the Camp Roosevelt shell bed at Plum Point (sections 11 and 121, Figure 4) and the Baltimore Gas and Electric Company well core taken at Flag Pond (near section 163, Figure 4). He also records Zone III from the Plum Point Member along the Pamunkey River in Virginia (Abbott, 1982).

ECDZ IV, V, and VI characterize the studied strata above the PP-1 disconformity, and their boundaries coincide with disconformities recognized by physical stratigraphic evidence (Figure 12a, 12b, and 12c). The base of Zone IV, defined by the last appearance of *Delphineis ovata*, lies at or less than 1 m (3.3 ft) above the PP-1 disconformity within

the stratigraphically condensed Camp Roosevelt shell bed. Zone IV ranges through the PP-1 and PP-2 sequences and is replaced at the PP-3 disconformity by Zone V. It is also found in the *Chione* facies (PP-2 sequences) at Paris (section 122).

Defined by the first appearance of two species, the base of Zone V tracks the PP-3 surface in the Calvert Cliffs and Potomac region. However, in the Patuxent River area where the PP-2/PP-3 sequence transition is gradational, the base of diatom Zone V lies within the PP-2 sequence (Figure 12b). The last appearance of *D. penelliptica* defines the top of Zone V and the base of Zone VI, and coincides with the highly irregular CT-0 surface. This perfect coincidence demonstrates both the reality of the unconformity and the relative independence of the diatom biostratigraphic scheme to lithofacies variation.

Elsewhere in the Salisbury Embayment, diatoms have been found in Fairhaven-type lithologies near Oak Grove (ECDZ IV) and at Fones Cliffs (ECDZ V), both on the Rappahannock River (Abbott, 1982). Fairhaven-like sediments cropping out near Richmond contain diatom species younger than Zone VI (that is, above the last appearance of *Distephanus stauracanthus*; Abbott, 1982). These data indicate that the Fairhaven paleoenvironment migrated southward with time, being replaced in Maryland by Plum Point and Choptank lithologies.

Chronostratigraphic Utility

The disconformities provide a reasonable basis for chronostratigraphic subdivision and correlation within the study area. Available biostratigraphic data indicate that the disconformity-bounded depositional sequences are not diachronous within the limits of resolution of diatom zonation, although the top of each depositional sequence must vary somewhat in age locally owing to erosional relief (Figure 13). Depositional sequences, or sets of two depositional sequences, occupy unique time intervals (that is, correspond to single biozones). Inasmuch as exposures limit the evaluation of diachroneity over down-dip distances of less than 30 km (18.6 mi) in a structural embayment of very slight initial dip, this result should not be too surprising. The fortunate consequence for regional geologic interpretation, however, is that the depositional sequences delimit a series of facies tracts that can be used to reconstruct depositional systems for six relatively brief time periods (see Kidwell, 1982a).

Ages and Magnitudes of Hiatuses

Each of the disconformities formed over a period of time less than the duration of a single diatom zone (Figure 13). Locally and in updip areas, however, erosion enlarged some breaks in the record to equal or exceed one zone in duration. The CT-0 disconformity, for example, encompasses all or almost all of Zone V time where it cuts deeply into the PP-3 depositional sequence in the Calvert Cliffs and along the Potomac River (Figures 12a and 12c). Traced updip, the CT-1 hiatus expands to represent late Zone IV and all of Zone V time as well as Zone VI time by merging with the CT-0 and PP-3 disconformities in northern Calvert County; and along the Potomac River, the SM-0 hiatus progressively encompasses the larger part of Zone VI time by merging with the CT-1 surface.

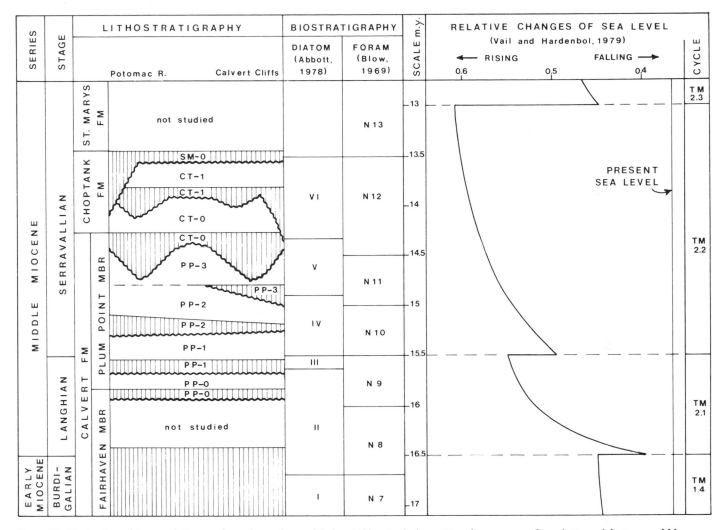

Figure 13: Biostratigraphic correlation and geochronology of Calvert-Choptank depositional sequences. Correlation of diatom and Neogene foraminiferal zones from Abbott (1978, 1982); geochronologic scale in m.y. (van Couvering and Berggren, 1977) from Vail and Hardenbol (1979). The PP-1 disconformity of the Calvert Formation approximates the Langhian-Serravallian boundary, where Vail and Hardenbol (1979) have recognized a minor, Type 2 interregional unconformity (15.5 m.y. event). A major hiatus recognized by Andrews (1978) and Abbott (1978, 1982) within the Fairhaven Member (shown in part) of the Calvert Formation brackets the Burdigalian-Langhian boundary, where Vail and Hardenbol (1979) identified another minor Type 2 unconformity (16.5 m.y. event).

To evaluate the disconformities in terms of absolute age, the diatom zones are correlated (Abbott, 1978, 1982) to the standard Neogene foraminiferal scheme of Blow (1969), which in turn has recently been calibrated to a geochronologic scale (van Couvering and Berggren, 1977; Vail and Hardenbol, 1979). As summarized in Figure 13, several diatom zones span one or more erosional surface, thus limiting the accuracy of age estimates. For example, the PP-0 disconformity lies somewhere within diatom Zone II, the PP-2 surface within Zone IV, and the CT-1 surface within Zone VI. Also diatom zone boundaries coinciding with erosional surfaces do not all coincide with well-dated boundaries of standard foraminiferal zones for the Neogene; for example, the bases of diatom Zones V and VI fall within foraminiferal zones N11 and N12. Perhaps the best constrained erosional surface is the PP-1 surface, which truncates diatom Zone II and lies very near the top of Zone III, the shortest of all of Abbott's (1978) East Coast Diatom Zones. The diatom

information places the Langhian-Serravallian stage boundary just above the PP-1 surface within the Camp Roosevelt shell bed. The age of the SM-0 surface is not as well-constrained owing to the absence of diatoms in the overlying St. Marys Formation. However, it must lie near the foraminiferal zone N12-N13 boundary (Figure 13) based on diatoms found in the underlying CT-1 sequence (Abbott, 1982).

Both Andrews (1978) and Abbott (1978, 1982) have recognized a major hiatus between Abbott's Zones I and II within the Fairhaven Member of the Calvert Formation (Figure 13). This hiatus, which includes the Burdigalian-Langhian stage boundary, has a minimum duration of 2.5 m.y. (in the work cited). In contrast, disconformities in the Plum Point Member, Calvert Formation, all represent hiatuses less than 0.5 m.y. in magnitude. This includes the PP-1 surface and associated Camp Roosevelt shell bed, which coincide with the Langhian-Serravallian stage boundary.

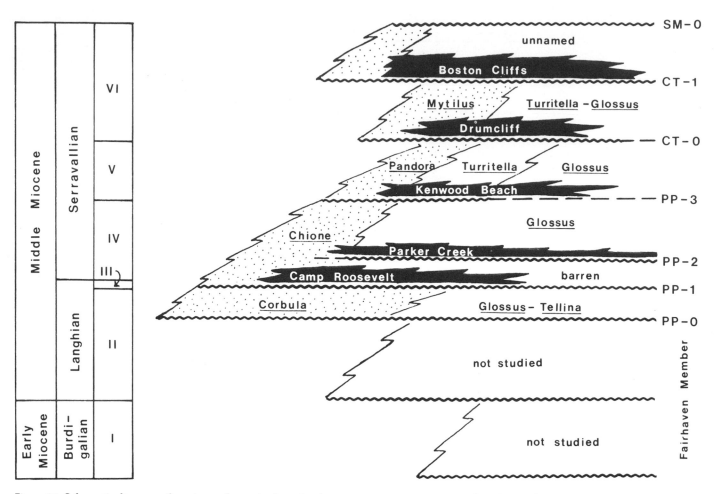

Figure 14: Schematic diagram of marine and marginal marine facies in transgressive-regressive cycles, and occurrence of stratigraphic discontinuities within the lower Chesapeake Group of southern Maryland. Fairhaven lithologies recur in the Serravallian Stage further south along structural strike in Virginia. Diagram not to scale.

Finally, disconformities within the Choptank Formation and at the base of the St. Marys Formation represent at least 0.5 m.y. each over most of their geographic extent, and considerably more locally (Figure 13).

Origin

Paleoenvironmental interpretation of the Plum Point and Choptank strata indicates a series of transgressive-regressive cycles with asymmetric facies records (Figure 14). This evidence, along with limited available information on onlapping relations between Burdigalian, Langhian, and Serravallian strata in the Salisbury Embayment, can be used to infer the origin and stratigraphic significance of the described surfaces. Unfortunately, published information on the mid-Fairhaven hiatus, not included in this study but probably the most significant break in the Maryland Miocene section, is insufficient to permit interpretation of its origin.

The lower Plum Point Member records a transgressive phase of deposition, with maximum open-shelf water depths attained in the massive barren interval of the PP-1 depositional sequence. Above the PP-2 surface and Parker Creek bone bed, the Plum Point and Choptank strata record an overall regressive phase punctuated by transgressive pulses

and condensed shell deposits. Within this context, most of the disconformities represent ravinement surfaces formed erosionally during early transgression or late regression.

The PP-0 firmground, dominated by crustacean burrows including the typical marginal-marine form *Gyrolithes*, records the transgression of a marginal marine shoreface across older Fairhaven strata. The most conservative interpretation, consistent with the associated fauna, considerable erosional relief, and placement within a single biozone, is that the PP-0 surface is a submarine ravinement surface. However, because the paleoenvironmental significance of the underlying Fairhaven is unknown and owing to the limited exposures of the PP-0 surface, an earlier, possibly prolonged interval of subaerial exposure cannot be ruled out. The burrowed shoreface was followed in transgression by shallow-water oyster-bearing sands, including small reef bodies on paleohighs (Holland Cliffs), soft-bottom muds dominated by deposit-feeding communities and storm-transported skeletal debris from shallower areas (species-rich facies), and episodically prograding, polyhaline disturbed mud and sand habitats dominated by *Corbula*.

The PP-1 surface records rapid transgression of an open-marine shoreface across the prograding marginal marine system. Shutdown or slowdown of terrigenous sediment supply to the study area, in response to rapid base-level rise, fostered the accumulation of the Camp Roosevelt shell bed, a condensed deposit of open marine sand-bottom and shell-gravel faunas (Kidwell 1982a; Kidwell and Jablonski, 1983). Eventual resumption of a sediment supply found the study area a distal and relatively deep-water open-shelf environment, resulting in the barren, sand-poor mud of the upper PP-1 sequence.

In contrast, the PP-2 surface is a mid-cycle omission horizon. Rather than forming through erosion and reduced net sedimentation during rapid transgression, the surface and its associated Parker Creek bone bed accumulated under conditions of sediment starvation during maximum transgression or early regression. This origin is underscored by the nature of the vertebrate assemblage: although the assemblage is condensed (Myrick, 1979), elements are not heavily worn and disarticulated as is typical of erosional bone and teeth lags. The stratigraphic relations of the horizon are also consistent with this origin, in which the hiatus should have a maximum value in the most distal regions. Throughout most of the study area, the PP-2 surface and Parker Creek bone bed form a thin, distinctive horizon, but in extreme updip exposures in the Calvert Cliffs, they grade into a sequence of bone-rich sandy layers lacking a well-defined burrowed horizon. Eventual resumption of terrigenous sediment supply to the study area resulted in accumulation of the regressive *Glossus* (soft mud) and *Chione* (silty sand) facies, both open-marine subtidal deposits.

In the Calvert Cliffs, where it is clearly erosional in origin, the PP-3 surface marks the passage of a marginal-marine shallow-water environment, either a shoreface or perhaps a bay-mouth shoal. Its simplest interpretation is a part of the regression initiated during PP-2 time, although the coincidence of the erosional PP-3 surface with a zone boundary may indicate a more significant break. The mixed polyhaline and open-marine fauna of the Kenwood Beach shell bed is succeeded by a regressive series of subtidal facies (*Glossus*, *Turritella*, and *Isognomon*) culminating in the very shallow subtidal to intertidal *Pandora* facies.

The CT-0 disconformity marks the termination of the Plum Point transgressive-regressive cycle and initiation of a new pulse of transgression within the study area. Much of the topographic relief on this surface, as much as 14 m (46 ft) locally, probably dates to late, erosional stages of the Plum Point regression—sediments immediately below the disconformity accumulated in intertidal flat and marsh environments (*Pandora* facies of the Calvert Cliffs, and *Ocypode* facies along the Potomac River, Virginia) characterized by channel systems, and it is not improbable that they were succeeded by fully supratidal coastal conditions. Topographic relief of this magnitude is not uncommon along modern coastlines and in Holocene transgressive records (Reineck and Singh, 1975). The intertidal origin of sediments resting directly upon the CT-0 disconformity (basal *Mytilus* assemblages within the Drumcliff shell bed and mussel-barnacle-echinoid assemblages and algal mats of the Governor Run sand (Kidwell, 1982a) underscores the probability of subaerial conditions during some period of CT-0 formation, and that much of the erosional paleotopography is regressive rather than transgressive in origin. Direct evidence of subaerial exposure such as paleosol development

and root casts is lacking. Rapid transgression is consistent with the condensed, deepening-up sequence of marine faunal assemblages within the Drumcliff shell bed. The shell bed grades upward rapidly into a bathymetrically symmetrical series of fossil-poor regressive facies that prograded when the sediment supply system re-equilibrated to the new base level.

Stratigraphic relations within the CT-1 depositional sequence are less well-documented and understood than other sequences due to the inaccessibility of the sequence in most cliff sections and the smaller outcrop belt. However, the CT-1 surface has a burrowed appearance everywhere it is examined and exhibits considerable topographic relief, often mimicking that of the underlying CT-0 although offset up- or downdip by some distance (Figures 7, 8, and 9); fauna of the Boston Cliffs shell bed are dominated by subtidal species, although polyhaline as well as fully marine taxa are included. It thus appears to represent a second transgressive pulse within the Choptank Formation. Unlike the history of the CT-0 sequence, sediment supply was resumed before maximum transgression was attained— facies above the Boston Cliffs shell bed record relatively deep subtidal and fully marine benthic habitats (*Glossus* and *Turritella* facies).

The SM-0 disconformity is the subject of ongoing research. It bevels the Choptank Formation throughout the study area, especially in downdip exposures, and is marked locally by an erosional lag of phosphatic material and comminuted vertebrate debris; the upper part of the Boston Cliffs shell bed is indurated where truncated by the SM-0 surface and is marked by a ferricrete crust. Although sediments of the overlying St. Marys Formation describe an overall regressive sequence that could be interpreted as a continuation of the Choptank sequence, erosional beveling and local channeling of Choptank strata and the development of a phosphatic lag suggest disconformity and a significant episode of submarine and possibly subaerial erosion and non-deposition.

The physical features and paleontology of the Calvert-Choptank section are consistent with deposition along a non-deltaic coastline that experienced several cycles of rapid, erosional and non-depositional transgression and progradational regression (see Kraft, 1978). The features do not fit a deltaic model (see Miall, 1979). Facies sequences consistently fine rather than coarsen upward, and the section lacks distributary channel and bar sands, delta front sands, and delta-plain deposits even though several depositional sequences shallow-up into intertidal and possibly subaerial environments. The sedimentary structures and context of sub-wave base marine deposits are inconsistent with accumulation on prodeltaic slopes. It is thus unlikely that the disconformity-bounded depositional sequences owe their origin to alternating conditions of erosion and progradation related to delta-lobe switching. Moreover, the condensed shell beds associated with the erosional surfaces accumulated in open-marine, subtidal environments under conditions of prolonged low net sedimentation rather than erosional reworking (Kidwell, 1982a, 1982b). They are thus better explained in terms of marine transgression of non-deltaic coastlines following base-level rise, than by the erosional reworking of delta-front and plain environments in response to sediment cutoff alone.

Relations to Interregional Unconformities

In the absence of seismic profiles for direct tracing of off-shore subsurface reflectors into the Salisbury Embayment, determining the relationship of recognized interregional unconformities to the basin margin disconformities described here is limited to similarities in geologic age and onlap relations. Possible correlations are discussed without presupposing any eustatic or relative sea level significance for the interregional unconformities.

Biostratigraphic correlation places a 2.5 m.y. hiatus within the Fairhaven Member of the Calvert Formation at the Burdigalian-Langhian stage boundary (Abbott, 1978, 1982; Andrews, 1978), approximately 16.5 m.y. ago (Figure 13). In seismic records, Vail and Hardenbol (1979) have dated a minor, Type 2 unconformity at this same boundary (Type 2 unconformities are thought to result from a rapid rise in sea level following a period of stillstand or slow fall, inferred from onlap over a progradational sequence). Younger, Langhian aged strata of the Fairhaven Member should significantly onlap older strata if this major hiatus is related to this unconformity. Unfortunately, information on onlap-offlap relations, physical characteristics, and paleoenvironmental context for the mid-Fairhaven unconformity is unavailable for evaluation.

Within the studied Plum Point-Choptank section, which is Langhian to Serravallian in age, several laterally persistent erosional surfaces have been identified. Elsewhere, Vail and Hardenbol (1979) have identified only two unconformities of interregional significance, at the Langhian-Serravallian boundary at 15.5 m.y., and within the Serravallian stage at approximately 13.0 m.y. (Figure 13). Both are minor, Type 2 unconformities.

The 15.5-m.y.-ago event correlates closely with the PP-1 disconformity and its basal condensed shell bed (Figure 13). The PP-1 disconformity, however, is a marine ravinement surface rather than a major unconformity. Its stratigraphic position 10 m (33 ft) or less above the geologically more significant PP-0 surface, however, is beyond the resolution of typical seismic reflection data, and so the disconformable base of the Plum Point Member (PP-1 to PP-2 package) might represent the basin margin extension of Vail and Hardenbol's (1979) 15.5 m.y. sequence boundary. Available data for Langhian and Serravallian strata do indicate onlapping relations. Langhian-age strata, represented in Maryland by the upper part of the Fairhaven Member (Abbott, 1978; 1982; Figure 13), have not yet been recognized in Virginia, but Serravallian age strata extend throughout southern Maryland and as far south as Richmond, Virginia. In Maryland they are represented by the Plum Point Member of the Calvert Formation and the Choptank Formation (Figure 13); in Virginia they consist of Fairhaven-type diatomaceous sediments referred to as the Calvert Formation (Abbott, 1982).

The mid-Serravallian, 13.0 m.y. Type 2 unconformity is too young to correspond to any of the Plum Point or Choptank erosion surfaces, but may correlate with the SM-0 surface at the Choptank-St. Marys formational contact. Additional biostratigraphic evidence will be required, however, to determine whether the SM-0 hiatus was sufficiently

long to encompass this event and whether regional stratigraphic relations are consistent with an onlapping record. Marine depositional environments do regress during St. Marys accumulation, but marginal marine strata can be traced further updip even in the Calvert Cliffs than previously reported (Kidwell, work in progress), and the extent of non-marine coastal onlap has not yet been established. Blackwelder (1981) and Newell and Rader (1982) have discussed unconformities and onlap relations for Chesapeake Group strata younger than the St. Marys Formation and their possible sea level significance.

CONCLUSIONS

Three hiatuses within the Maryland Miocene correlate with minor, Type 2 interregional unconformities recognized in seismic sections elsewhere by Vail and Hardenbol (1979). These are found: (1) within the Fairhaven Member of the Calvert Formation (16.5 m.y. event at Burdigalian-Langhian stage boundary); (2) at the approximate base of the Plum Point Member of the Calvert Formation (15.5 m.y. event at Langhian-Serravallian stage boundary); and (3) at the contact of the Choptank and St. Marys formations (13.0 m.y. event within the Serravallian stage). The disconformable contact of the Calvert and Choptank formations (CT-0 surface) also indicates a period of subaerial erosion and possibly onlap. It dates at approximately 14.5 m.y. (Figure 13), but does not correlate with any published interregional reflector. Correlation of Calvert-Choptank disconformities with offshore subsurface reflectors using geologic age and correspondence in onlap relations should be relatively straightforward, since the complications of delta-lobe switching that characterize some basin margins are absent in this situation.

The other disconformities of the Plum Point-Choptank section are transgressive or regressive ravinement surfaces or distal omission horizons associated with stratigraphically condensed fossil accumulations. These relations are an expression of stratigraphic complexity within larger-scale, interregionally significant depositional sequences, and would be beyond the resolution of most seismic reflection data. The isochroneity of the Plum Point-Choptank disconformities over down-dip distances on a regional scale requires further documentation, but the surfaces should grade into conformable sections within the shallow subsurface given their origins, and within the outcrop belt the disconformities are not measurably diachronous. They thus provide a valuable basis for chronostratigraphic subdivision and correlation within the study area and demonstrate the applicability of depositional sequence analysis to outcrop scale features. The practical consequences for the Maryland Miocene include the identification of facies tracts for paleogeographic reconstruction of the Salisbury basin margin (see Kidwell, 1982a) and a chronostratigraphic framework for paleobiologic analysis of its diverse and historically important faunas.

ACKNOWLEDGMENTS

This paper is based on portions of a dissertation submitted to Yale University to which I am grateful for a Graduate

Fellowship. I thank my advisors, Karl M. Waage and Donald C. Rhoads, as well as Robert E. Gernant and Clarissa (Mrs. A.L.) Dryden for their advice and interest, and W.H. Abbott, T.G. Gibson, D. Jablonski, G.H. Johnson, P.H. Kelley, W.L. Newell, W.C. Poag, E.K. Rader, and J.S. Schlee, volume editor, for valuable reviews. I am particularly grateful to William H. Abbott, then of the South Carolina Geological Survey, for micropaleontologic analysis of samples. Field work was supported by the Yale Schuchert Fund (1977), the Geological Society of America (1978, 1979), the Society of Sigma Xi (1978-1980), and the Woman's Seamen's Friend Society of Connecticut (1978-1980). Grateful acknowledgment is also made to the Donors of the Petroleum Research Fund, administered by the American Chemical Society, for continuing support of this research. J. Overs carefully typed the manuscript. This is a contribution of the Laboratory of Geotectonics, University of Arizona.

REFERENCES CITED

Abbott, W.H., 1978, Correlation and zonation of Miocene strata along the Atlantic margin of North American using diatoms and silicoflagellates: Marine Micropaleontology, v. 3, p. 15-34.

——, 1982, Diatom biostratigraphy of the Chesapeake Group, Virginia and Maryland, in T.M. Scott and S.B. Upchurch, eds., Miocene of the southeastern United States: Special Publication, n. 25, Florida Bureau of Geology, p.

Andrews, G.W., 1978, Marine diatom sequence in Miocene strata of the Chesapeake Bay region, Maryland: Micropaleontology, v. 24, p. 371-406.

Blackwelder, B.W., 1981, Late Cenozoic marine deposition in the United States Atlantic Coastal Plain related to tectonism and global climate: Palaeogeography, Palaeoclimatology, Palaeoecology, v. 34, p. 87-114.

——, and L.W. Ward, 1976, Stratigraphy of the Chesapeake Group of Maryland and Virginia: Geological Society of America Southeast Section Field Trip Guidebook, 55 p.

Blow, W.H., 1969, Late middle Eocene to Recent planktonic foraminiferal biostratigraphy: Geneva, Proceedings, First International Conference on Planktonic Microfossils, p. 199-421.

Calver, J.L., and C.R.B. Hobbs, Jr., eds., 1963, Geologic Map of Virginia: Virginia Division of Mineral Resources, Scale 1:500,000.

Cleaves, E.T., J. Edwards, Jr., and J.D. Glaser, eds., 1968, Geologic Map of Maryland: Maryland Geological Survey, Scale 1:250,000.

Dryden, A.L., Jr., 1930, Stratigraphy of the Calvert Formation at the Calvert Cliffs, Maryland: Unpublished Ph.D. Dissertation, John Hopkins University, 235 p.

——, 1933, Xenohelix in the Maryland Miocene: Proceedings, National Academy of Science, v. 19, p. 139-143.

——, 1936, The Calvert Formation in southern Maryland: Proceedings, Pennsylvania Academy of Science, v. 10, p. 42-51.

Frey, R.W., and A. Seilacher, 1980, Uniformity in marine invertebrate ichnology: Lethaia, v. 13, p. 183-207.

Gazin, C.L., and R.L. Collins, 1950, Remains of land mammals from the Miocene of the Chesapeake Bay region: Smithsonian Miscellaneous Collections, v. 16, n. 2, 21 p.

Gernant, R.E., 1970, Paleoecology of the Choptank Formation (Miocene) of Maryland and Virginia: Maryland Geological Survey Report of Investigations, v. 12, 90 p.

——, 1971, Invertebrate biofacies and paleoenvironments, in R.E. Gernant, T.G. Gibson, and F.C. Whitmore, Jr., eds., Environmental history of Maryland Miocene: Maryland Geological Survey Guidebook, v. 3, p. 19-30.

——, 1972, The paleoenvironmental significance of Gyrolithes (Lebensspur): Journal of Paleontology, v. 46, p.735-741.

Gibson, T.G., 1962, Benthonic foraminifera and paleoecology of the Miocene deposits of the middle Atlantic Coastal Plain: Unpublished Ph.D. Dissertation, Princeton University, 198 p.

——, 1971, Miocene of the middle Atlantic Coastal Plain, in R.E. Gernant, T.G. Gibson, and F.C. Whitmore, Jr., eds., Environmental history of Maryland Miocene: Maryland Geological Survey Guidebook, v. 3, p. 1-15.

Glenn, L.C., 1904, Systematic paleontology, Miocene Pelecypoda: Maryland Geological Survey, Miocene Volume, p. 274-401.

Harris, G.D., 1893, The Tertiary geology of Calvert Cliffs, Maryland: American Journal of Science, 3rd Series, v. 45, p. 21-31.

Howard, J.D., and H.-E. Reineck, 1981, Depositional facies of high-energy beach-to-offshore sequence: Comparison with low-energy sequence: AAPG Bulletin, v. 65, p. 807-830.

Kellogg, R., 1965-1969, Fossil marine mammals from the Miocene Calvert Formation of Maryland and Virginia: U.S. National Museum Bulletin, n. 247, parts 1-9, 197 p.

Kellogg, R., 1969, Cetothere skeletons from the Miocene Choptank Formation of Maryland and Virginia: U.S. National Museum Bulletin, n. 294, parts 1 and 2, 39 p.

Kidwell, S.M., 1979, Stratigraphic condensation and the formation of major shell beds in the Miocene Chesapeake Group: Geological Society of America Abstracts, v. 11, p. 457.

——, 1982a, Stratigraphy, invertebrate taphonomy, and depositional history of the Miocene Calvert and Choptank formations, Atlantic Coastal Plain: Unpublished Ph.D. Dissertation, Yale University, 514 p.

——, 1982b, Time scales of fossil accumulation—— patterns from Miocene benthic assemblages: Proceedings, Third North America Paleontology Convention, v.1, p. 295-300.

——, and D. Jablonski, 1983, Taphonomic feedback; ecological consequences of shell accumulation, in M.J.S. Teresz and P.L. McCall, eds., Biotic interactions in Recent and fossil benthic communities: New York, Plenum Press, p. 195-248.

Kraft, J.C., 1978, Coastal stratigraphic sequences, in R.A. Davis, Jr., ed., Coastal sedimentary environments: New York, Springer-Verlag, p. 361-383.

Malkin, D.S., 1953, Biostratigraphic study of Miocene Ostracoda of New Jersey, Maryland, and Virginia: Journal of Paleontology, v. 27, p. 761-799.

Martin, G.C., 1904, Systematic paleontology, Miocene Gastropoda: Maryland Geological Survey, Miocene Volume, p. 131-270.

Miall, A.D., 1979, Facies Models 5—deltas, in R.G. Walker, ed., Facies models: Geoscience Canada Reprint Series, v. 1, p. 43-56.

Myrick, A.C., Jr., 1979, Variation, taphonomy, and adaptation of the Rhabdosteidae (=Eurhinodelphidae) (Odontoceti, Mammalia) from the Calvert Formation of Maryland and Virginia: Unpublished Ph.D. Dissertation, University of California, 411 p.

Newell, W.L., and E.K. Rader, 1982, Tectonic control of cyclic sedimentation in the Chesapeake Group of Virginia and Maryland: Geological Society of America Northeastern-Southeastern Section Guidebook Field Trip No. 1, p. 1-27.

Payton, C.E., ed., 1977, Seismic stratigraphy -- applications to hydrocarbon exploration: AAPG Memoir 26, 516 p.

Poag, W.C., 1979, Stratigraphy and depositional environments of Baltimore Canyon Trough: AAPG Bulletin, v. 63, p. 1452-1466.

Radwanski, A., 1977, Burrows attributable to the Ghost Crab Ocypode from the Korytnica Basin middle Miocene (Holy Cross Mountains, Poland): Acta Geologica Polonica, v. 27, p. 217-225.

Reineck, H.-E., and I.B. Singh, 1975, Depositional sedimentary environments. Berlin, Springer-Verlag, 439 p.

Schoonover, L.M., 1941, A stratigraphic study of the mollusks of the Calvert and Choptank formations of southern Maryland: Bulletin of American Paleontology, v. 25, 94B, p. 169-199.

Shattuck, G.B., 1904, Geological and paleontological relations, with a review of earlier investigations: Maryland Geological Survey, Miocene Volume, p. 33-137.

Spangler, W.B., and J.J. Peterson, 1950, Geology of Atlantic Coastal Plain in New Jersey, Delaware, Maryland, and Virginia: AAPG Bulletin, v. 34, p.1-99.

Ulrich, E.O., and R.S. Bassler, 1904, Systematic paleontology, Miocene Bryozoa: Maryland Geological Survey, Miocene Volume, p. 404-429.

Vail, P.R., and J. Hardenbol, 1979, Sea level changes during the Tertiary: Oceanus 22, v. 71-79.

——, et al, 1980, Unconformities of the North Atlantic: Philosophical Transactions, Royal Society of London, v. A294, p. 137-155.

van Couvering, J.A., and W.A. Berggren, 1977, Biostratigraphical basis of the Neogene time scale, in E.G. Kaufman and J.E. Hazel, eds., Concepts and methods of biostratigraphy: Stroudsburg, PA, Dowden, Hutchinson, and Ross, p. 283-306.

Ward, L.W., and B.W. Blackwelder, 1980, Stratigraphic revision of upper Miocene and lower Pliocene beds of the Chesapeake Group, middle Atlantic Coastal Plain: U.S. Geological Survey Bulletin, v. 1482-D, 61 p.

Significant Unconformities and the Hiatuses Represented by Them in the Paleogene of the Atlantic and Gulf Coastal Province

Joseph E. Hazel
Amoco Production Company
Tulsa, Oklahoma

Lucy E. Edwards
Laurel M. Bybell
U.S. Geological Survey
Reston, Virginia

A biostratigraphic, chronostratigraphic, and magnetostratigraphic model has been calibrated to produce a new time scale for the Paleogene. The model gives the biostratigraphic position and duration represented by significant unconformities in three areas of the Atlantic and Gulf Coastal Province: 1) western and central Alabama; 2) South Carolina; and 3) central Virginia to southwestern Maryland. In these areas, the most significant unconformity, in terms of duration represented and lateral extent, is found in the lower Eocene. In Alabama, this unconformity centers around 51.4 m.y. and represents a hiatus of about 1.4 m.y. In South Carolina, this unconformity centers around 50.3 m.y. and represents a hiatus of about 10.0 m.y. In Virginia-Maryland, the lower Eocene unconformity centers around 49.0 m.y. and represents a hiatus of about 7.3 m.y. A significant unconformity exists between the Cretaceous and Tertiary in all three areas. On the Atlantic coast the Cretaceous-Tertiary unconformity represents some missing Danian and significant missing Maestrichtian. In Alabama, however, there is virtually a complete Danian section and it is only most of the upper Maestrichtian that is missing. There are significant regional unconformities in all three areas.

INTRODUCTION

During the past 2 years, data from nearly 100 Cretaceous and Paleogene measured sections and boreholes have been integrated into a biostratigraphic, chronostratigraphic, and magnetostratigraphic model. Our goal has been to produce a framework that incorporates and therefore calibrates first and last appearance datums (FAD's and LAD's) for fossil organisms such as foraminifers, calcareous nannofossils, and dinoflagellates, and the record of magnetic reversals. Such a model can be used to analyze the sediment record in various locations and structural settings to chart the stratigraphic position of significant unconformities and, if it can be tied to a timescale, to document the durations of the hiatuses represented by these unconformities.

In this paper, our most recent model for the Paleogene is used to analyze the sediment record in three settings in the Atlantic and Gulf Coastal Province of the United States: 1) western and central Alabama; 2) the Charleston area of South Carolina; and, 3) central and northern Virginia to southwestern Maryland (Figure 1).

METHODS

Graphic Correlation

The basic framework of our model was produced using Graphic Correlation (GC) (Shaw, 1964; Miller, 1977). GC is an iterative procedure using two-axis graphs. For any two sections, sediment thickness in one section may be plotted against sediment thickness in the other section, and locations of significant datums (usually first and last occurrences of fossils, but magnetostratigraphic and other events may also be used) may be plotted as points on the graph. As long as the two sections include sediment representing coincident or overlapping intervals of time, a line of correlation (LOC), which need not be straight, must exist relating the two sections. Once an LOC is positioned, the equation (or equations) for this line is used to convert levels of events in one section to levels in the other section.

Graphic correlation may be used to compare any two sections with one another, but more importantly, it may be used to combine data from all sections to produce a hypothesis concerning the position of all events. For multiple sections, a LOC exists between each section and a single composite section. The composite section is constructed using information from all sections and should represent the data for a study better than any individual section. Development of the composite section is accomplished by projecting points on the graph for the various individual sections onto the LOC and thence to the composite. Units on the composite axis now represent a composite of the feet or meters of the compared sections and are referred to as "composite units." The composite axis is a scale that measures time as expressed in terms of the rate of rock accumulation

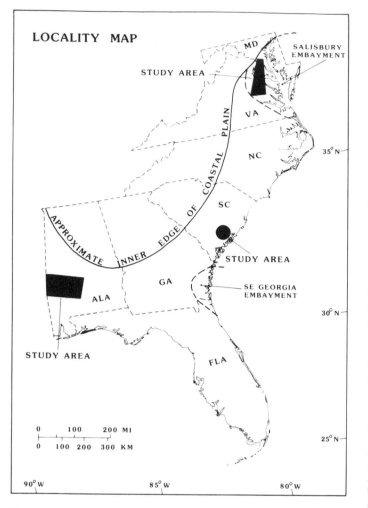

LOCALITY MAP

Figure 1: Locations of the three study areas discussed in the text.

of the section chosen to represent the first model.

ZONATIONS

Figure 2 gives a biostratigraphic zonal framework for the Paleocene through the middle Eocene. The planktic foraminifer zones of Toumarkine and Bolli (1970), for the middle Eocene, Stainforth et al (1975) for the Paleocene and lower Eocene, and Blow (1979) are calibrated to each other and to a nannofossil zonation based on the zones of Roth (1973), Bukry (1978), and Martini (1971) that empirically seem to be the most useful. The only defining nannofossil datum used that was not employed by Roth (1973), Bukry (1978), or Martini (1971) is the FAD of *Helicosphaera lophota* (Bramlette and Sullivan, 1961). Because of the recorded occurrence of *Tribrachiatus orthostylus (ref. Shamrai, 1963)* above the FAD of *Discoaster sublodoensis* (Bramlette and Sullivan, 1961) (see Davaud and Guex, 1978; Perch-Nielsen, 1977), its LAD cannot be used to mark the base of the *Discoaster lodoensis* Zone (Martini's NP13). The FAD of *Helicosphaera lophota* closely approximates the last abundant occurrence of *T. orthostylus*.

The zones were calibrated to magnetostratigraphic units primarily by graphic correlation modeling of the sections presented in Poore et al (in press) and Lowrie et al (1982).

TIME SCALE

As mentioned above, the result of graphic correlation is a scaled solution; that is, a scale along which datums are spaced. There is no *a priori* reason to conclude that there must be a linear relationship between composite units and time. However, this would be the simplest initial working hypothesis, one that must be tested in some manner. Linearity could be demonstrated if there existed radiometric dates on suitable minerals in rocks that could be directly or indirectly related with some accuracy to datums in the model. Unfortunately, although many Paleogene rocks have been dated (see Odin, 1982), too few of these rocks can be accurately related biostratigraphically to the model. Further, of those data that can be compared, most are based on glauconites and yield inconsistent results.

Time scales based on marine magnetic profiles and seafloor spreading rates provide an alternative method of testing linearity. Ness, Levi, and Couch (1980) presented such a time scale based on the South Atlantic that took into consideration the new internationally accepted decay constants. Figure 3 is a graphic correlation plot comparing the Ness, Levi, and Couch (1980) scale on the X-axis to our model on the Y-axis. In the body of the graph are positioned the tops and bottoms of the magnetic anomalies of the Paleogene (see LaBrecque, Kent and Cande, 1977; Ness, Levi and Couch, 1980; Lowrie et al, 1982).

The general linear arrangement of the anomalies indicates that the working hypothesis is acceptable. The Ness, Levi, and Couch (1980) scale shows the linearity of the graphic correlation model with a time scale based on spreading rates. However, we disagree with one of the calibration points used to construct the Paleogene part of the scale. We believe the 66.7 m.y. estimates for the Cretaceous-Tertiary boundary is too old and prefer 66.0 m.y. (from Obradovich and Cobban, 1975; recalculated according to Dalrymple, 1979). Further, preliminary work on spreading rates in different ocean basins suggests that the South Atlantic may not be the best of the ocean basins to use to derive a magnetic anomaly time scale for the early Paleocene and the later Late Cretaceous.

We conclude that with existing data, the best line of correlation between time and our biostratigraphic-magnetostratigraphic model is derived from using the 66.0 m.y. estimate for the Cretaceous-Tertiary boundary, which is fixed at 17.7 composite units (Cu) in the model, and we estimate 23.7 m.y. for the Oligocene-Miocene boundary as based on the LAD of *Cyclicargolithus abisectus* (ref. Muller, 1971; Wise, 1973) which has a value of 226.7 Cu in the model. The LAD of *C. abisectus* is associated with the Iversen Point basalts of California (Miller, 1981) dated by Turner (1970).

The composite unit values for fossil and magnetic datums in the model were converted to time using the formula for simple linear regression, $X = (Y-A) \div B$, and the values given above. Figure 2 shows the calibration of the zonal and magnetostratigraphic datums to time.

UNCONFORMITIES IN ALABAMA

Thierstein (1981) documented the calcareous nannofossil flora across the Cretaceous-Tertiary boundary in the Braggs

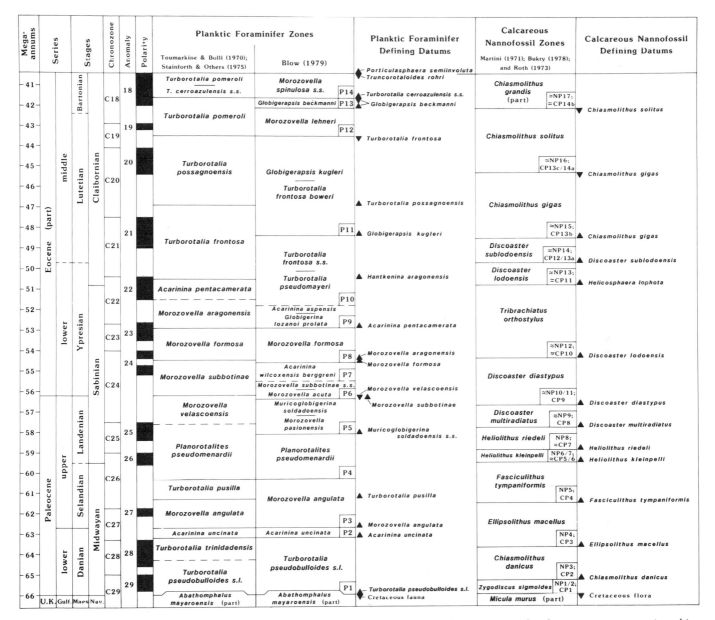

Figure 2: Calibration of Paleocene through middle Eocene planktic foraminifer zones, calcareous nannofossil zones, magnetostratigraphic units, and provincial and international stages. See text for discussion of the time scale. The figure is based on a model constructed by the U.S. Geological Survey.

section in central Alabama. Hay and Mohler (1967) also studied the nannofossils, and Olsson (1970) studied the planktic foraminifers from this locality. The basal part of the Pine Barren Member of the Clayton Formation here contains *Cruciplacolithus primus* (Perch-Nielsen, 1977) and is therefore within the *C. primus* zone of Romein (1979), which is equivalent to the middle part of the *Markalius inversus* zone (NP1) of Martini (1971). The FAD of *C. primus* in the model is calculated to be 65.8 m.y. The underlying Prairie Bluff Chalk contains *Micula murus* (ref. Martini, 1961; Bukry, 1973) apparently near its FAD. The FAD of *M. murus* at mid-latitudes is calculated to be at about 69.3 m.y. Thus, the unconformity between the Cretaceous and Tertiary represents a hiatus of about 3.5 m.y. Most of it represents missing Cretaceous.

Although transgressive and regressive phases are present

within the Midway Group, there seem to be no significant amounts of time unrepresented by sediment. The next major unconformity above the Cretaceous-Tertiary boundary lies between the Naheola and Nanafalia formations. Here, the short *Heliolithus kleinpelli* zone and perhaps some small parts of the underlying *Fasiculithus tympaniformis* and/or overlying *Heliolithus riedeli* zones are apparently missing. The Gravel Creek Sand Member, which is generally considered the lowest member of the Nanafalia Formation, but which does not appear to be genetically related to the Nanafalia, contains a pollen flora most similar to that of the Naheola Formation (Frederiksen, 1980). For the purposes of the present paper, the Gravel Creek is considered to be a part of the Naheola Formation and is assigned to the chronozone of the *Fasiculithus tympaniformis* zone. The Naheola-Nanafalia unconformity is calculated to represent a

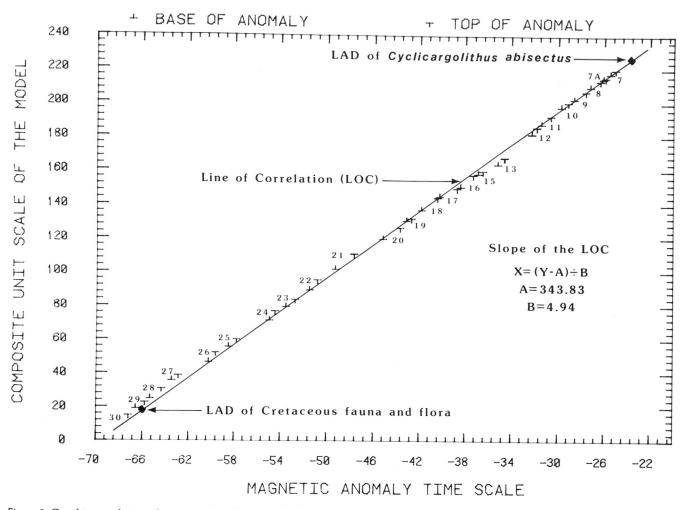

Figure 3: Graphic correlation plot comparing the ages calculated for Paleogene anomalies by Ness, Levi, and Couch (1980) on the X-axis to the position of these anomalies in a biostratigraphic-magnetostratigraphic model on the Y-axis. The position of the Cretaceous-Tertiary boundary (LAD of Cretaceous fauna and flora) and the LAD of the nannofossil *Cyclicargolithus abisectus*, which is used to mark the Oligocene-Miocene boundary, are indicated by black diamonds. These two points are used to calculate a line of correlation (LOC) between the model and time. See text for discussion.

hiatus of about 1.0 m.y. in Alabama.

The next significant hiatus is within the lower Eocene. In the shallow subsurface, clastic deposits referable to the *Tribrachiatus orthostylus* zone are present (Gibson, Mancini, and Bybell, 1982). These underlie lithologically similar deposits of the *Discoaster lodoensis* zone (Gibson and Bybell, 1981; Bramlette and Sullivan, 1961). Graphic correlation plots suggest that the contact between the *T. orthostylus* zone and the *Discoaster lodoensis*-zone deposits in this area is an unconformity representing a hiatus equivalent to the upper part of the *T. orthostylus* zone. We calculate the duration of this hiatus to be about 1.4 m.y. As mentioned, these two units are lithologically similar to each other and to the Tallahatta Formation.

The Tallahatta at the well-known Little Stave Creek locality contains a *Discoaster sublodoensis*-zone assemblage. We

consider the *Discoaster lodoensis*-zone deposits to also represent the Tallahatta. Until the regional stratigraphy is better understood, we refer to the *Tribrachiatus orthostylus*-zone deposits as "Tallahatta" Formation. The latter are very probably the marine equivalent of the Meridian Sand of Mississippi and far western Alabama.

The Tallahatta is overlain in sharp contact by the Lisbon Formation. Stenzel (1952) concluded that this contact represented a significant hiatus. However, we find no evidence to support this. The lower Lisbon contains *Chiasmolithus gigas* and *Lophodolithus mochlophorous* and other forms suggesting a normal succession from the *Discoaster sublodoensis*-zone flora of the Tallahatta.

Stenzel (1952) also concluded that a major hiatus was represented by an unconformity between the lower Lisbon (*Cubitostrea lisbonensis* zone) and the upper Lisbon (*C. sel-*

laeformis zone). Existing data suggest that the lower Lisbon is no younger than the top of the *Chiasmolithus gigas* zone, which we calibrate at 45.3 m.y. However, data are too poor for the lower part of the upper Lisbon for us to estimate its age. The flora of the upper part of the upper Lisbon is consistent with assignment to the *Chiasmolithus solitus* zone.

Our samples of the overlying Gosport Formation are also relatively poor in microfossils. The upper beds of the Lisbon contain *Campylosphaera dela*, the LAD of which we calculate at 42.7 m.y. Hardenbol and Berggren (1978) indicate that the Gosport can be assigned to the *Globigerapsis beckmanni* zone and/or the *Truncorotaloides rohri* zone. The base of the *G. beckmanni* zone is defined by the FAD of the nominate taxon, which we calculate to be at about 42.1 m.y. The Lisbon-Gosport unconformity may represent a hiatus of at least 0.6 m.y.

In addition to the references cited in the section above, valuable biostratigraphic data for the Alabama units were given by Loeblich and Tappan (1957a, 1957b), Berggren (1965), Mancini (1981), and Mancini and Oliver (1981). The positions of hiatuses in the Alabama section are shown in Figure 4.

UNCONFORMITIES IN SOUTH CAROLINA

The data in Figure 4 for the South Carolina section come primarily from the U.S. Geological Survey Clubhouse Crossroads Core Hole No. 1 in Dorchester County (Hazel et al, 1977). Major unconformities lie at the base of the Black Mingo Formation, the base of the Fishburne Formation (Gohn et al, 1983), and the base of the Santee Limestone.

The basal part of the Black Mingo Formation in the Clubhouse Crossroads Core contains ostracodes and calcareous nannofossils indicating a chronostratigraphic assignment no older than the middle of the *Chiasmolithus danicus* zone, about 64.6 m.y. In the underlying Cretaceous Peedee Formation, Hattner and Wise (1980) reported that the upper beds contain the nannofossil *Lithraphidites quadratus (ref. Bramlette and Martini, 1964)*. The top of the Peedee here is interpreted by the authors to be at least as young as the upper part of the *Lithraphidites quadratus* zone, or about 71.0 m.y. Thus, the Cretaceous-Tertiary unconformity here may represent a hiatus of about 6.4 m.y.

Four transgressive-regressive cycles can be recognized within the Black Mingo, but the data are insufficient to determine the hiatuses represented by observed contacts. The FAD of *Discoaster multiradiatus (ref. Bramlette and Riedel, 1954)* was encountered about 1 m below the top of the Black Mingo. We place the top of the Black Mingo at about 57.6 m.y. A thin calcareous unit, the Fishburne Formation, unconformably overlies the Black Mingo. It contains a calcareous nannofossil, dinoflagellate, ostracode, and foraminifer assemblage indicative of the middle part of the chronozone of the *Discoaster diastypus* zone, about 55.4 m.y. The unconformity, therefore, represents a hiatus of about 2.2 m.y.

Overlying the Fishburne disconformably is the Santee Limestone. Although the preservation of the microfauna and microflora is poor, the available macrofossil and microfossil data suggest that the Santee correlates with the *Cubi-*

tostrea sellaeformis-zone part of the Lisbon Formation of the Gulf Coast and the Gosport Formation. The unconformity between the Fishburne and the Santee represents a hiatus of about 10.0 m.y.

The upper part of the Santee (the Cross Member) was assigned to the upper Eocene by Hazel et al (1977). This is in error. The assignment was based on the presence of *Sphenolithus pseudoradians (ref. Bramlette and Wilcoxon, 1967)* the first occurrence of which defines the base of Martini's (1971) *S. pseudoradians* zone, and the secondary markers *Helicosphaera bramlettei (ref. Muller, 1970; Jafar and Martini, 1975), H. euphratis (ref. Haq, 1966), H. intermedia* Martin, 1965 and *Sphenolithus obtusus* Bukry, 1971, which were thought to have evolved in the late Eocene. *Sphenolithus pseudoradians* is now known to range into the middle Eocene and Martini (1976) abandoned the zone. Subsequent work on the calcareous nannofossils of the Clubhouse Crossroads Core Hole No. 1 has resulted in the identification of *Helicosphaera seminulum* (Bramlette and Sullivan, 1961) from the upper part of the Cross Member. This species does not range above the middle Eocene (for example, see Okada and Thierstein, 1979; Roth, 1973). *Sphenolithus obtusus* occurs in the middle Eocene (Okada and Thierstein, 1979). Because of taxonomic problems and preservation we are no longer confident of our identification of *Helicosphaera euphratis* and *H. intermedia*. We have observed *H. bramlettei* in middle Eocene rocks elsewhere in the Coastal Province.

The presence of *Helicospharea seminulum* 1.0 m below the top of the Cross in the Clubhouse Crossroads Corehole No. 1 and *Truncorotaloides rohri* in the upper 0.33 m indicates that the top of the Santee is no younger than middle Eocene, which differs from the interpretation of Powell and Baum (1982) who placed the Cross Member in the Priabonian Stage (upper Eocene). We estimate the age of the top of the Santee Formation to be about 40.5 m.y.

Figure 4 gives the positions of hiatuses in the South Carolina section.

UNCONFORMITIES IN VIRGINIA-MARYLAND

The section shown in Figure 4 is composited from sections and wells studied from the area of the Pamunkey River in central Virginia to Prince Georges County, Maryland. Significant unconformities occur at the Cretaceous-Tertiary boundary, the Brightseat-Aquia contact, and above and below the Piney Point Formation.

The oldest Tertiary unit in the area is the Brightseat Formation. The Brightseat contains an ostracode, dinoflagellate, foraminifer, and calcareous nannofossil assemblage indicating placement in the *Chiasmolithus danicus* zone. It appears to correlate with the McBryde Limestone Member of the Clayton Formation in Alabama. A maximum duration for deposition of the Brightseat is about 65.4 m.y. to about 63.8 m.y. (time of the *Chiasmolithus danicus* zone). If the Brightseat accounts for only part of the zone, it probably represents an even shorter period of time. In Prince Georges County, the Brightseat overlies the Severn Formation. The Severn contains the LAD of the distinctive ostracode *Fissocarinocythere pidgeoni (ref. Berry, 1925)*, which is

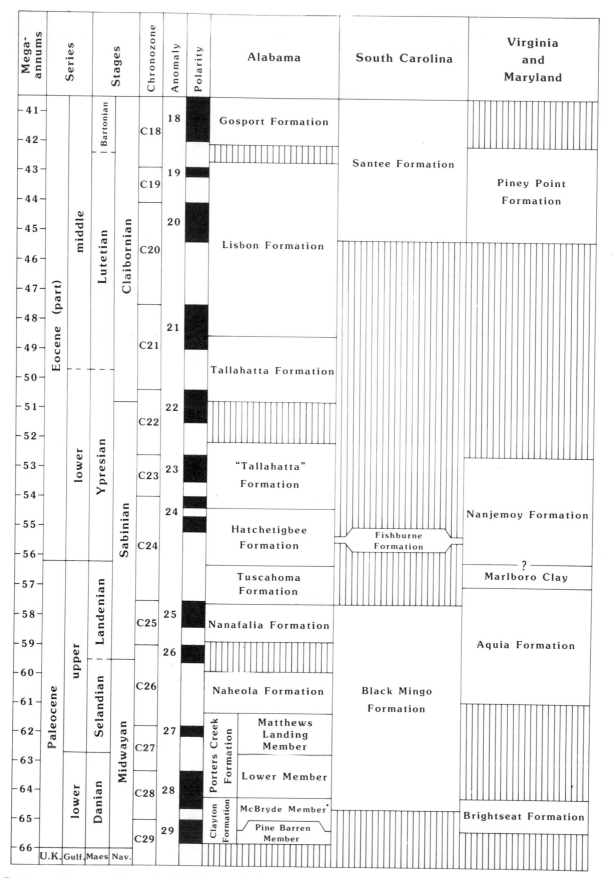

Figure 4: Positions of important hiatuses in the Paleocene through middle Eocene in Alabama, South Carolina, and Virginia - Maryland. See Figure 2 for relationships to zonations.

calculated at about 70.7 m.y. Thus, the Cretaceous-Tertiary unconformity in this area represents a hiatus of about 5.3 m.y.

The Brightseat is overlain disconformably by the Piscataway Member of the Aquia Formation. The lower part of the Piscataway contains a *Fasciculithus tympaniformis*-zone nannofossil assemblage. We calculate the age of the base of the Aquia at about 60.9 m.y. and the duration of the Brightseat-Aquia hiatus as 2.9 m.y.

Regressive and transgressive cycles can be recognized in the Aquia, Marlboro Clay, and Nanjemoy formation sequences, but none of the observed contacts seems to represent a significant hiatus. The top of the Nanjemoy in two coreholes in Virginia lies within the *Tribrachiatus orthostylus* zone at about 52.6 m.y. In the Oak Grove core, Westmoreland County (Gibson et al, 1980; Reinhardt, Newell and Mixon, 1980), the lower Eocene Nanjemoy is overlain by the Miocene Calvert Formation. In the Putneys Mill core in New Kent County and in outcrop along the Pamunkey River, the Nanjemoy is overlain disconformably by the Piney Point Formation. The Piney Point contains a middle Eocene, Claibornian fauna and flora that correlates the Piney Point with the upper part of the Lisbon Formation. At Putneys Mill, the unconformity between the Nanjemoy and the Piney Point represents a hiatus of about 7.3 m.y.

Latest middle Eocene (Gosport equivalent) is seemingly absent in this region. In the southern part of the Salisbury Embayment in exposures and in the shallow subsurface, the Piney Point is overlain by a lithologically distinct, thin unit of latest Oligocene or earliest Miocene age. The name Piney Point has been erroneously used for this unit (for example, Olsson, Miller, and Ungrady, 1980). Brown, Miller and Swain (1972) correctly point out that the type Piney Point Formation contains a Claibornian (middle Eocene) ostracode assemblage.

Figure 4 gives the positions of hiatuses in the Virginia-Maryland section.

SUMMARY

In the sections studied, the most significant unconformity within the Paleogene, in terms of hiatus and lateral extent, is in the lower Eocene. In South Carolina, the time not represented by deposits centers around about 50.3 m.y. In the Salisbury Embayment, deposition seems to have continued later (or sediments have not been eroded away), and the hiatus centers around 49.0 m.y. In Alabama, the major unconformity detected is in the lower Eocene within a series of similar, Tallahatta-like lithologies. The hiatus here centers around 51.4 m.y.

A significant unconformity exists in all three areas between the Cretaceous and Tertiary. This unconformity is much more extensive along the middle and southern Atlantic Coast where Tertiary sedimentation apparently did not begin until *Chiasmolithus danicus*-zone time (ca 65 m.y.). In the three areas, the top of the Cretaceous ranges from the lower part of the upper Maestrichtian to the upper part of the middle Maestrichtian. The presence of a significant unconformity at the Cretaceous-Tertiary boundary in the sections studied appears to be the general rule for the Atlantic and Gulf Coastal Province, but there may be exceptions.

For example, Ganapathy, Gartner and Jiang (1981) documented what seems to be a continuous section across the Cretaceous-Tertiary boundary in Texas.

REFERENCES CITED

Berggren, W.A., 1965, Some problems of Paleocene-lower Eocene planktonic foraminiferal correlations: Micropaleontology, v. 11, p. 278-300.

Blow, W.H., 1979, The Cainozoic Globigerinida: Leiden, E.J. Brill, v. 1-3, p. 1-1413, pls. 1-264.

Bramlette, M.M., and F.R. Sullivan, 1961, Coccolithophorids and related nannoplankton of the Early Tertiary in California: Micropaleontology, v. 7, p. 129-174.

Brown, P.M., J.A. Miller, and F.M. Swain, 1972, Structural and stratigraphic framework, and spatial distribution of permeability of the Atlantic Coastal Plain, North Carolina to New York: U.S. Geological Survey Professional Paper 796, 79 p.

Bukry, D., 1978, Biostratigraphy of Cenozoic marine sediment by calcareous nannofossils: Micropaleontology, v. 24, p. 44-60.

Dalrymple, G.B., 1979, Critical tables for conversion of K-Ar ages from old to new constants: Geology, v. 7, p. 558-560.

Davaud, E., and J. Guex, 1978, Traitement analytique "manuel" et algorithmique de problèmes complexes de corrélations biochronologiques: Eclogae Geologicae Helvetiae, v. 71, p. 581-610.

Frederiksen, N.O., 1980, Lower Tertiary sporomorph biostratigraphy: Preliminary report, in Juergen Reinhardt and T.G. Gibson, eds., Upper Cretaceous and lower Tertiary geology of the Chattahoochee River Valley, western Georgia and eastern Alabama, in R.W. Frey, ed., Excursions in southeastern geology, v. 2: Atlanta, Geological Society of America, 93rd Annual Meeting Field Trip Guidebooks, p. 422-423.

Ganapathy, R., S. Gartner, and M.J. Jiang, 1981, Iridium anomaly at the Cretaceous-Tertiary boundary in Texas: Earth and Planetary Science Letters, v. 54, p. 393-396.

Gibson, T.G., et al, 1980, Biostratigraphy of the core, in Geology of the Oak Grove core: Virginia Division of Mineral Resources Publication 20, p. 14-30.

Gibson, T.G., and L.M. Bybell, 1981, Facies changes in the Hatchetigbee Formation in Alabama-Georgia and the Wilcox-Claiborne Group unconformity: Gulf Coast Association of Geological Societies, Transactions, v. 31, p. 301-306.

Gibson, T.G., E.A. Mancini and L.M. Bybell, 1982, Paleocene to middle Eocene stratigraphy of Alabama: Gulf Coast Association of Geological Societies, Transactions, v. 32, p. 449-458.

Gohn, G.S., et al, 1983, The Fish-Bunne Formation (Lower Eocene) a newly defined subsurface unit in the South Carolina Coastal Plain: U.S. Geological Survey Bulletin 1537-C, p. C1-C16.

Hardenbol, J., and W.A. Berggren, 1978, A new Paleogene numerical time scale, in G.V., Cohee, M.F. Glaessner, and H.D. Hedberg, eds., Contributions to the geologic time scale: AAPG, Studies Geology, v. 6, p. 213-234.

Hattner, J.G., and S.W. Wise, Jr., 1980, Upper Cretaceous

calcareous nannofossil biostratigraphy of South Carolina: South Carolina Geology, v. 24, p. 41-117.

Hay, W.W., and H.P. Mohler, 1967, Calcareous nannoplankton from early Tertiary rocks at Pont Labau, France, and Paleocene-early Eocene correlations: Journal of Paleontology, v. 41, p. 1505-1541.

Hazel, J.E., et al, 1977, Biostratigraphy of the deep corehole (Clubhouse Crossroads Corehole 1) near Charleston, South Carolina: U.S. Geological Survey Professional Paper 1028-F, p. 71-89.

LaBrecque, J.L., D.V. Kent, and S.C. Cande, 1977, Revised magnetic polarity for Late Cretaceous and Cenozoic time: Geology, v. 5, p. 330-335.

Loeblich, A.R., Jr., and Helen Tappan, 1957a, Planktonic foraminifera of Paleocene and early Eocene age from the Gulf and Atlantic Coastal Plains: U.S. National Museum Bulletin 215, p. 173-198.

———, and ———, 1957b, Correlation of the Gulf and Atlantic Coastal Plain Paleocene and lower Eocene formations by means of planktonic foraminifera. Journal of Paleontology, v. 31, p. 1109-1137.

Lowrie W., et al, 1982, Paleogene magnetic stratigraphy in Umbrian pelagic carbonate rocks: The Contessa sections, Gubbio: Geological Society of America, Bulletin, v. 93, p. 414-432.

Mancini, E.A., 1981, Lithostratigraphy and biostratigraphy of Paleocene subsurface strata in southwest Alabama: Gulf Coast Association of Geological Societies, Transactions, v. 31, p. 359-367.

Mancini, E.A., and G.E. Oliver, 1981, Planktic foraminifers from the Tuscahoma Sand (upper Paleocene) of southwest Alabama: Micropaleontology, v. 27, p. 204-225.

Martini, E., 1971, Standard Tertiary and Quaternary calcareous nannoplankton zonation, in A. Farinacci, ed., Proceedings, Second Planktonic Conference: Rome, Edizioni Tecnoscienza, v. 2, p. 739-785.

Martini, E., 1976, Cretaceous to Recent calcareous nannoplankton from the central Pacific Ocean (DSDP Leg 33), in Initial reports of the deep sea drilling project: Washington, D.C., U.S. Government Printing Office, v. 33, p. 838-423.

Miller, P.L., 1981, Tertiary calcareous nannoplankton and benthic foraminifera biostratigraphy of the Point Arena area, California: Micropaleontology, v. 27, p. 419-443.

Miller, F.X., 1977, The graphic correlation method, in E.G. Kauffman, and J.E. Hazel, eds., Concepts and methods in biostratigraphy: Stroudsburg, Pennsylvania, Dowden, Hutchinson and Ross, Inc., p. 165-186.

Ness, G., S. Levi, and R. Couch, 1980, Marine magnetic anomaly time scales for the Cenozoic and Late Cretaceous: a précis, critique and synthesis: Review of Geophysics and Space Physics, v. 18, p. 753-770.

Obradovich, J.D., and W.A. Cobban, 1975, A time scale for the Late Cretaceous of the Western Interior of North America: Geological Association of Canada, Special Paper 13, p. 31-54.

Odin, G.S., ed., 1982, Numerical dating in stratigraphy: Chichester, John Wiley and Sons, parts 1 and 2, p. 1-630.

Okada, H., and H.R. Thierstein, 1979, Calcareous nannoplankton - leg 43, Deep Sea Drilling Project, in Initial reports of the deep sea drilling project: Washington, D.C., U.S. Government Printing Office, v.43, p. 507-573.

Olsson, R.K., 1970, Planktonic foraminifera from base of Tertiary, Millers Ferry, Alabama: Journal of Paleontology, v. 44, p. 598-604.

Olsson, R.K., K.G. Miller, and T.E. Ungrady, 1980, Late Oligocene transgression of Middle Atlantic Coastal Plain: Geology, v. 8, p. 549-554.

Perch-Nielsen, K, 1977, Albian to Pleistocene calcareous nannofossils from the western South Atlantic, DSDP Leg 39, in Initial reports of the deep sea drilling project: Washington, D.C., U.S. Government Printing Office, v. 39, p. 147-176.

Poore, R.Z., et al, in press, Late Cretaceous-Cenozoic magnetostratigraphic and biostratigraphic correlations of the South Atlantic Ocean, DSDP Leg 73, in Initial reports of the deep sea drilling project: Washington, D.C., U.S. Government Printing Office, v. 73.

Powell, R.J., and G.R. Baum, 1982, Eocene biostratigraphy of South Carolina and its relationship to Gulf Coastal Plain zonations and global changes of coastal onlap: Geological Society of America, Bulletin, v. 93, p. 1099-1108.

Reinhardt, J., W.L. Newell, and R.B. Mixon, 1980, Tertiary lithostratigraphy of the core, in Geology of the Oak Grove core: Virginia Division of Mineral Resources Publication 20, p. 1-13.

Roth, P.H., 1973, Calcareous nannofossils - Leg 17, Deep Sea Drilling Project, in Initial reports of the deep sea drilling project: Washington, D.C., U.S. Government Printing Office, v. 17, p. 695-795.

Romein, A.J.T., 1979, Lineages in Early Paleogene calcareous nannoplankton: Utrecht Micropaleontology Bulletin, v. 22, p. 5-231.

Shaw, A.B., 1964, Time in stratigraphy: New York, McGraw-Hill Book Co., 365 p.

Stainforth, R.M., et al, 1975, Cenozoic planktonic foraminiferal zonation and characteristics of index forms: Kansas University Paleontology Contributions, Article 62, 425 p.

Stenzel, H.B., 1952, Correlation chart of Eocene at outcrop in eastern Texas, Mississippi, and western Alabama: Mississippi Geological Society Guidebook, 9th Field Trip, p. 32-33.

Thierstein, H.R., 1981, Late Cretaceous nannoplankton and the change at the Cretaceous-Tertiary boundary: Society of Economic Paleontologists and Mineralogists, Special Publication 32, p. 355-394.

Toumarkine, M., and H.M. Bolli, 1970, Evolution de Globorotalia cerroazulensis (Cole) dans l'Eocène moyen et supérieur de Possagno (Italie): Revue de Micropaléontologie, v. 13, p. 131-145.

Turner, D.L., 1970, Potassium-argon dating of Pacific coast Miocene foraminiferal stages, in O.L. Bandy, ed., Radiometric dating and paleontologic zonation. Geological Society of America, Special Paper 124, p. 91-129.

Regional Unconformities and Depositional Cycles, Cretaceous of the Arabian Peninsula

P.M. Harris
Gulf Research and Development Company
Houston, Texas

S.H. Frost
G.A. Seiglie
N. Schneidermann
Gulf Oil Exploration and Production Company
Houston, Texas

Three regional unconformities are recognized during stratigraphic studies of Cretaceous sedimentary rocks of the Arabian Peninsula. The Cretaceous sequence can be divided into early (Neocomian through early Aptian), middle (late Aptian through early Turonian), and late (Coniacian through Maestrichtian) units by three unconformities of major importance: (1) an unconformity that represents erosion or nondeposition of the middle Aptian section and marks a significant lithological change from shallow shelf limestones of the Shuaiba Formation to overlying shallow shelf clastics of the Nahr Umr Formation; (2) an unconformity that reflects erosion or nondeposition of Turonian strata and is a break between shallow shelf limestones of the Mishrif Formation and overlying deep-water beds of the Fiqa Formation; and (3) an unconformity that separates the Cretaceous and Tertiary formations capping shallow shelf limestones of the Simsima Formation. A regional unconformity at the Jurassic-Cretaceous boundary has not been recognized.

The subaerial exposure and erosion represented by each of the three unconformities is of great economic importance due to the development of reservoir-quality secondary porosities in rudist reef-bearing limestones. When the ages of these unconformities and the intervening depositional cycles in the Arabian Peninsula are compared with published eustatic curves, several differences are evident.

INTRODUCTION

The Arabian Peninsula has been the focus of considerable geological research because it contains the majority of the world's giant petroleum reservoirs. Hydrocarbons have accumulated in carbonate and siliciclastic rocks of several ages, but Mesozoic carbonates contain most of the reservoirs. This paper describes and interprets the regional stratigraphic changes in Cretaceous formations of the Arabian Peninsula. In terms of petroleum reserves, the most important of these changes were caused by intermittent subaerial erosion, marine transgressions and attendant deposition of carbonate sediments, and marine regressions and the associated deposition of clastic sediments across a vast carbonate platform.

The most significant variations in the regional stratigraphy, from a hydrocarbon exploration standpoint, were caused by changes in relative sea level over the carbonate platform. These changes in water depth have been recognized throughout the region and, furthermore, have been used to interpret individual stratigraphic sections and identify localized depositional events. Our interpretation of changes of relative sea level and associated sedimentation on the Arabian Peninsula benefitted this study by: (1) helping to determine the position of prominent seismic reflectors within the Cretaceous sequence; (2) providing insight into regressive and transgressive stratigraphic sequences between seismic reflectors; (3) focusing interest on stratigraphic units in which subaerial exposure and resultant formation of attractive moldic porosities are more likely; (4) refining our understanding of the burial history of various stratigraphic units and thus the timing of thermal maturation and possible migration of hydrocarbons; and, (5) forcing the development of a regional Cretaceous model for the deposits providing some predictability in the area before drilling and a guide for interpreting the drilled section.

REGIONAL SETTING

The geological setting of the Arabian Peninsula can be greatly simplified into three major elements (Figure 1): (1) the Arabo-Nubian Shield, a land area several hundreds of kilometers in width in Saudi Arabia and narrowing toward the northwest and south; (2) a belt of clastic sediments also hundreds of kilometers in width, that rims the eastern margin of the shield and comprises both continental and shallow marine deposits derived from the shield; and, (3) a carbon-

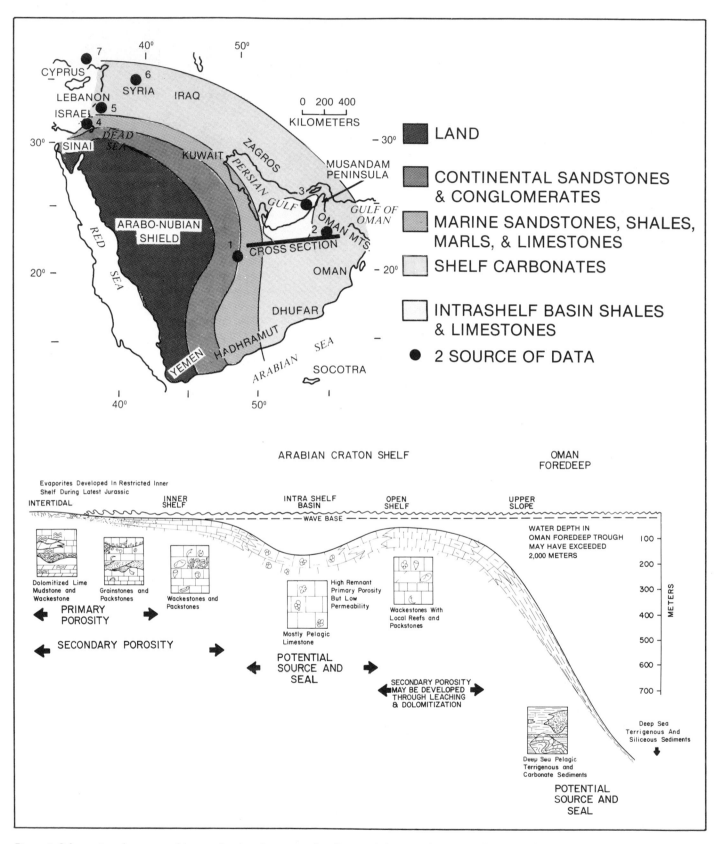

Figure 1: Schematic paleogeographic map (top) and cross section (bottom) showing depositional setting of Cretaceous lithologies. As shown on the map, siliciclastic and carbonate sediments form broad, continuous belts that rim the land area formed by the Arabo-Nubian Shield. Deeper-water carbonate sediments formed within intrashelf basins. Numbers approximately locate our main sources of data: (1) eastern Saudi Arabia; (2) northern Oman; (3) eastern offshore Persian Gulf; (4) Israel; (5) Lebanon; (6) Syria; and (7) Turkey. Cross section relates lithofacies and depositional environments and suggests the importance of the lithofacies to reservoir development.

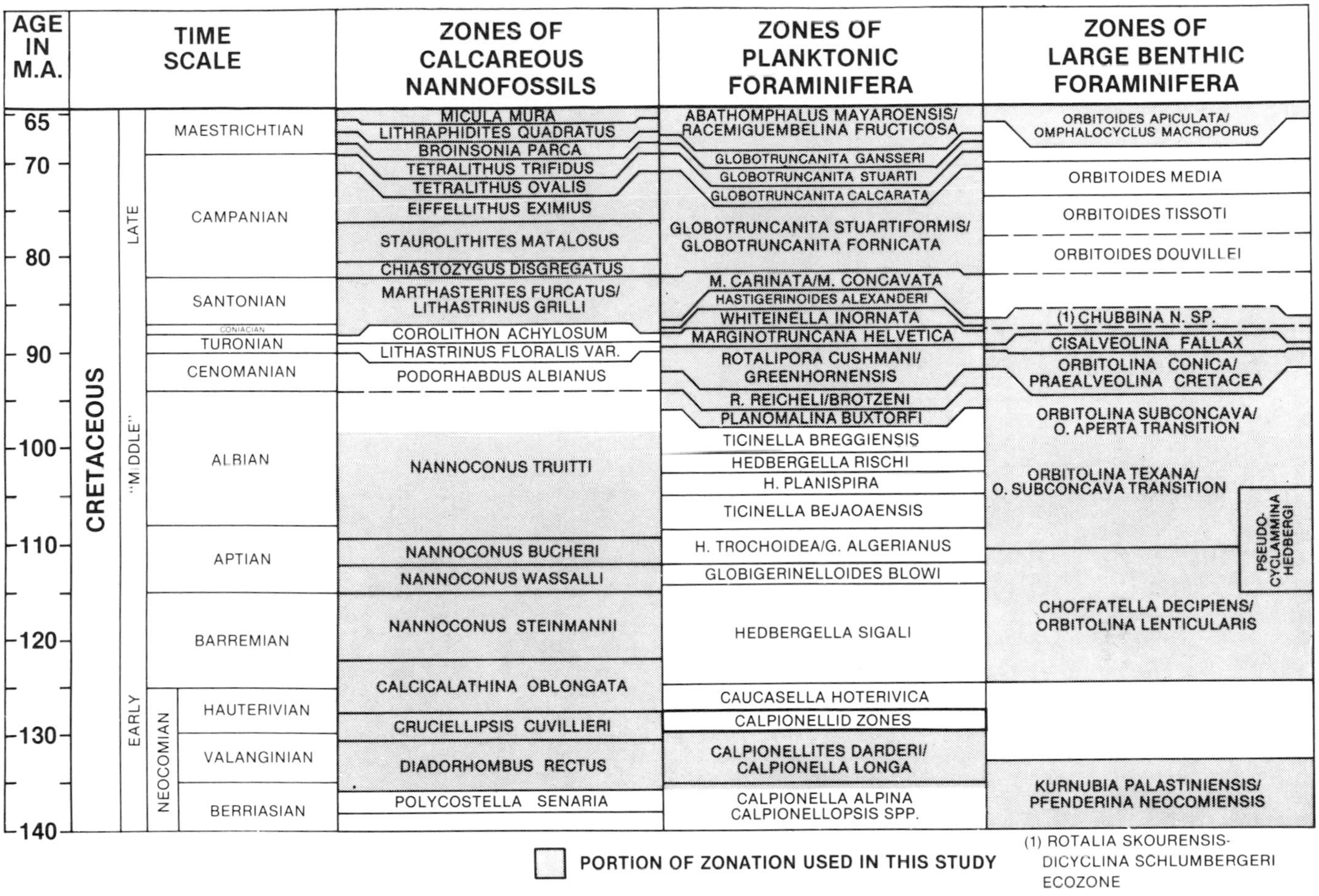

AGE IN M.A.	TIME SCALE			ZONES OF CALCAREOUS NANNOFOSSILS	ZONES OF PLANKTONIC FORAMINIFERA	ZONES OF LARGE BENTHIC FORAMINIFERA
65	CRETACEOUS	LATE	MAESTRICHTIAN	MICULA MURA / LITHRAPHIDITES QUADRATUS / BROINSONIA PARCA	ABATHOMPHALUS MAYAROENSIS/RACEMIGUEMBELINA FRUCTICOSA	ORBITOIDES APICULATA/OMPHALOCYCLUS MACROPORUS
70			CAMPANIAN	TETRALITHUS TRIFIDUS / TETRALITHUS OVALIS / EIFFELLITHUS EXIMIUS	GLOBOTRUNCANITA GANSSERI / GLOBOTRUNCANITA STUARTI / GLOBOTRUNCANITA CALCARATA	ORBITOIDES MEDIA
80				STAUROLITHITES MATALOSUS / CHIASTOZYGUS DISGREGATUS	GLOBOTRUNCANITA STUARTIFORMIS/GLOBOTRUNCANITA FORNICATA	ORBITOIDES TISSOTI / ORBITOIDES DOUVILLEI
		"MIDDLE"	SANTONIAN	MARTHASTERITES FURCATUS/LITHASTRINUS GRILLI	M. CARINATA/M. CONCAVATA / HASTIGERINOIDES ALEXANDERI / WHITEINELLA INORNATA	(1) CHUBBINA N. SP.
90			CONIACIAN / TURONIAN	COROLITHON ACHYLOSUM / LITHASTRINUS FLORALIS VAR.	MARGINOTRUNCANA HELVETICA	CISALVEOLINA FALLAX
			CENOMANIAN	PODORHABDUS ALBIANUS	ROTALIPORA CUSHMANI/GREENHORNENSIS / R. REICHELI/BROTZENI / PLANOMALINA BUXTORFI	ORBITOLINA CONICA/PRAEALVEOLINA CRETACEA / ORBITOLINA SUBCONCAVA/O. APERTA TRANSITION
100			ALBIAN	NANNOCONUS TRUITTI	TICINELLA BREGGIENSIS / HEDBERGELLA RISCHI / H. PLANISPIRA / TICINELLA BEJAOAENSIS	ORBITOLINA TEXANA/O. SUBCONCAVA TRANSITION (PSEUDOCYCLAMMINA HEDBERGI)
110			APTIAN	NANNOCONUS BUCHERI / NANNOCONUS WASSALLI	H. TROCHOIDEA/G. ALGERIANUS / GLOBIGERINELLOIDES BLOWI	CHOFFATELLA DECIPIENS/ORBITOLINA LENTICULARIS
120			BARREMIAN	NANNOCONUS STEINMANNI	HEDBERGELLA SIGALI	
		EARLY	HAUTERIVIAN	CALCICALATHINA OBLONGATA / CRUCIELLIPSIS CUVILLIERI	CAUCASELLA HOTERIVICA / CALPIONELLID ZONES	
130		NEOCOMIAN	VALANGINIAN	DIADORHOMBUS RECTUS	CALPIONELLITES DARDERI/CALPIONELLA LONGA	KURNUBIA PALASTINIENSIS/PFENDERINA NEOCOMIENSIS
140			BERRIASIAN	POLYCOSTELLA SENARIA	CALPIONELLA ALPINA CALPIONELLOPSIS SPP.	

▢ PORTION OF ZONATION USED IN THIS STUDY

(1) ROTALIA SKOURENSIS-DICYCLINA SCHLUMBERGERI ECOZONE

Figure 2: Successions of index fossils for the Cretaceous. Time scale is based on Van Hinte (1976), Kauffman (1977), Childs (1981; COSUNA), and internal data.

ate shelf hundreds of kilometers in width and lying seaward of the clastic belt, upon which predominantly shallow shelf carbonates were deposited. The shelf contains intrashelf basins filled by shales and deeper-water limestones.

These broadly-defined lithologic belts are of course more complexly interrelated than shown. Their distribution was examined in detail by Saint Marc (1978) and Murris (1980) and is discussed later in this paper. It is largely the relationships through time between these lithologic belts that provide data important to interpreting the depositional cycles and relative sea-level changes in the area. For example, the significant transgressions of shelf carbonates onto the clastic sediments and shield signifies increasing water depth, due either to eustatic rise or to subsidence caused by tilting of the shield. Conversely, the progradation of clastic deposits over long distances onto the carbonate shelf signifies decreasing water depth, due either to an eustatic drop or uplift of the shield. These generalized relations hold true for most of the geological section; however, departures are observed. In cases where both shallow- and deeper-water shales are present, we have relied on fossil assemblages to interpret paleobathymetry. In general, regional changes in sedimentation patterns during the Jurassic and most of the Cretaceous appear to have been controlled more by sea-level fluctuation than by tectonics.

BIOSTRATIGRAPHY

Biostratigraphic studies of the Cretaceous are instrumental in allowing us to determine the age of various formations, rates of sedimentation, the amount of time represented by unconformities, and paleodepths. The fossils most useful are calcareous nannofossils, planktonic foraminifers, and large benthic foraminifers (Figure 2). Each microfossil group has great utility but they are limited in some lithologic units due to unfavorable depositional environments. For example, large benthic foraminiferal zones are the basis for our biostratigraphic studies of shallow shelf carbonates of the Early and middle Cretaceous (Barremian through Turonian), whereas calcareous nannofossils and planktonic foraminifers permit a detailed zonation of marls and shales of the Late Cretaceous.

The sequence of planktonic foraminifers is based in part on the work of Bolli (1957) and Van Hinte (1976). Not all of the zones are represented in the area: (1) the upper *Globotruncanita? gansseri* zone and the *Abathomphalus mayaroensis* zone are not represented because the waters were too shallow during the deposition of the Simsima Formation; (2) the *Planoglobulina glabrata* ecozone of the late Santonian has only a local character and its age was determined with nannofossils; (3) during the Coniacian, the *Hastigeri-*

noides alexanderi ecozone has only a local but constant character; (4) the Turonian biostratigraphic zones are truncated by erosion; and (5) below the *Planomalina buxtorfi* zone no planktonic foraminiferal zone was recognized because waters were too shallow.

The late Maestrichtian benthic foraminiferal zone, *Orbitoides apiculata-Omphalocyclus macroporus* zone, is well represented in the shallow-water deposits of the Simsima Formation. No benthic-foraminiferal zone is present down to the early Coniacian where a *Rotalia skourensis - Dicyclina schlumbergeri* ecozone is represented locally. Below, because the Turonian is truncated, only the Cenomanian part of the *Cisalveolina fallax* zone is present.

DEPOSITIONAL CYCLES

The stratigraphy of the Cretaceous, as summarized on Figure 3, is based on literature study, on outcrop samples collected in Oman and the Musandam Peninsula, and on cuttings, conventional cores, and wire-line logs from more than 30 wells in the region. We do not recognize a regional unconformity at the base of the Cretaceous section (see also Soliman and Al Shamlan, 1982). Although relative sea level appears to have been rising through the Late Jurassic, the sedimentation rates on the carbonate shelf kept pace, maintaining shallow shoal to arid supratidal depositional conditions over much of the area. Representative stratigraphic sections of the Cretaceous from the localities numbered in Figure 1 are each segmented into three distinct large-scale units by regional unconformities of major importance: (1) an unconformity that reflects erosion or nondeposition of the middle Aptian section and marks a major change in depositional facies from shallow shelf limestones below to shallow shelf clastic sediments above; (2) an unconformity, during which Turonian deposits were removed, representing a break between shallow shelf limestones and the overlying deep-water deposits; and (3) an unconformity that defines the boundary between the Cretaceous and Tertiary and caps a shallow shelf limestone deposit.

The three large-scale lithic units of the Cretaceous stratigraphic section bounded by these regional unconformities are Early, middle, and Late Cretaceous in age. The position of the unconformities in the Cretaceous and the paleodepth interpretations of the sequences between the breaks in deposition are summarized by a paleodepth curve (Figure 4). Deposition occurred in three large-scale cycles, each with its own characteristics. The Upper Cretaceous cycle represents the most extensive transgression of the Arabo-Nubian Shield, whereas the middle Cretaceous cycle appears to have been a minimal transgression onto the craton. The Lower Cretaceous cycle has an intermediate position.

Lower Cretaceous Deposits

The Lower Cretaceous Thammama Group, approximately 2,000 ft (610 m) thick, includes beds of Neocomian (Berriasian through Hauterivian), Barremian, and Aptian ages. These strata were deposited during an extensive marine incursion over the Arabian Peninsula when carbonate shelf to shallow basin conditions were established over a

time span of nearly 30 m.y. Although the Lower Cretaceous sequence varies regionally, it consists of deep-shelf carbonate deposits for most of the lower two-thirds of its thickness and in its upper one-third changes to a shallowing upward sequence. Deep-water carbonates occur locally in the upper one-third of the sequence where intrashelf basins were developed. Clastic sediments are more prevalent in eastern Saudi Arabia (Powers et al, 1966; Soliman and Al Shamlan, 1982) than they are in central Oman or offshore Abu Dhabi; they become even more important to the west and northwest. References concerning the Thammama Group are by Hudson and Chatton (1959); Harris, Hay, and Twombley (1968); Hassan, Mudd, and Twombley (1975); Johnson and Budd (1975); Twombley and Scott (1975); Hood and Downey (1975); Sethudehnia (1978); Saint Marc (1978); Murris (1980); and Hassan and Wada (1981).

The Lower Cretaceous lithic sequence is three-fold (Figure 4). The lowermost cycle is terminated by a hiatus during the late Valanginian. Much of the Valanginian section is missing in eastern Saudi Arabia (Soliman and Al Shamlin, 1982), though the section appears to be complete in both central Oman and offshore Abu Dhabi. The section is composed of the carbonates of the Habshan Formation (Figure 3) that developed as a shoaling upward cycle. The Habshan Formation formed during the initial Cretaceous flooding of the stable cratonal platform and embodies the progradation and subsequent filling of the old Middle to Late Jurassic cratonal margin depression. Deeper-water deposits of the Rayda, Sulaiy, and Salil formations were formed when water depth increased due to downwarp of the Oman carbonate-ramp margin that lasted through the early Valanginian (Figure 4). These deposits are primarily hemipelagic wackestones and occasional packstones rich in small miliolid forams, planktonic foraminifera, calcispheres, calpionellids, micropellets, and calcitized sponge spicules. The basal beds of the Rayda may in fact be equivalent to the upper part of the latest Jurassic Hith evaporites further west on the craton.

Water depth decreased late in the early Valanginian due to an eustatic drop or tectonic uplift. Sediments of the Yamama Formation reflect this change as they are shallow-shelf limestones that prograded eastward across the area as a bank margin carbonate shoal. Hassan, Mudd, and Twombley (1975) provide an example of the thickness and lithofacies distribution of the Yamama Formation for Abu Dhabi. Continued decrease in water depth caused deposition in Saudi Arabia to cease; locally a significant amount of section may be missing due to subaerial erosion. Outcrops in the central Oman Mountains indicate this event by the presence of shallow-water carbonate facies which were locally deposited on agitated, current-swept sea bottoms and which cap the shallowing upward cratonal margin sequence. There is no evidence of a substantial break in deposition.

The middle cycle of the Early Cretaceous formed during the next increase and subsequent decrease in water depth over the shelf (Figure 4). The cycle beginning during the early Hauterivian and ending in the early Barremian precipitated the Lekhwair Formation (Figure 3). It consists of carbonate-ramp sediments that are alternating cycles of shallow subtidal turbulent grain-rich limestones with somewhat deeper-water subtidal mud-rich deposits. There is little

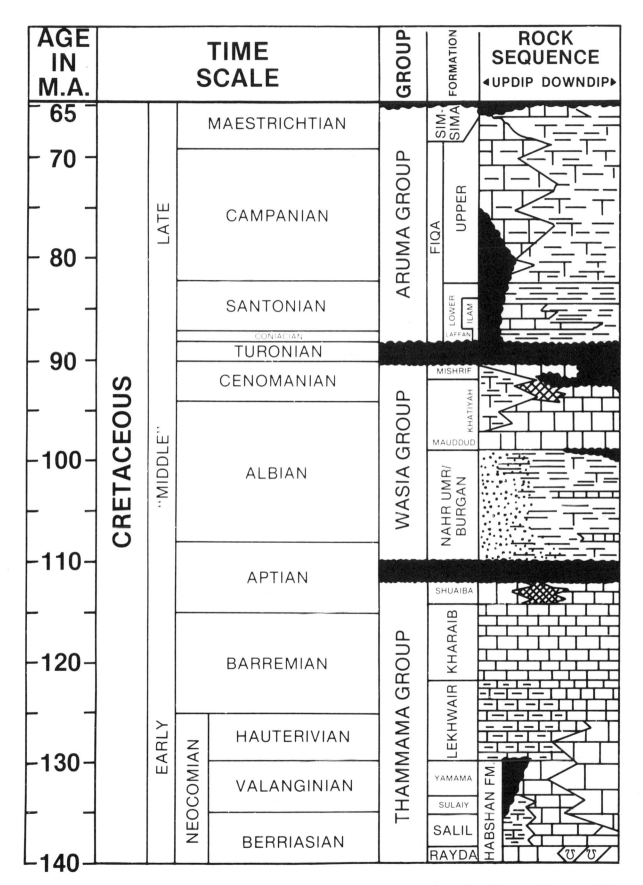

Figure 3: Cretaceous stratigraphic units recognized in the subsurface of the eastern Arabian Peninsula. The Lower Cretaceous Thammama Group, Middle Cretaceous Wasia Group, and Upper Cretaceous Aruma Group are separated by regional unconformities.

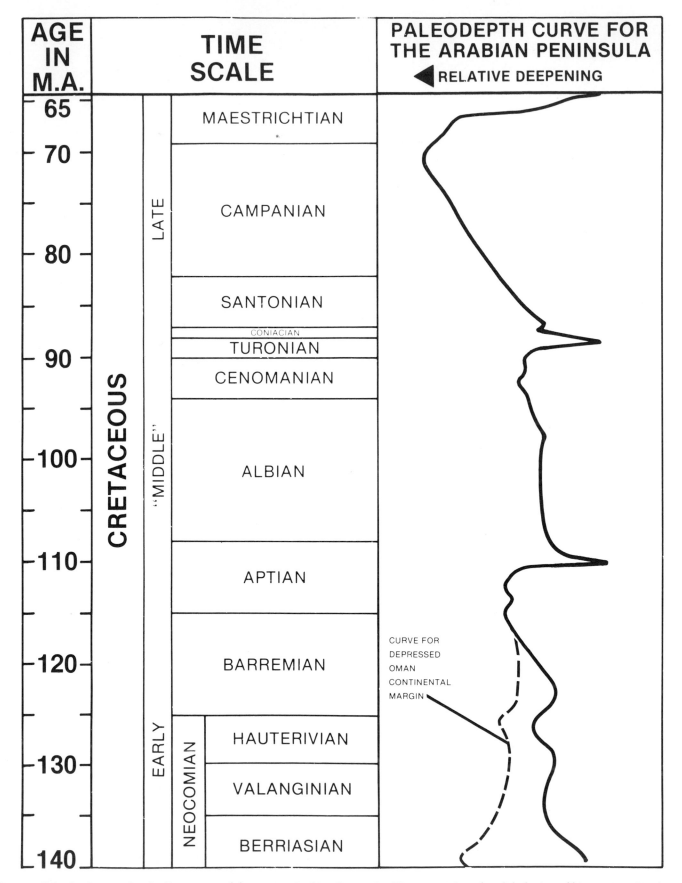

Figure 4: Paleodepth curve for the Cretaceous of the eastern Arabian Peninsula. The curve is based on lithologic and biostratigraphic data as well as regional geologic studies.

evidence in the biota that suggests water depth changes during deposition, but the lithofacies in the upper one-third of the Lekhwair are shallower water deposits and are less argillaceous than those of the lower part. The end of the cycle brought clastic sediments from the Arabo-Nubian Shield which were deposited in portions of eastern Saudi Arabia (Powers et al, 1966), but neither the significant influx of clastics nor evidence of exposure is present further east on the carbonate shelf.

The third and final cycle of the Lower Cretaceous sequence is the most interesting because it contains a significant reservoir-bearing carbonate termed the Shuaiba Formation. The cycle, beginning in the early Barremian and ending in the early middle Aptian (Figure 4), is capped by a significant regional unconformity. Both the Kharaib and Shuaiba formations (Figure 3) have more terrigenous sediments in eastern Saudi Arabia; the Kharaib is a sandstone (Soliman and Al Shamlan, 1982) and the Shuaiba is a dolomitic and shaley limestone. The Kharaib Formation is essentially the continuation of ramp-style carbonate deposition from the Lekhwair Formation. Kharaib cycles display the same shoaling-upward carbonate lithofacies as do those of the Lekhwair Formation, but are thicker and represent longer durations. Examples of cycles within the Kharaib Formation in Abu Dhabi are described in reports by Hassan, Hay, and Twombley (1975); Johnson and Budd (1975); and Harris, Hay, and Twombley (1968).

Numerous significant reservoirs are developed in buildups of caprinid rudists within the Shuaiba Formation. The shelf deposits are commonly bioturbated wackestones and packstones, with the caprinid rudist buildups occurring in belts of variable width which partially rim intrashelf basins. Commonly, the bases of the rudist buildups formed in slightly deeper water, and the representative vertical section is one showing diversification of growth forms of the caprinids and the development of a more grain-rich matrix associated with these shells (Frost, Bliefnick, and Harris, 1983). The changes are undoubtedly due in part to localized shoaling conditions, but the common shoaling sequence observed in numerous Shuaiba buildups suggests that a stillstand or lowering of water depth also played an important role in sequence development. Pelagic biogenic sediments formed in intrashelf basins of a few tens to exceptionally hundreds of meters depth (Figure 1).

Middle Cretaceous Deposits

The middle Cretaceous sequence, the Wasia Group, formed during the late Aptian through the latest Cenomanian and/or earliest Turonian. It represents an upward change from shallow-shelf clastics to shallow-shelf carbonates, is as much as 2,000 ft (610 m) thick, and represents deposition during a time span of 21 m.y. Again, there was localized development of deeper-water carbonates within intrashelf basins. Clastic sediments are more important in eastern and northeastern Saudi Arabia and the northwestern Persian Gulf where siltstones, sandstones, and even conglomerates occur, in contrast to shales and shaley limestones that formed to the east (Powers et al, 1966; Soliman and Al Shamlan, 1982). Alluvial plain to marine sandstones of the Burgan Formation form prolific reservoirs in Kuwait and

southern Iraq. References describing the Wasia Group include those by Tschopp (1967); Glennie et al (1973, 1974); Reulet (1977); Saint Marc (1978); Murris (1980); Harris and Frost (in press); and Jordan, Connally, and Vest (in press).

The middle Cretaceous deposits are not as complex depositionally as those of the Lower Cretaceous. Deposition of the Burgan sandstones grading eastward into shales or shaley limestones of the Nahr Umr Formation began in the late Aptian and continued into the late Albian (Figure 3). We established the stratigraphic subdivision of Wasia Group carbonates on a regional basis according to the usage of El-Naggar and Al Rifaiy (1973). They subdivide Wasia Group carbonates into the Mauddud Formation and the overlying Mishrif Formation. Shallow-shelf, reservoir-bearing carbonates of the Mauddud and Mishrif formations (Figure 3; Tschopp, 1967; Harris and Frost, in press; Jordan, Connally, and Vest, in press) formed during the late Albian, Cenomanian, and earliest Turonian. Complicating the subdivision of El-Naggar and Al Rifaiy (1973) is the occurrence of local facies of hemipelagic carbonates of intraplatform basinal origin. These are known as the Rumaila Formation in the northern Gulf area (El-Naggar and Al Rifaiy, 1972), and as the Khatiyah Formation in the southeastern Gulf area (Murris, 1980; Jordan, Connally, and Vest, in press).

The Burgan and Nahr Umr formations comprise the initial deposits of the middle Cretaceous cycle of flooding after a period of subaerial exposure (Figure 4). They embody a major incursion of terrigenous clastic sediments from the shield eastward onto the former carbonate platform. The depositional setting contains terrestrial and nearshore marine coarse to medium grained sandstone, grading into a vast offshore shale belt up to 311 mi (500 km) wide. The shales grade eastward to impure limestones exposed in the Musandam Peninsula and Oman Mountains. This limestone is dominated by the large benthic foraminifers *Orbitolina texana* and *Orbitolina subconcava* in the upper part of the unit. The Nahr Umr sedimentary cycle lasted a relatively long time. Sedimentation on the outer part of the Nahr Umr shelf required more than 12 m.y. to deposit 500 to 700 ft (152 to 213 m) of sediments, in contrast to the overlying Mauddud and Mishrif carbonates that are more than twice as thick and were deposited in about 8 m.y. The outer Nahr Umr shelf was not a starved basin, but is best interpreted as a marine shelf where the shallow setting led to continued reworking of the fine clastic sediments during storms to develop many local hiatuses of short duration. A color change from grayish rocks in the lower part of the Nahr Umr to lighter-colored, more oxidized units in the upper part of the sequence may represent a trend toward shallowing during deposition.

A late Albian transgression initiated deposition of the shallow-water carbonates of the Mauddud Formation (Figure 3). Thus began a cycle of carbonate deposition which lasted until the latest Cenomanian or earliest Turonian. The transgression is indicated in the Khatiyah Formation basinal facies by the appearance of a planktonic foraminiferal fauna dominated by *Planomalina buxtorfi* and *Hedbergella* spp. During the Cenomanian, the foraminiferal assemblage increases its specific diversity with species of the genera *Rotalipora*, *Whiteinella*, *Heterohelix*, and *Hedbergella*. A regression closes this sedimentation cycle and is indicated by the presence of only two species of planktonic foraminifers: *Heterohelix moremani* and *H.* cf. *globulosa*.

The Khatiyah Formation (Figure 3) is typically a laminated pelagic lime mudstone containing abundant *Oligostegina* calcispheres. Planktonic foraminifera and calcareous nannofossils are abundant; benthic foraminifera and molluscs are sparse. Beds in the deeper parts of the intrashelf basin are dark-colored, potential hydrocarbon source rocks, suggesting that although the basins were large enough to have had substantial circulation in the upper part of the water column they were also deep enough to have developed an oxygen minimum zone at or near the sediment-water interface.

The lowermost deposits of the Mauddud Formation exhibit a sharp decrease in clastic content from the underlying Nahr Umr Formation (Figure 3). At the same time, the biota of both units are nearly identical. One possible explanation is that flooding of the Arabo-Nubian Shield covered some of the early Paleozoic formations that may have been contributing clastic sediments. Another is that there may be a hiatus of short duration between the two formations. The expansion of the clastic-free shelf in the late Albian generated a widespread lithic sequence in the lowermost Mauddud Formation that consists of a basal *Orbitolina*-rich wackestone overlain transitionally by deep, open shelf, hemipelagic lime mudstones, characterized by calcispheres and planktonic foraminifera. The thicker part of the Mauddud Formation is composed of relatively uniform shallow shelf skeletal and peloidal packstones and wackestones. These deposits are of generally consistent composition over broad areas, suggesting that the shelf had a low depositional gradient and that facies belts were broad. Radiolitid rudist-reef buildups are abundant in the outer shelf and intrashelf basin slope facies. Several contain large hydrocarbon accumulations.

The top of the Mauddud Formation is defined herein on a regional basis to coincide with the uppermost stratigraphic occurrence of orbitolinid large benthic foraminifera. This datum includes the overlap between *Orbitolina conica* and the lowest stratigraphic appearance of *Praealveolina cretacea*. The most characteristic fossils of the shallow shelf facies of the Mauddud Formation are large benthic foraminifera, including *Orbitolina concava*, *Orbitolina conica*, and *Chrysalidina gradata*. Alveolinid foraminiferans include *Ovalveolina* which first appears in the lower Mauddud and more diverse assemblages in the upper Mauddud which are characterized by *Praealveolina*, *Pseudedomia*, and *Taberina*. Outcrops show that the most important megafossils are chondrodonts and scattered radiolite rudists with few corals or stromatoporoids. In portions of northwestern Oman, extension block faulting in the Cenomanian created horsts and grabens which influenced sedimentation patterns during deposition of the Mishrif Formation (Harris and Frost, in press; Figure 3). As a result, sedimentation of Mishrif carbonates is more heterogeneous than that of the Mauddud Formation and most of the Thammama Group. Porous grainstone facies of the Mishrif Formation commonly occur in elongate strips which parallel the eroded margins of tilted upthrown fault blocks.

The middle Cretaceous lithic section was terminated by a major break in deposition during the Turonian-Coniacian

(Figure 3). No index foraminifers of Turonian age occur in this sequence. This unconformity commonly serves as a distinct break between shallow-platform carbonates of the middle Cretaceous and deeper-water deposits of the Upper Cretaceous section. Significant removal of sedimentary rocks associated with the unconformity also complicates the regional picture greatly by totally removing Mishrif deposits and masking an underlying hiatus that locally occurs between the Mauddud and Mishrif formations.

Upper Cretaceous Deposits

Both the Lower and middle Cretaceous lithic sequences formed during periods of relative regional tectonic quiescence. In contrast, the Upper Cretaceous sequence was deposited during significant tectonic activity and, consequently, is much more variable in regional thickness and facies (Glennie et al, 1973; 1974). The tectonism, resulting in large part from the collision and partial subduction of a margin of the eastern Arabian plate with a spreading ridge whose axis was probably centered in the present-day Gulf of Oman (Wilson, 1969; Glennie et al, 1973, 1974; Gealey, 1977; Pearce et al, 1981), was of two types: (1) latest Cenomanian and Turonian tilting of the shelf and the depression of the craton to form the Oman Foredeep trough that trended parallel and just westward of the present-day Oman Mountains; and, (2) Campanian-Maestrichtian thrusting toward the west, mainly evident in the area of the present-day northern Oman Mountains and the Musandam Peninsula.

The Upper Cretaceous lithologies of the Aruma Group (Figure 3) reflect this tectonic instability (Glennie et al, 1973, 1974; Murris, 1980; Ricateau and Riche, 1980; Tippit, Pessagno, and Smewing, 1981). They are predominantly deep-water deposits, commonly 1,500 to 3,000 ft (457 to 914 m) thick, including chalks and marls and containing turbidites, debris flows, and radiolarian oozes in the deepest parts of the foredeep basin. Deposits in the foredeep basin may be in excess of 10,000 ft (3,048 m) thick. Deep-shelf deposits extend into eastern Saudi Arabia (Powers et al, 1966; Soliman and Al Shamlan, 1982) and even farther to the west, beyond the area of the foredeep formation

Middle to late Turonian sediments are apparently absent over much of the eastern Arabian Peninsula, although reworked planktonic foraminifera and calcareous nannofossils occur in the basal Coniacian in certain wells of the northern United Arab Emirates area. Well developed Turonian carbonates, some of them pelagic in nature, are present along the northern rim of the Arabo-Nubian Shield.

The Laffan shale, the basal Coniacian transgressive deposit of the Aruma Group, contains thin beds of limestone, dolomite, and marl. The overlying Ilam Formation and its lateral equivalents consist of two Coniacian-Santonian facies: shelf limestone and platform-slope marl. Shelf limestones are typically wackestones and packstones dominated by small miliolid foraminifers, as well as large benthic foraminifera including *Dicylina schlumbergeri*, *Cuneolina pavonia*, *Rotalia skourensis*, and *Chubbina* sp. Radiolite and caprinid rudists and other molluscs are common. Dasycladacean green algae and *Oligostegina*-type calcispheres characterize the deeper shelf facies of the Ilam.

Platform-slope marls contain abundant calcareous nanno-fossils and planktonic foraminifera.

The lower Fiqa Formation was deposited during the Coni-acian and Santonian (Figure 3) as dominantly hemipelagic fine terrigenous fill in the subsiding Omani Foredeep trough and as the westward extension of these fine clastic sedi-ments. Non-calcareous shales, with occasional interbeds of marl, typify the lower Fiqa. The number and diversity of planktonic foraminifers increase and reach a maximum dur-ing the early part of the Santonian. At the end of the Santo-nian, the specific diversity decreases and is represented by the *Planoglobulina glabrata* ecozone, which closes the lower Fiqa cycle of deposition. Deep-sea clastic facies equivalent to the lower Fiqa and basal upper Fiqa are termed the Muti Formation. They are a mixture of autochthonous planktonic foraminiferal/nannofossil hemipelagic sediment and turbi-dites of coarse limestone lithoclasts. The carbonate litho-clasts typically contain shallow shelf fossils which are as old as Permian. The Muti probably represents erosion of uplifted carbonates on an older cratonal margin to the east.

The upper Fiqa of Campanian to middle Maestrichtian age consists of pelagic biogenic marls and calcareous shales (Figure 3). An eustatic rise appears to have reinforced the effect of regional tilt and depression of a portion of the cra-tonal margin so that deep shelf or even slope conditions formed over a substantial portion of the Arabian Peninsula (Figure 4). Species diversity increased during the Campanian reaching a peak during the *Globotruncanita calcarata* chron. Local variations in the lithology of the upper Fiqa include turbidite sequences of marls, calcareous shales, and grain-rich carbonates, as well as dolomite or sandstone. Other facies equivalents of the upper Fiqa are sandstones, con-glomerates, and shales (Wilson, 1969; Glennie et al, 1974). The sandstones are commonly fine-grained graywacke with shale, dolomite, or calcite matrix; the conglomerates con-tain clasts of quartz, igneous and metamorphic rocks, chert, and limestone, set in a shaley matrix with planktonic fora-minifera and calcareous nannofossils.

At the onset of the Maestrichtian, a regression took place as indicated by the low specific diversity in the *Globotrun-canita stuarti* zone and in the lower part of the *G.? gansseri* zone. The Simsima Formation of late Maestrichtian age is formed of shelf limestone and a deeper-water marly facies. The shelf limestones are a basal lithoclast conglomerate overlain by a shoaling carbonate cycle which becomes increasingly restricted, and is capped by shallow subtidal and tidal-flat lime mudstones and wackestones. Shoal con-ditions are denoted by buildups of corals and rudists and abundant large benthic foraminifera which include the index species *Orbitoides apiculata*, *Omphalocyclus macroporus*, and *Siderolites calcitrapoides*. Locally, upper Maestrichtian offshore equivalents of the Simsima consist of interbedded planktonic foraminiferal/nannoplankton marls, pelagic mudstones, and relatively shallow-water skeletal wacke-stones.

Much of the Upper Cretaceous sequence is missing in the updip section of eastern Saudi Arabia (Figure 3; Powers et al, 1966) due to a combination of basement uplift plus regional tilting and slow transgression by the rising Creta-ceous sea. Even in central Oman, portions of the Campan-ian section are missing. The section appears to be missing in

areas where continuous submarine currents eroded the sea bottom for long periods of time, though the missing section observed in wells could also be in part produced by faulting.

A regional unconformity, which caps the Cretaceous lith-ologic sequence, was formed in the Upper Maestrichtian and marks the boundary between the Cretaceous and Tertiary formations. Porosity developed within the Simsima Forma-tion is partly moldic and was formed during this exposure. At some localities the upper Simsima contains large collapse breccias that apparently formed through evaporite solution.

REGIONAL RELATIONSHIPS

The major lithic packages and their confining unconformi-ties, as described here, are also recognized in Syria and Tur-key (Ala and Moss, 1979), Lebanon (Beydoun, 1977), Israel, and as far west as Egypt, northern Libya, and possibly Morocco. The major clastic-to-carbonate cycles and shelf-to-basin transitions, interpreted often as a result of local tec-tonics or sea-level changes, seem to be correlative over the entire southern margin of the Tethys. To fully evaluate the possible regional extent and distribution of the lithic units and their bounding unconformities, we conducted an exten-sive review of publications on the Middle East. We concen-trate here on the eastern Mediterranean area, where the more extensive lithologic and biostratigraphic studies of Israel allow detailed comparisons with our area.

In Israel, a hinge belt (Gvirtzman and Klang, 1972) sepa-rates Mesozoic shallow-marine shelf deposits on the east from basinal and continental slope deposits to the west (Ginsburg et al, 1975; Bein and Gvirtzman, 1977). The carbonate-shelf deposits interfinger in the Sinai and Negev areas with marginal marine and continental deposits extend-ing toward the Arabo-Nubian Shield (Freund, 1978). Our depositional curve for the Arabian Peninsula (Figure 4) is based on interpretation of the shelf sequence for that area since the effects of sea-level changes are more difficult to rec-ognize in the basinal sediments.

Although locally interrupted by tectonics or erosion, the majority of the cycles defined on the eastern Arabian Penin-sula are also recognized in the stratigraphic column of Israel. The Lower Cretaceous cycles are best illustrated on the recently published stratigraphic table of Israel (Bartov et al, 1981), with three marine incursions recognized in the south. These are interpreted as major sea-level rises (Bein and Gvirtzman, 1977; Rosenfeld and Raab, 1981). The sedi-ments of the eastern Arabian Peninsula depicted on Figure 3 were deposited relatively further from the shield area, and thus relate better to northern Israeli-Lebanese and Syrian sequences.

In Israel, the transgressive carbonates of the Albian-Turonian Judea Group (Arkin and Hamaoui, 1967) have been studied extensively. They are characteristically lime-stones and dolomites with occasional rudist buildups and other indicators of locally very shallow marine deposition (Schneidermann, 1970; Sass and Bein, 1978, 1982; and many others). A minor eustatic event interpreted in early Cenomanian sediments (Sass and Oppenheim, 1965; Rosen-feld and Raab, 1974) is most likely responsible for differenti-ation between Albian-early Cenomanian *Eoradiolites* and

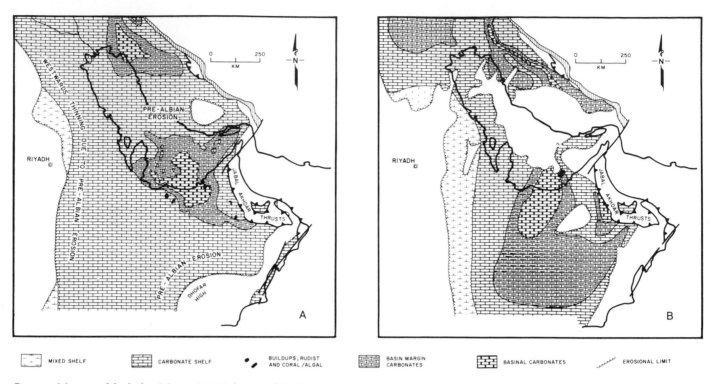

MIXED SHELF   CARBONATE SHELF   BUILDUPS, RUDIST AND CORAL/ALGAL   BASIN MARGIN CARBONATES   BASINAL CARBONATES   EROSIONAL LIMIT

Figure 5: Maps modified after Murris (1980) showing lithofacies and environments of deposition for strata of early Aptian age (A) and late Cenomanian age (B). In both cases, these rocks were subsequently above sea level and subjected to erosion and freshwater leaching, as now represented by regional unconformities.

the late Cenomanian *Caprina* and *Caprinula* rudist reefs (Bein, 1976). During the Turonian, differentiation of shelf deposits into intrashelf basins, similar to the Khatiyah facies of the eastern Arabian Peninsula, is also observed in Israel (Freund, 1961; Sass and Bein, 1982). Toward the shield area, clastic components increase dramatically (Bartov and Steinitz, 1977). A major sea-level drop is identified in the middle to late Coniacian by drastic facies changes, major paleodepth variations and karst development (Picard, 1959; Weiler, 1968; Magaritz, 1974; Lewy, 1975; Buchbinder and Magaritz, in press; and many others).

The Upper Cretaceous strata of the Mount Scopus Group in Israel (Flexer, 1968) are chalky and marly and contain a few layers of chert and phosphate. They represent open-marine, relatively deep environments of deposition (Flexer, 1971). The clastic source from the Arabo-Nubian Shield was still active (Bartov et al, 1972; Garfunkel, 1978). A minor shallowing is recognized in the late Santonian, preceding a second eustatic event in the late Campanian. These events are not recognized by us in the Arabian Peninsula, but they have been observed elsewhere (Kauffman, 1977). An upper Maestrichtian unconformity is also recognized in Israel.

Two examples from the eastern Arabian Peninsula illustrate the regional relationships between lithologies that formed as a result of flooding and subsequent exposure: (1) widespread shallow-water deposition of carbonate sediments during the early Aptian, terminated by a regional unconformity in the middle Aptian; and, (2) deposition of carbonate sediments during the Cenomanian, locally interrupted by exposure during the Cenomanian and regionally terminated by exposure during the Coniacian. In both cases,

the exposure that ended the flooding depositional cycles is of great importance due to the development of hydrocarbon reservoir-quality secondary porosities.

During the early Aptian an onlap of shallow-marine carbonate sediments occurred as limestones of the Shuaiba Formation were deposited well onto the craton (Figure 5A). Nearly all of the rudist buildups of the Shuaiba Formation producing hydrocarbons do so from moldic porosity formed by fresh-water leaching of rudists and other skeletal fragments during formation of the middle Aptian unconformity. Precise biostratigraphic analyses across the early middle Aptian unconformity indicate that the hiatus represents approximately 1.5 million years. The prospectiveness of Shuaiba rudist buildups as hydrocarbon exploration targets results from the interaction of three main factors: (a) the development of an intraplatform basin by differential subsidence, establishing deep open-marine conditions on what had formerly been a shallow carbonate platform during deposition of the Kharaib Formation and localizing a buildup-prone belt around its rim; (b) a rapid eustatic rise in the early Aptian causing the buildups to grow vertically in contrast to the sheet-like biostromes of rudists or coral/algal communities which typify a low relief platform on which the Kharaib Formation or Lekhwair Formation were deposited; and, (c) the stratigraphic position of these buildups in the upper Shuaiba (perhaps resulting in part from the initiation of a sea-level drop in late Shuaiba time) positioning them favorably for subsequent fresh-water leaching.

Giant oil fields formed at Bab and Bu Hasa in Abu Dhabi (Twombley and Scott, 1975) and Shaybah Field in eastern Saudi Arabia (Soliman and Al Shamlan, 1982) produce

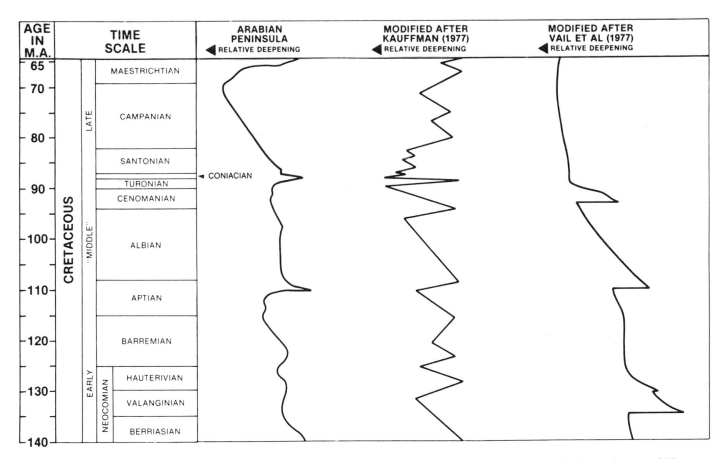

Figure 6: Comparison of paleodepth curve for eastern Arabian Peninsula with curves of Kauffman (1977) and Vail, Mitchum, and Thompson (1977).

from moldic porosity formed within rudist mounds and associated sediments. Shale seals for these significant reservoirs were formed during deposition of the overlying Nahr Umr Formation (Figure 3).

During mid-Albian, the Arabian craton was the site of deposition of terrigenous clastic sediments. The shelf area became increasingly carbonate dominated during the Cenomanian as deposits of the Mishrif and Mauddud formations of the Wasia Group onlapped the craton (Figure 5B). Large-scale facies relations are similar to those recognized for the Shuaiba Formation. Carbonate sands formed in association with buildups of caprinid and radiolite rudists that rimmed intrashelf basins or developed on upthrown fault blocks. Submarine fans of shell debris were distributed over the adjacent shelf and basin. Primary interparticle porosities in the rudist sands were high, but were commonly greatly reduced by pore-filling calcite cements. As shown by the giant Fateh Field in offshore Dubai (Jordan, Connally, and Vest, in press), the reservoir-quality porosity in the fore reef debris aprons and rudist core facies was formed principally by fresh-water leaching associated with a Turonian-Coniacian exposure. Biostratigraphic estimates on the age of the missing section at the unconformity range from 3 to 3.5 million years, in portions of offshore Abu Dhabi and to more than 15 million years in parts of Saudi Arabia bordering the Arabo-Nubian Shield.

COMPARISONS WITH OTHER SEA-LEVEL CURVES

Our Middle East paleodepth curve exhibits a strong regional character. The curve is constructed from lithostratigraphic and biostratigraphic data derived from a mid-shelf position, reflecting local tectonic activity, sedimentation patterns, and eustatic overprint. Any single curve is biased toward the regional noise induced by local tectonism, and only by comparing a number of such curves can the significance of the differences be suggested and worldwide events be related to eustacy. Several differences are evident when our curve is compared with the published coastal onlap curves of Kauffman (1977) and Vail, Mitchum, and Thompson (1977) (Figure 6). Differences are of two types: (1) the timing of relative transgressions, regressions or unconformities; and, (2) relative amount of transgression or regression. It is not surprising that differences exist between the various curves considering that the data was collected from different geographic locations and different geologic foundations and reflect discrepancies in absolute age curves, position of various biostratigraphic markers, and overall degree of biostratigraphic resolution.

Our curve suggests that the maximum marine transgression occurred during the Late Cretaceous with an intermediate position in the Early Cretaceous. The generalized curve of Vail, Mitchum, and Thompson (1977) closely parallels our curve except that the relative amounts of transgression for the Early and middle Cretaceous are reversed. Kauffman's (1977) curve has greater differences. For instance, it suggests that the greatest transgression occurred during the middle Cretaceous; the Upper Cretaceous sequence formed during increasing regression. Hancock and Kauffman (1979) interpret the transgression in the Western Interior of the United States during the later part of the Late Cretaceous as not as great as that visualized on either the curve of Vail, Mitchum, and Thompson (1977) or our curve. They relate this difference to uplift during the Sevier-Laramide orogenies. By contrast, Schmid (1959), Hancock (1961), Reyre (1966), Christensen (1975), and Reyment and Morner (1977) suggest that the greatest transgression in northern Europe and west Africa occurred during the Late Cretaceous.

The major breaks in deposition used to segment the Cretaceous deposits of the Arabian Peninsula into three lithic sequences seem to correlate very well with regressions and breaks depicted on the curve of Kauffman (1977). Discrepancies exist with Vail, Mitchum, and Thompson (1977) in that they assigned the major break during the Early Cretaceous to the basal Valanginian and that a major break terminating the middle Cretaceous deposits occurs during the Cenomanian and not the Coniacian. Details of the Cretaceous curve of Vail, Mitchum, and Thompson (1977) have not been published so it is difficult to determine if there are any other significant differences. Because the curve is generalized for the Cretaceous, short-term fluctuations of only a few million years are not seen. Differences do exist in the timing of the detailed variations depicted on our curve and those shown by Kauffman (1977), although differences in the magnitude of the variations are harder to compare because magnitude is not emphasized on the Kauffman (1977) curve. Kauffman's (1977) curve indicates more variation in the Coniacian and Santonian; the middle Cretaceous segments are similar although we have not emphasized the magnitude of the Cenomanian regression and subsequent transgression in the manner Kauffman (1977) shows. Early Cretaceous fluctuations are similar in number but their exact timing and relative magnitudes are not the same on the two curves.

ACKNOWLEDGMENTS

Foraminifera identifications and interpretation were made by Frost and Seiglie. Identification of calcareous nannofossils was by B.L. Shaffer. We thank Z. Lewy, Y. Mimran, and C.G. Kendall for critically reviewing various drafts of the manuscript. Gulf Oil Exploration and Production Company provided the opportunity to publish this paper.

REFERENCES CITED

Ala, M.A., and B.J. Moss, 1979, Comparative petroleum geology of southeast Turkey and northeast Syria: Journal of Petroleum Geology, v. 1, p. 3-27.

Arkin, Y. and M. Hamaoui, 1967, The Judea Group (Upper Cretaceous) in central and southern Israel: Israel Geological Survey Bulletin, v. 42, 17 p.

Bartov, Y., and G. Steinitz, 1977, The Judea and Mount Scopus Group in the Negev and Sinai with trend surface analysis of thickness data: Israel Journal of Earth-Science, v. 26, p. 119-148.

——— , et al, 1972, Late Cretaceous and Tertiary stratigraphy and paleogeography of Southern Israel: Israel Journal of Earth-Science, v. 21, p. 69-97.

——— , et al, 1981, Regional stratigraphy of Israel, A guide to geological mapping: Geological Survey of Israel Curr. Research, 1980, p. 38-41.

Bein, A., 1976, Rudistid fringing reefs of Cretaceous shallow carbonate platform of Israel : AAPG Bulletin, v. 60, p. 258-272.

——— , and G. Gvirtzman, 1977, A Mesozoic fossil edge of the Arabian Plate along the Levant coastline and its bearing on the evolution of the eastern Mediterranean, in B. Biju-Duval and L. Montadert, eds., International symposium on the structural history of the Mediterranean basins (Yugoslavia): Edi-Technie, p. 95-110.

Beydoun, Z.R., 1977, Petroleum prospects of Lebanon, Reevaluation: AAPG Bulletin, v. 61, p. 43-64.

Buchbinder, B., and M. Magaritz, in press, Senonian to Neogene paleokarst in Israel, 31 p.

Bolli, H.M., 1957, Planktonic foraminifera from the Oligocene-Miocene Cipero and Lengua formations of Trinidad, B.W.I.: in A.R. Loeblich, and H. Tappan, ed., Studies in foraminifera, U.S. National Museum Bulletin, v. 215, p. 97-124.

Childs, O.E., 1981, Fifth annual progress report of correlation of stratigraphic units of North America (COSUNA).

Christensen, W.K., 1975, Upper Cretaceous belemnites from the Kristianstad area in Scania: Fossils and Strata 7, 69 p.

El-Naggar, Z.R., and I.A. Al-Rifaiy, 1972, Stratigraphy and microfacies of type Magwa Formation of Kuwait, Arabia, part 1, Rumaila limestone member: AAPG Bulletin, v. 56, p. 1464-1493.

——— , and I.A. Al-Rifaiy, 1973, Stratigraphy and microfacies of type Magwa Formation of Kuwait, Arabia, part 2, Mishrif limestone member: AAPG Bulletin, v. 57, p. 2263-2279.

Flexer, A., 1968, Stratigraphy and facies development of the Mount Scopus Group (Senonian-Paleocene) in Israel and adjacent countries: Israel Journal of Earth-Science, v. 17, p. 85-114.

——— , 1971, Late Cretaceous paleogeography of northern Israel and its significance for the Levant geology: Paleogeography, Paleoclimatology, Paleoecology, v. 10, p. 293-316.

Freund, R., 1961, Distribution of lower Turonian ammonites in Israel and neighboring countries: Bulletin of the Research Council of Israel, 8G, p. 79-100.

——— , 1978, Judean Hills and Galilee regional synthesis of sedimentary basins: Jerusalem, Pre-Congress Excursion Guidebook, Tenth International Congress of Sedimentology, p. 5-31.

Frost, S.H., D.M. Bliefnick, and P.M. Harris, 1983, Deposition and porosity evolution of a Lower Cretaceous rudist buildups, Shuaiba Formation of eastern Arabian Peninsula, in P.M. Harris, ed., Carbonate buildups: Society of Economic Paleontologists and Mineralogists, Core Workshop No. 4, p. 381-410.

Garfunkel, Z., 1978, The Negev-Regional synthesis of sedimentary basins: Jerusalem, Pre-Congress Excursion Guidebook, Tenth International Congress on Sedimentology, p. 35-73.

Gealey, W.K., 1977, Ophiolite obduction and geological evolution of the Oman Mountains and adjacent areas: Geological Society of American Bulletin, v. 88., p. 1183-1191.

Ginsburg, A., et al, 1975, Geology of Mediterranean shelf of Israel: AAPG Bulletin, v. 59, p. 2142-2160.

Glennie, K.W., et al, 1973, Late Cretaceous nappes in the Oman Mountains and their geologic evolution: AAPG Bulletin, v. 57, p. 5-27.

—— , et al, 1974, Geology of the Oman Mountains: Verhandelingen Van Het Koninklijk Nederlands Geologisch Mijnbouwkundig, Genootschap, 423 p.

Gvirtzman, G., and A. Klang, 1972, A structural and depositional hinge-line along the coastal plain of Israel evidenced by magneto tellurics: Geological Survey of Israel Bulletin, v. 55, 18 p.

Hancock, J.M., 1961, The Cretaceous system in Northern Ireland: Quarterly Journal of the Geological Society of London, v. 117, p. 11-36.

—— , and E.G. Kauffman, 1979, The great transgressions of the Late Cretaceous: Quarterly Journal of the Geological Society of London, v. 136, p. 175-186.

Harris, P.M. and S.H. Frost, in press, Middle Cretaceous carbonate reservoirs, northwestern Oman: AAPG Bulletin.

Harris, T.J., J.T.C. Hay, and B.N. Twombley, 1968, Contrasting limestone reservoirs in the Murban Field Abu Dhabi: Dhahran, Saudi Arabia, in Proceedings, Second AIME Regional Technical Symposium, p. 149-187.

Hassan, T.H. and Y. Wada, 1981, Geology and development of Thammama Zone 4, Zakum Field: Journal of Petroleum Technology, v. 33, p. 1327-1337.

—— , G.C. Mudd, and B.N. Twombley, 1975, The stratigraphy and sedimentation of the Thammama Group (Lower Cretaceous) of Abu Dhabi: Dubai, Ninth Arab Petroleum Congress, Paper 107 (B-3), p. 1-11.

Hood, B.F., and W.A. Downie, 1975, Development of the Umm Shaif Field in Abu Dhabi: Dubai, Ninth Arab Petroleum Congress, Paper 130 (B-1), p. 1-7.

Hudson, R.G.S, and M. Chatton, 1959, The Musandam Limestone (Jurassic to Lower Cretaceous) of Oman, Arabia: Paris, Notes et mem. Moyen-Orient, v. 3, Museum of Natural History, p. 69-92.

Johnson, J.A.D. and S.R. Budd, 1975, The geology of the zone B and zone C Lower Cretaceous limestone reservoirs of the Asab Field, Abu Dhabi: Dubai, Ninth Arab Petroleum Congress, Paper 109 (B-3), p. 1-24.

Jordan, C.F., R.C. Connally, and H.A. Vest, in press, Upper Cretaceous carbonates of the Mishrif Formation, Fateh Field, Dubai, U.A.E., in P.O. Roehl, and P.W. Choquette, ed., Carbonate petroleum reservoirs, Springer-Verlag.

Kauffman, E.G., 1977, Geological and biological overview, Western Interior Cretaceous Basin: Mountain Geologist, v. 14, p. 75-99.

Lewy, Z., 1975, The geological history of southern Israel and Sinai during the Coniacian: Israel Journal of Earth-Science, v. 24, p. 19-43.

Magaritz, M., 1974, Lithification of chalky limestones: A case study in Senonian rocks from Israel: Journal of Sedimentary Petrology, v. 44, p. 947-954.

Murris, R.J., 1980, Middle East, Stratigraphic evolution and oil habitat: AAPG Bulletin, v. 64, p. 597-618.

Pearce, J.A., et al, 1981, The Oman Ophiolite as a Cretaceous arc-basin complex, evidence and implications: Philosophical Transactions of the Royal Society of London, Series A, v. 3000, n. 1454, p. 299-317.

Picard, L., 1959, Geology and oil exploration of Israel: Bulletin of Resource Council of Israel, v. 68, p. 1-30.

Powers, R.W., et al, 1966, Geology of the Arabian Peninsula, Sedimentary geology of Saudi Arabia: U.S. Geological Survey, Professional Paper 560D, 147 p.

Reulet, J., 1977, Carbonate reservoir of a marine shelf, Mishrif Formation (Middle East), in Essai de Caracterisation Sedimentologique des Depots Carbonates: Centres de Recherches de Boussens et de Pau, Elf Aquitaine p. 169-177.

Reyment, R.A., and N.A. Morner, 1977, Cretaceous transgressions and regressions exemplified by the South Atlantic: Special Paper Palaeontological Society of Japan, v. 21, p. 247-261.

Reyre, D., 1966, Particularites geologiques des bassins cotiers de L'ouest Africain, in Bassins sedimentaires du littoral Africain, symposium, I. Littoral Atlantique: Union Int. Sci. Geol., Ass. Serv. Geol. Africain, Int. Geol. Cong. 22, p. 253-301.

Ricateau, R. and P.H. Riche, 1980, Geology of the Musandam Peninsula (Sultanate of Oman) and its surroundings: Journal of Petroleum Geology, v. 3, p. 139-152.

Rosenfeld, A., and M. Raab, 1974, Cenomanian-Turonian ostracodes from the Judea Group in Israel: Geological Survey of Israel Bulletin, v. 62, 64 p.

—— , and M. Raab, 1981, Lower Cretaceous ostracodes from Israel: Geological Survey of Israel, 1980, p. 62-65.

Saint Marc, P., 1978, Arabian Peninsula, in M. Moullade, and A.E.M. Nairn, eds., The Phanerozoic geology of the World II, the Mesozoic, Amsterdam, Elsevier, p. 435-462.

Sass, E. and M.J. Oppenheim, 1965, The petrology of some Cenomanian sediments from the Judean Hills, Israel, and the paleo-environmental break of the Motza Marl: Israel Journal of Earth-Science, v. 14, p. 91-107.

—— , and A. Bein, 1978, Platform carbonates and reefs in the Judean Hills, Carmel, and Galilee: Jerusalem, Post Congress Excursion Guidebook, Tenth International Congress on Sedimentology, p. 239-274.

—— , and A. Bein, 1982, The Cretaceous carbonate platform in Israel: Cretaceous Research, v. 3, p. 135-144.

Schmid, F., 1959, Biostratigraphie du Campanien-Maastrichtien du NE de la Belgique sur la base des belemnites: Annales, Societe Geologique de Belgique, v. 82, p. B235-256.

Schneidermann, N., 1970, Genesis of some Cretaceous carbonates in Israel: Israel Journal of Earth-Science, v. 19, p. 97-115.

Sethudehnia, A., 1978, The Mesozoic sequence in southwest Iran and adjacent areas: Journal of Petroleum Geology, v. 1, p. 3-42.

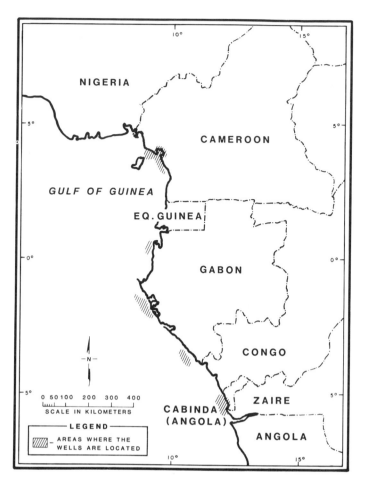

Figure 1: Location of wells, Zaire to Cameroon.

planispira is extremely abundant in the deepest part of the cycle, which is inner neritic or less probably middle neritic. At the end of the cycle the foraminifers are again rare. In some wells, an unconformity separates this cycle from the overlying cycle.

Second middle Cretaceous minor cycle: early and middle Cenomanian. This cycle comprises the *Lenticulina* cf. *ouachensis - Favusella washitensis* and *Textularia praelonga* zones. The fauna of the lower part of this cycle is difficult to determine because of uphole cavings in all wells. However, *Thomasinella* sp. is more frequently found in the lower part of the zone, and it is characteristic of the shallowest marine Cenomanian paleoenvironments in the area. Other common foraminifers include: *Favusella washitensis*, *Hedbergella* cf. *plainspira*, *Neobulimina* sp. D, *Lenticulina* cf. *ouachensis* and *Textularia praelonga*. Several species of *Rotalipora* are less common. The end of this sedimentary cycle is indicated by the abundance of glauconitized gastropods with a few specimens of *Textularia praelonga*.

Third middle Cretaceous minor cycle: late Cenomanian to middle Turonian. The *Whiteinella brittonensis*, *Praeglobotruncana helvetica* and *Whiteinella inornata - Dicarinella imbricata* zones correspond to this cycle. The most common foraminifer during the late Cenomanian is *Whiteinella brittonensis*, indicating an inner to middle neritic paleoenvironment. No keeled planktonic foraminifers are present in the *W. brittonensis* zone. The specific diversity increases during

the early Turonian, and the keeled planktonic foraminifer *Praeglobotruncana helvetica* appears. The specific diversity and the paleodepths continue to increase through the middle Turonian. A profound unconformity truncates the upper Turonian.

The last major transgressive-regressive cycle of sedimentation we observe ranges from Coniacian to Maestrichtian and is subdivided into three minor transgressive-regressive cycles of sedimentation (Figure 2) as follows:

The first Late Cretaceous minor cycle: Coniacian and Santonian. In most wells of the studied area, the earliest foraminiferal fauna of this cycle cannot be separated from the uphole cavings of deep water fauna. The fauna *in situ* probably represents a shallow paleoenvironment. The water depth increased at the end of the Coniacian and during the early and middle Santonian. A large number of keeled planktonic foraminifers occur in the deepest paleoenvironments, including: *Marginotruncana sigali*, *M. schneegansi*, *M. manaurensis*, *Globotruncanita elevata* and *Dicarinella asymetrica*. At the end of the Santonian, ostracodes (mainly *Cythereis* sp. A and *C. reticulata*) are the dominant microfauna, indicating a regression at the end of this cycle.

Second Late Cretaceous minor cycle: Campanian and earliest Maestrichtian(?). The fauna of the early part of the cycle consists mainly of planktonic foraminifers with *Globotruncana fornicata* as the dominant species. Other significant species are *Globotruncanita stuartiformis*, *G. elevata* and *Globotruncana lapparenti*, indicating a deep water paleoenvironment, the deepest water for the Late Cretaceous of central West Africa. The sea in Zaire and Cabinda (Angola) shallowed near the end of the cycle so that several shallow water species of the genus *Gabonella* (mainly *G. lata* and *G. spinosa*) dominated. This regression is the last episode of the cycle.

Third Late Cretaceous minor cycle: Maestrichtian. The Maestrichtian transgression is indicated by the abundance of planktonic foraminifers. *Globotruncana contusa*, *G. arca*, *Globotruncanita stuarti*, *G? gansseri*, *Racemiquembelina fructicosa*, *Pseudoguembelina excolata*, *Platystaphyla brazoensis*, *Pseudotextularia elegans*, *P. deformis*, *Heterohelix* spp., *Rugoglobigerina rotundata*, *R. rugosa*, *R. macrocephala*, *Plummerita reicheli*, *P. hantkeninoides*, *Globotruncanella havanensis*, *G. petaloidea*, and *Trinitella scotti* occur in the transgressive episode. Shallowing of the sea near the end of the Maestrichtian is indicated by the decreasing number of keeled planktonic foraminifers and the abundance of *Orthokarstenia clavata* and *Afrobolivina afra*. Above, a thin layer occurs consisting only of *A. afra* and *Orthokarstenia* cf. *laevigata*. The cycle ends with the deposition of a thin layer of oolitic marly limestone, found in some of the Zaire wells. An unconformity separates this cycle of sedimentation from the Paleocene.

RELATIVE SEA-LEVEL CHANGES, OFFSHORE CENTRAL GABON

Flysch-type agglutinate foraminifers are the most common fauna of the deep-water paleoenvironments of the Gabon Upper Cretaceous. Gradstein and Berggren (1980) consider that flysch-type foraminiferal faunas lived at water depths below 200 m (656 ft). The shelf-break of West Africa

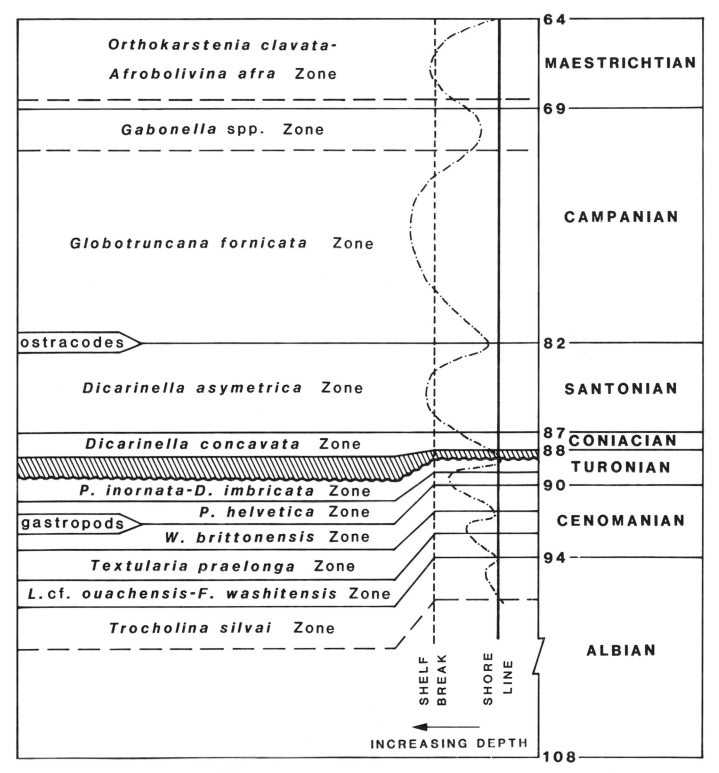

Figure 2: Transgressive-regressive cycles of sedimentation and biostratigraphic zones of Zaire and Cabinda (P.R. of Angola).

was probably above 200 m (656 ft) depth during the Cretaceous. Foraminifers are rare in the well samples because of high sedimentation rates from turbidity currents. Because this fauna lived between bathyal and abyssal depths, it is difficult to clearly define the Late Cretaceous transgressive-regressive cycles of sedimentation in Gabon.

Based mainly on seismic data, Vidal (1980) indicated that

the Campanian and Maestrichtian paleoenvironment offshore central Gabon was a shallow-marine platform where sediments were dispersed by waves and tidal action. Foraminiferal faunas do not support Vidal's assumption. The Campanian and Maestrichtian flysch-type fauna of Gabon includes the genera: *Ammobaculites, Ammodiscus, Ammovertellina, Bathysiphon, Clavulina, Dorothia, Gaudryina,*

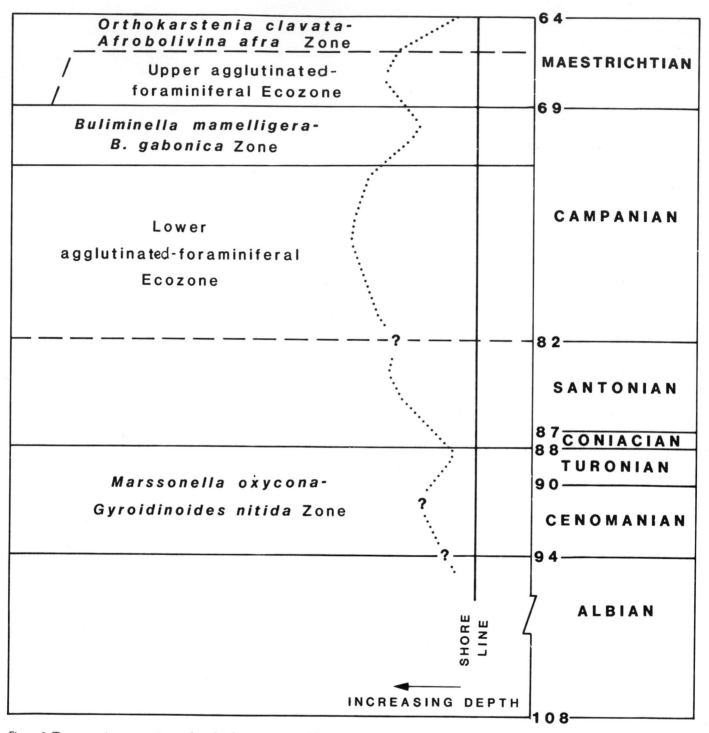

Figure 3: Transgressive-regressive cycles of sedimentation and biostratigraphic zones of Gabon.

Haplophragmoides, Hormosina, Plectina, Psammosphaera, Recurvoides, Reophax, Rhizammina, Rzehakina, Saccammina, Spiroplectammina, and *Trochammina.*

The tentative transgressive-regressive cycles of sedimentation and the foraminiferal zonation of Gabon are shown in Figure 3.

Middle Cretaceous major cycle: probably late Albian to Turonian. No subdivision into minor cycles is possible because of the rarity of foraminifers. *Epistomina* sp. and a

few poorly-preserved foraminifers characterize the Albian shallow-water faunas off Gabon. *Thomasinella* is representative of the Cenomanian inner neritic paleoenvironments of northern Gabon. *Marssonella oxycona* and *Gyroidina nitida* are the dominant species of an outer neritic fauna occurring in rocks of Cenomanian-Turonian age off central Gabon. The *Marssonella oxycona - Gyroidinoides nitida* zone occurs in this cycle. The upper boundary of the zone is determined by the uppermost occurrence of *Marssonella oxycona* or

Gyroidina nitida or both; the lower boundary is not well defined. Above this zone, foraminifers are absent and the electric logs show an abrupt change, suggesting an unconformity.

Late Cretaceous major cycle: Coniacian to Maestrichtian. Three minor cycles occur in the Late Cretaceous of Gabon, as follows:

First Late Cretaceous minor cycle: In the southern wells, ostracodes *Cythereis* sp. A. and *C. reticulata* appeared during the early Coniacian-Santonian. The number and specific diversity of planktonic foraminifers (mainly *Dicarinella asymetrica*, *Marginotruncana manaurensis* and *M. angusticarinata*) increases upward as the number of ostracodes decreases. The top of the Santonian, which corresponds to the regressive sequence, is truncated by an unconformity or a fault. In the western wells, the rocks of Coniacian-Santonian age are indicated by the presence of *Hastigerinoides watersi*. The end of the Santonian is suggested by the rarity of foraminifers in northern and central Gabon.

Second Late Cretaceous minor cycle: Campanian. Campanian age rocks are indicated in some wells by the occurrence of *Pseudotextularia plummerae*. A flysch-type fauna characterizes the lower agglutinate-foraminiferal ecozone (Figure 3). Its lower boundary coincides approximately with the Santonian-Campanian boundary. The number and diversity of foraminifers increases upward, and represents a change from a bathyal to abyssal paleoenvironment. The upper boundary of the ecozone corresponds to the uppermost abundance of agglutinate foraminifers.

A regression occurred near the end of the cycle. It is represented by an outer neritic assemblage which constitutes the *Buliminella mamelligera* - *B. gabonica* zone. The most significant species are *Buliminella mamelligera*, *B. gabonica* and *Orthokarstenia dentata*. Their uppermost occurrence determines the upper boundary of the zone. The lower boundary of this zone coincides with the upper boundary of the lower agglutinate-foraminiferal ecozone. The end of the cycle coincides approximately with the end of the Campanian.

Third Late Cretaceous minor cycle: Maestrichtian: Off central Gabon this cycle was initiated by an early Maestrichtian transgression. It is represented by the upper agglutinate-foraminiferal ecozone, the upper boundary of which is defined by the uppermost abundance of the Maestrichtian agglutinate foraminifers. The lower boundary of the zone coincides with the upper boundary of the Campanian.

The last Cretaceous regression during the late Maestrichtian is indicated by a decrease in species diversity and by the dominance of the shallow-water foraminifers *Orthokarstenia clavata* and *Afrobolivina afra*. The *Orthokarstenia clavata* - *Afrobolivina afra* zone is found in the entire Maestrichtian stage of Zaire, Cabinda, and Cameroon. However, the two index species of this zone occur only in the late Maestrichtian shallow-water facies off central Gabon.

RELATIVE SEA-LEVEL CHANGES, DOUALA BASIN, CAMEROON

As in Zaire, Cabinda (Angola), and offshore Gabon, two major transgressive-regressive sedimentation cycles occur in the Douala basin, Cameroon, from late Albian to Maes-

trichtian. The ages were determined by the foraminiferal zonation shown in Figure 4 and by stratigraphic position. The zones are discussed below the cycles of sedimentation.

Middle Cretaceous major transgressive-regressive cycle of sedimentation: late Albian to Turonian. The oldest age determined with foraminifers (*Globigerinelloides algerianus*) in the Douala basin was late Albian; however, most of the sediments of Albian age are composed of unfossiliferous terrestrial sands. Algal (mainly *Parachaetetes*) and oolitic limestones constitute the Cenomanian-Turonian part of the transgressive sequences. Subdivision of this major cycle is not possible because of the scarcity of biostratigraphic and paleoenvironmental data. The absence of foraminifers in the clastic sediments above the Cenomanian-Turonian limestones indicates the end of the cycle and possibly the presence of an unconformity.

Late Cretaceous major transgressive-regressive cycle of sedimentation: Coniacian? to Maestrichtian. Only two minor cycles were observed in the Upper Cretaceous of the Douala basin, as compared to three in Zaire and Cabinda. This is probably due to the scarcity of foraminifers and other fossils.

Combined first and second late Cretaceous minor cycle: Coniacian, Santonian, and Campanian. Foraminifers were absent during the early part of this cycle. Fayose (1976) described a rock sequence (from Coniacian to Campanian) in Calabar, eastern Nigeria, which corresponds in age to this cycle. He indicated that the lowest part of the sequence is cross-stratified and contains coal measures, but that it becomes marine in the younger parts of the sequence. No Santonian index foraminifers were found in the Douala basin wells, and no regression was observed until the end of the Campanian.

The uppermost occurrence of *Clavulina gabonica* is of Campanian and Santonian age and indicates the upper boundary of the *C. gabonica* zone. This boundary approximately indicates the peak of the Campanian transgression. The lower boundary of this zone cannot be determined. *C. gabonica* is associated with a deep-water assemblage of flysch-type agglutinate foraminifers.

The *Buliminella mamelligera* - *B. gabonica* zone extends between the uppermost occurrence of *Clavulina gabonica* at the lower boundary and the uppermost occurrence of *Buliminella mamelligera*, *B. gabonica* or *Orthokarstenia dentata* at the upper boundary of the zone. The extinction of these foraminifers indicates the end of the Campanian. These species were abundant in shallower waters than the flysch-type assemblages and mark the terminal regression of the cycle.

Third Late Cretaceous minor cycle: Maestrichtian. The *Orthokarstenia clavata* - *Afrobolivina afra* zone indicates the age of this cycle (Seiglie and Baker, 1982). Early Maestrichtian deep-water agglutinate foraminifers are followed by an abundance peak of small specimens of *Praebulimina* (*P.* cf. *baccata*, *P. kickapooensis*, *P. prolixa* and *P. exiqua*). This fauna suggests on outer neritic or uppermost bathyal paleoenvironment. A decrease in specific diversity of foraminifers and the dominance of two shallow-water species, *Orthokarstenia clavata* and *Afrobolivina afra*, indicate that a regression ended this sedimentation cycle. It is separated from the Paleocene by an unconformity.

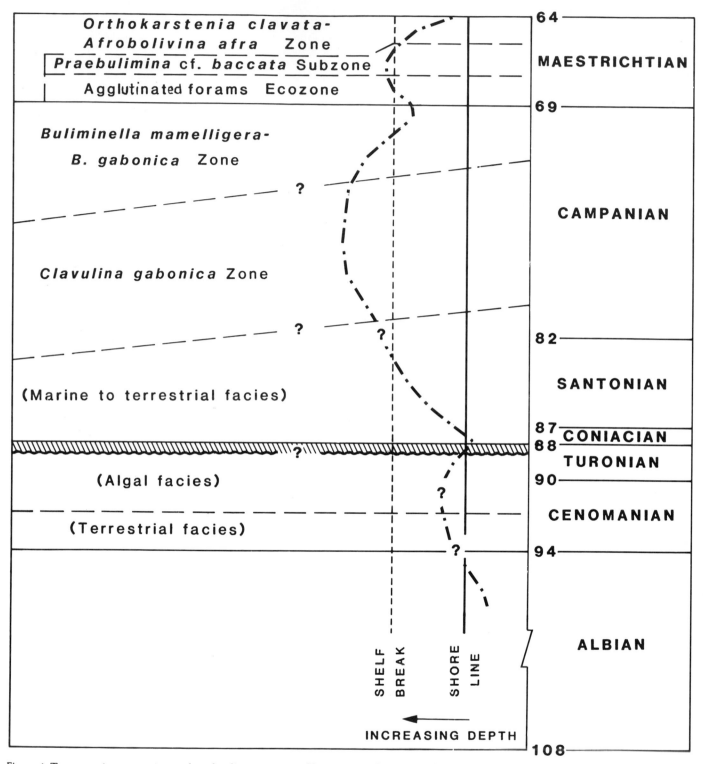

Figure 4: Transgressive-regressive cycles of sedimentation and biostratigraphic zones of Douala basin, Cameroon.

COMPARISON WITH OTHER AREAS OF THE WORLD

Vail et al (1977) published a simplified curve for global cycles of sea-level changes in the Cretaceous. It is, however, too simplified to be compared with the West African cycles of sedimentation. Matsumoto (1980) also summarized the transgressions-regressions of the middle and Upper Creta-

ceous for many areas of the world; to a certain degree, there is correlation among the relative sea-level changes in all the areas that he presented. Some of the differences in correlation may be attributed to local tectonics or sedimentation rate. Other differences may be attributed to: (1) interpreting the ages of the fossils and (2) duration of stages, in terms of millions of years. The central West African cycles of sedi-

LEGEND

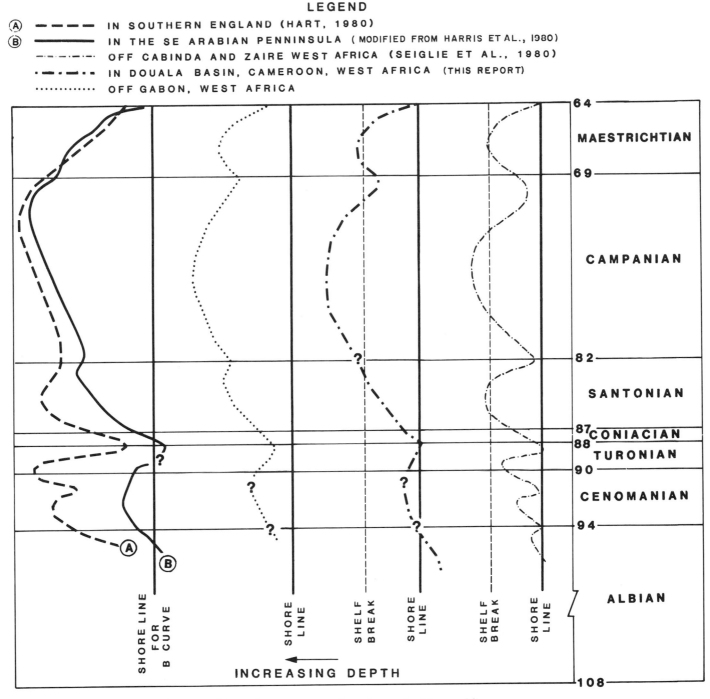

(A) — — — — — IN SOUTHERN ENGLAND (HART, 1980)
(B) ————— IN THE SE ARABIAN PENNINSULA (MODIFIED FROM HARRIS ET AL., 1980)
—··—··— OFF CABINDA AND ZAIRE WEST AFRICA (SEIGLIE ET AL., 1980)
—·—·—· IN DOUALA BASIN, CAMEROON, WEST AFRICA (THIS REPORT)
············· OFF GABON, WEST AFRICA

Figure 5: Cretaceous sea-level changes in West Africa as compared to other parts of the world.

mentation, compared with cycles in other parts of the world, show a closer degree of correlation.

For southern England, Hart (1980a) showed relative sea-level changes as they relate to paleodepths determined from foraminiferal assemblages. With one exception, these curves approximately coincide with the West African transgressive-regressive cycles of sedimentation (Figure 5). Hart's relative sea-level changes show only one cycle of sedimentation from Campanian through Maestrichtian, while we determined two cycles in West Africa, one Campanian and another Maestrichtian. This difference may be caused

by the upwarp of southeastern England (Pegrum et al, 1975). Hart (1980b) also showed middle Cretaceous sea-level changes based on foraminifers. Hart's (1980b) curve of sea-level change has two minor differences with the one of West Africa: (1) a small regression during the early Cenomanian; and, (2) the middle Cenomanian regression appears to be slightly older than the middle Cenomanian regression of West Africa.

Hancock and Kauffman (1979) described the transgressions and regressions of northern Europe. They are closer to the Zaire and Cabinda (Angola) sedimentation cycles than are those of southeastern England by Hart (1980a). There are two minor differences: (1) the late Santonian regression is not so strong in northern Europe as in Zaire and Cabinda (Angola); and, (2) the Maestrichtian transgression is deeper in northern Europe. The relative depths of the other transgressions and regressions are similar in both areas.

Naidin et al (1980) present a generalized summary of the Russian platform transgressive-regressive cycles of sedimentation. Differences among the Russian cycles are probably a consequence of local tectonics and different amounts of sedimentation. North central Asia cycles (Naidin et al, 1980) are closer to the central West African cycles. A generalization of these curves shows three cycles during the Middle Cretaceous and three during the Late Cretaceous, all roughly similar in age to those of central West Africa.

The relative sea-level changes of the eastern Arabian Peninsula (Curve B Figure 5) discussed by Harris et al (1980) were compared with those of West Africa from Seiglie and Baker (1982). The minor differences in the curves for the two areas are probably caused either by the local tectonics or sedimentation rate, or by the poor paleodepth resolution in some of the Cretaceous foraminiferal assemblages of the eastern Arabian Peninsula.

The transgressive-regressive cycles of sedimentation in the Western Interior of North America (Kauffman, 1977) do not coincide with the central West African cycles. However, those presented by Hancock and Kauffman (1979) are closer to the West African cycles. Kauffman (personal communication) showed us a more recent, unpublished study on the Western Interior cycles, where they are relatively similar to those of central West Africa.

CONCLUSIONS

We have not considered subsidence rates or uplift in relation to the West African cycles of sedimentation, which would be necessary to convert them to eustatic sea-level changes. However, these cycles approximately coincide in age and relative depths with those of other areas of the world. The effects of the local salt tectonics and the sedimentation rate were not significant enough to change the age in which the West African cycles occurred. Therefore, they coincide in age with the eustatic sea-level changes, but not in absolute depths.

ACKNOWLEDGMENTS

Thanks to N. Schneidermann, S. H. Frost, P. M. Harris, and K. E. Scott from Gulf Oil Exploration and Production Company, Houston Technology Center, for their suggestions and to Gulf Oil Corporation for permission to publish this paper.

REFERENCES CITED

Fayose, E.A., 1976, Cretaceous micropaleontology of the Calabar flank, southeastern Nigeria: 7th African Micropaleontological Colloquium, lle-lfe, abstracts, Nigeria, p. 41.

Gradstein, F.M., and W.A. Berggren, 1980, Flysch-type agglutinated foraminifera and the Maestrichtian to Paleogene history of the Labrador and North Seas: Marine Micropaleontology, v. 6, n. 3, p. 211-269.

Hancock, J.M., and E.G. Kauffman, 1979, The great transgressions of the Late Cretaceous: Journal of the Geological Society of London, v. 136, p. 175-186.

Harris, P.M., et al, 1980, Cretaceous sea level and stratigraphy eastern Arabian Peninsula, in Denver, abstract, 1980 AAPG-SEPM-EMD Annual Convention: AAPG Bulletin, v. 64, n. 5, p. 719 (unpublished figure part of oral presentation).

Hart, M.B., 1980a, A water depth model for the evolution of planktonic Foraminierida: Nature, v. 286, p. 252-254.

——— , 1980b, The recognition of mid-Cretaceous sea-level changes by means of foraminifera: Cretaceous Research, v. 1, p. 289-297.

Kauffman, E. G., 1977, Geological and biological overview: Western Interior Cretaceous Basin: Mountain Geologist, v. 14, n. 3-4, p. 75-99.

Matsumoto, T., 1980, Inter-regional correlation of transgressions and regressions in the Cretaceous Period: Cretaceous Research, v. 1, p. 359-373.

Naidin, D.P., et al, 1980, Cretaceous transgressions and regressions on the Russian platform, in Crimea and Central Asia: Cretaceous Research, v. 1, p. 375-387.

Obradovich, J.D., and W.A. Cobban, 1975, A time-scale for the Late Cretaceous of the western interior of North America, in W.G.E. Caldwell, ed., Cretaceous system in the western interior of North America: Geological Association of Canada, Special Paper 13, p. 31-34.

Pegrum, R.M., G. Rees, and D. Naylor, 1975, Geology of the North-West European continental shelf, v. 2, The North Sea: London, Graham Trotman Dudley Ltd, 269 p.

Petters, S.W., 1981, Late Cretaceous-Tertiary marine cycles and paleoenvironments in Gulf of Guinea, eastern equatorial Atlantic margin: AAPG Bulletin, v. 65, n. 6, p. 1668-1669 (abstract).

Reyment, R.A., 1979, The Mid-Cretaceous of the Nigerian coastal basin. Annales Musee d' Histoire Naturelle de Nice, v. 4 (for 1976), p. 1-15.

——— , P. Bengtson, and E.A. Tait, 1976, Cretaceous transgressions in Nigeria and Sergipe-Alagoas (Brazil): Anais Academia Brasileira Ciencias, v. 48, p. 253-2664, (supplemental).

Robaszinski, F., et al, 1979, Atlas de Foraminiferes planktoniques du Cretace Moyen (mer Boreal et Tethys): Cahiers Micropaleontologies, v. 1, 185 p., v. 2, 181 p.

Seiglie, G.A., and M.B. Baker, 1982, Foraminiferal zonation of the Cretaceous off Zaire and Cabinda, West Africa, and its geological significance, in J. Watkins, ed., Studies in Continental Margin Geology: AAPG Memoir 34, p. 651.

Vail, P.R., et al, 1977, Seismic stratigraphy and global changes of sea level, in C.E. Payton, ed., Seismic stratigraphy-applications to hydrocarbon exploration: AAPG Memoir 26, p. 49-212.

Vidal, J., 1980, Geology of Grondin field, in M.T. Halbouty, ed., Giant oil and gas fields of the decade 1968-1978: AAPG Memoir 30, p. 577-590.

Late Oligocene-Pliocene Transgressive-Regressive Cycles of Sedimentation in Northwestern Puerto Rico

George A. Seiglie
Mounir T. Moussa
Gulf Oil Exploration and Production Company
Houston, Texas

Several transgressive-regressive sedimentation cycles were recognized in the upper Oligocene to lower Pliocene in northwestern Puerto Rico. These are represented by the upper San Sebastian-Lares, "lower Montebello," Los Puertos, "Aymamon," and Quebradillas formations. The inner neritic to terrestrial paleoenvironments of the Cibao Formation do not allow the recognition of cycles.

Correlation of the northern Puerto Rican cycles of sedimentation with global sea-level changes is possible because tectonic movements in the Oligocene to Pliocene did not affect the water paleodepths significantly. The cycles of sedimentation were used to correlate some of the lithostratigraphic units of northern Puerto Rico.

INTRODUCTION

The setting for the development of Puerto Rico during the middle and late Tertiary was determined by the major tectonic events that occurred in the Caribbean region from the early Mesozoic to late Eocene. Since the beginning of the Oligocene, Puerto Rico has been relatively stable. The entire Oligocene-to-Pliocene sequence forms a homoclinal dip with dips ranging from almost horizontal to a maximum of 4° to the north. No evidence of any major tectonic disturbance is recognized in these rocks and the attitude apparently approximates the original depositional dip.

Most of the upper Oligocene to upper Miocene in northern Puerto Rico consists of shallow-water limestones. Terrigenous sediments were deposited during shorter periods of time.

This study describes the late Oligocene-to-Pliocene transgressive-regressive cycles of sedimentation of northwestern Puerto Rico using biostratigraphic and paleoenvironmental data. These cycles are comparable to the eustatic sea-level changes recognized by Vail, Mitchum, and Thompson (1977).

Figure 1 shows the study area, surface-sample localities, and well locations. More than 2,500 thin sections of surface samples, ditch cuttings and cores were examined in this study.

STRATIGRAPHY AND PALEOENVIRONMENTS

Most of the formations in northern Puerto Rico were originally described by Hubbard (1923) and were later redescribed and mapped by Meyerhoff (1933). More recently, Monroe (1980) described the Tertiary rock sequences of northern Puerto Rico. His age determinations, however, are based on data that are obsolete and thus any correlation based on that evidence is not satisfactory. Also, his descriptions of the paleoenvironments are too generalized to provide meaningful data for interpreting any change in the eustatic sea level.

The stratigraphy of Puerto Rico has been discussed by Moussa and Seiglie (1975) and Meyerhoff (1975). They recognized a sequence of six formations of Oligocene to early Pliocene age. From oldest to youngest these are: San Sebastian Formation, Lares Limestone, Cibao Formation, Los Puertos Limestone, "Aymamon" limestone, and Quebradillas Limestone.

TRANSGRESSIVE-REGRESSIVE CYCLES IN NORTHWESTERN PUERTO RICO

Five transgressive-regressive cycles of sedimentation were recognized. The faunas on which recognition of these cycles is based are listed in Figure 2.

1. *Upper San Sebastian-Lares transgressive-regressive cycle.* The San Sebastian Formation consists mostly of terrigenous sediments (conglomerates, sandstones, and shales). After these were deposited, a transgression began as manifested by the appearance of mangroves (Habib, 1971) and marine fossils. Sedimentation gradually changed to the carbonates of the Lares Limestone. The lower Lares is late Oligocene to earliest Miocene in age, as shown by the presence

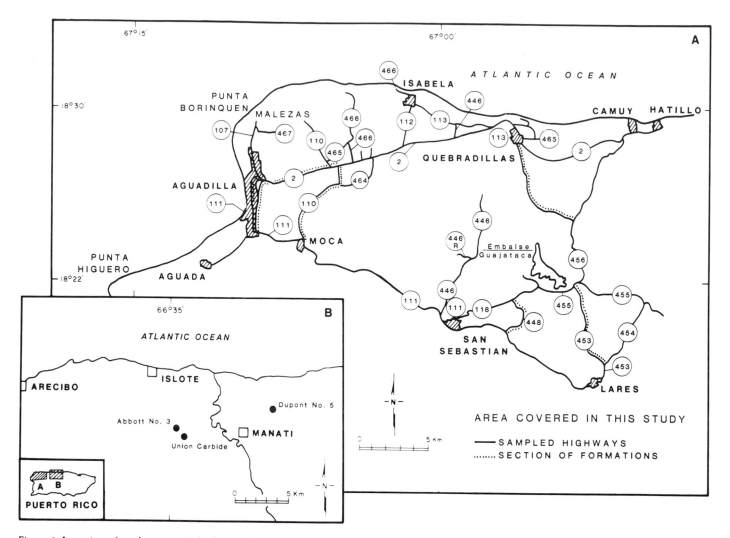

Figure 1: Location of study areas: A, highways from which samples were taken and localities in which samples in continuous sequences were taken; B, location of water wells and the area where the nine Islote wells were drilled.

of *Miogypsinoides thalmanni, Miogypsina panamensis, Lepidocyclina undosa* and *L. mantelli.* The basal part of the Lares is characterized by back-reef debris or oysters, or both. The maximum depth during Lares time is represented by parts of the back-reef complexes and by coral bioherms, both of which can be seen in outcrop. Regression is reflected in the presence of oyster-barnacle "reefs," oyster "reefs," and shales containing echinoids and pectens.

In the Dupont No. 5 well the Lares consists of deeper water deposits (Figure 3). A transgression is represented by an increase in the percentage of planktonic foraminifers in relation to the total number of foraminifers. The maximum water paleodepth during this transgression is shown by a maximum of 48 percent of planktonic foraminifers. The occurrence of planktonic foraminifers seaward of the Lares reef, their high percentage, and their relatively high specific diversity are interpreted as indicative of an outer neritic paleoenvironment. Water masses surrounding Puerto Rico have had a low organic productivity from the late Oligocene to recent. There is no known evidence of upwelling or large rivers that supplied nutrients, and thus organic matter is mostly absent in sediments. The Puerto Rican shelf has been

generally devoid of planktonic foraminifers since the late Oligocene. The presence of *Lepidocyclina* and *Heterostegina* characterized the Lares Limestone in this well. A decrease in the percentage of foraminifers represents the final regression.

In the Union Carbide No. 1 well (Figure 4) the transgression of the San Sebastian-Lares cycle is represented by increased number of specimens and percentage of *Amphistegina.* A high percentage of *Amphistegina* is considered representative of fore-reef paleoenvironments (Seiglie, 1968, 1971). A decrease in the percentage of *Amphistegina* indicates the regression. Also oysters characterized the close of the cycle. The fauna of this cycle in the Union Carbide No. 1 well is distinguished from the overlying "lower Montebello" cycle by the relative abundance of several species of *Lepidocyclina* and *Heterostegina antillea.*

2. *"Lower Montebello" transgressive-regressive cycle.* This cycle is represented by the "lower Montebello limestone" which is the upper member of the Lares Limestone in central-northern Puerto Rico. The transgressive phase of the "lower Montebello" cycle began after the deposition of the oyster-barnacle and oyster "reefs" of the underlying Lares.

Age	Paleoenvironmental index fossils and paleoenvironments		Lithostratigraphic units
PLIOCENE	(4) Cross-stratified calcarenites (3) Amphistegina assemblages (2) Operculinoides cojimarensis assemblage (1) Globigerine ooze	(beach) (inner to middle neritic) (middle to outer neritic) (bathyal)	QUEBRADILLAS LS.
LATE MIOCENE			
MIDDLE MIOCENE	(3) Encrusting coralline algae — Amphistegina (2) Coral heads, gastropod-bivalve bed, reef debris (1) Branching coralline algae-miliolids-soritids	(fore reef) (reef & back-reef) (back-reef lagoon)	"AYMAMON LS."
	Cross-stratified calcarenites	(beach)	— — — — ? — — — —
	Branching coraline algae-miliolids-soritids	(back-reef lagoon)	LOS PUERTOS LS.
EARLY MIOCENE	(Terrestrial, back-reef, littoral, coastal lagoons)		CIBAO LS.
	Montebello Ls Mbr: Lepidocyclina and Heterostegina biostromes	Mudstone unit: outer-middle neritic	LOWER "MONTEBELLO"equiva- lents in time
	Oyster-barnacle"reefs", oyster"reefs", echinoid-pecten sh.		
OLIGOCENE	Coral bioherms, Lepidocyclina Oysters, pectens	(back-reef) (reef) (back-reef)	LARES LS.
	Oysters, pectens Mangrove, coastal lagoons Terrestrial		SAN SEBASTIAN

Figure 2: Ages, index fossils, and paleoenvironments of the lithologic units on which the transgressive-regressive cycles of sedimentation are based.

In outcrop, the deepest part of the transgression is indicated by *Lepidocyclina undosa, L. mantelli,* and *Heterostegina antillea* biostromes. A gastropod limestone containing *Miogypsina* cf. *tani* cropping out north of the town Lares represents the end of this cycle. It represents a middle early Miocene paleoenvironment close to the shore.

In the Dupont No. 5 well, a mudstone unit is the time equivalent of the "lower Montebello limestone." Increase in the percentage of planktonic foraminifers in the lower part of the unit indicates the transgression; decrease in the upper part of the unit characterizes the regression which closes this cycle (Figure 3).

The "lower Montebello" transgression in the Union Carbide No. 1 well (Figure 4) is indicated by an increased number of specimens and percentages of *Miogypsina* and *Amphistegina.* The onset of deposition of the land-derived sediments containing inner neritic soritids of the overlying Cibao Formation is evidence for the regression and end of the cycle. In this well, the "lower Montebello" cycle is distinguished from the underlying Lares cycle by abundance of the early Miocene *Miogypsina tani.*

From Monroe's (1980) paleontologic data, the "lower Montebello" cycle is probably equivalent in age to the "Rio Indio limestone."

3. *Deposition of the Cibao Formation.* The Cibao deposition in northwestern Puerto Rico began with the accumulations of oysters and coastal lagoonal deposits interbedded with terrestrial sediments. *Miosorites americanus* (see Seiglie, Grove, and Rivera, 1977) appeared in the basal part of the Cibao following the extinction of *Lepidocyclina undosa* which is characteristic of the Lares and "lower Montebello." This succession of foraminifers occurs throughout the Caribbean (Frost and Langenheim, 1974). Inner neritic, marginal, and terrestrial (calcareous and clastic) sediments occur throughout the Cibao Formation and the cycle of sedimentation is difficult to determine.

The upper part of the "Montebello limestone" is probably equivalent in age to the Cibao Formation and the Quebrada Arenas Limestone Member. We need further studies to confirm this relationship.

4. *Los Puertos transgressive-regressive cycle.* A gradual increase in marine carbonates indicates the transgression from the Cibao marginal deposits to the back-reef lagoons of the Los Puertos Limestone in northwestern Puerto Rico. The fossil assemblage that characterizes the latter paleoenvironment consists of branching coralline algae, miliolids, and soritids (foraminifers). *Borelis melo* appears in the lower Los Puertos Limestone, suggesting an early middle Miocene age for that formation. The end of the Los Puertos cycle coincided with the deposition of a unit of cross-stratified calcarenite consisting of fragments of back-reef organisms; this calcarenite is probably a beach deposit.

5. *"Aymamon" transgressive-regressive cycle.* The "Aymamon" transgression followed the deposition of the cross-

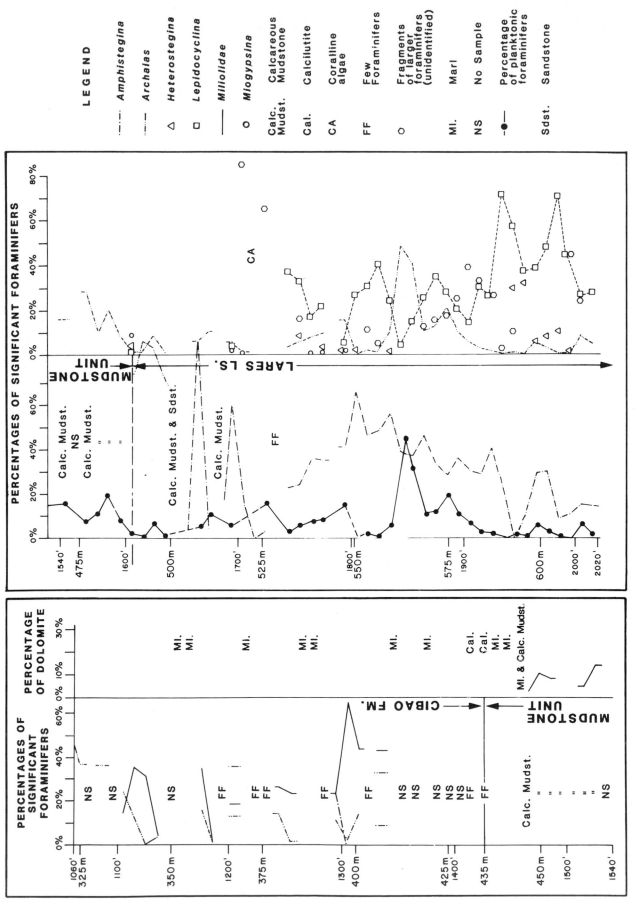

Figure 3: Percentages of the most significant foraminifers of the Lares Limestone and the "mudstone unit" (age equivalent of the Montebello Limestone) in the Dupont No. 5 well. Part of the Cibao Formation is included for comparison. Data on the lithology of the formations is included.

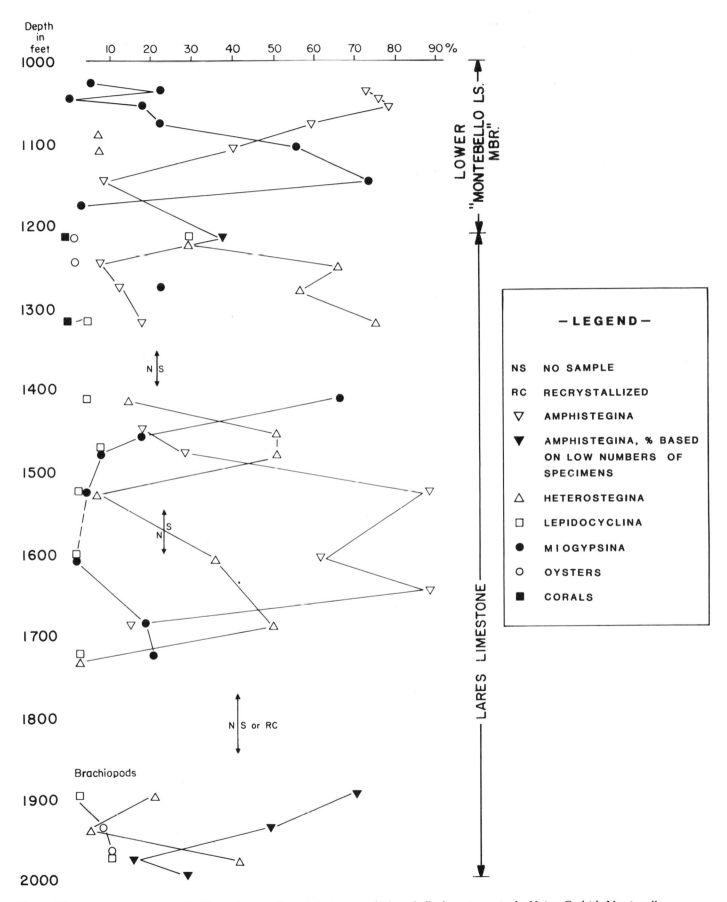

Figure 4: Percentages of the most significant foraminifers of the Lares and Montebello limestones in the Union Carbide No. 1 well.

stratified beach calcarenites of the Los Puertos Limestone. In northwestern Puerto Rico, the transgression is represented by a succession of three different paleoenvironments: (1) a back-reef lagoon that contained branching coralline algae, miliolids, and soritids; (2) open-shelf coral "reef" debris and gastropod-bivalve deposits; and (3) a fore reef containing an encrusting coralline algae-*Amphistegina angulata* assemblage and, rarely, planktonic foraminifers. The water depths must have been between 30 and 50 m. The "Aymamon" limestone is truncated by an erosional unconformity which separates it from the overlaying Quebradillas Limestone.

No index fossils are known from the "Aymamon"; however, its stratigraphic position suggests a middle or late Miocene age.

6. *Quebradillas transgressive-regressive cycle.* The "Aymamon" subaerial erosional surface is overlain by 1 to 2 cm of shallow-water mollusks followed by several meters of sediments with deep-water planktonic foraminifers. This succession is evidence for the sudden Quebradillas transgression. The early Pliocene planktonic foraminifers of the Quebradillas Limestone were described by Seiglie (1978) from the Islote borings (Figure 1). They indicate a bathyal paleoenvironment deeper than 200 m. This and other Quebradillas paleoenvironments were described by Seiglie and Moussa (1975). A regression followed as manifested by a decreased number of specimens and percentage of planktonic foraminifers, and by the increase in benthonic foraminifers, mainly *Operculinoides cojimarensis*. *O. cojimarensis* flourished in an outer-to middle-neritic paleoenvironment and it was followed by an *Amphistegina* assemblage that included inner neritic foraminifers. In the Antilles, this assemblage is indicative of middle-to inner-neritic paleoenvironments of unprotected, narrow shelves. This cycle closes with the deposition of cross-stratified beach calcarenites and calcareous sandstones. An unconformity separates these from the overlying Pleistocene sediments. The Pliocene-Pleistocene unconformity is marked by an erosional surface and an ancient paleosol followed by sediments containing a Pleistocene foraminiferal fauna.

COMPARISON WITH THE EUSTATIC SEA-LEVEL CYCLES

A comparison of the eustatic sea-level cycles described by Vail, Mitchum, and Thompson (1977) and the Puerto Rican cycles (Figure 5) shows that both are similar. In northwestern Puerto Rico, no foraminifers are known that can be used to determine precise ages of the Tertiary rocks, with the exception of the early Pliocene *Globorotalia margaritae* Zone (Bolli and Bermudez, 1965) or the *Globorotalia tumida* s.s. - *Sphaeroidinellopsis subdehiscens paenedehiscens* Zone (Blow, 1969) of the Quebradillas Limestone. Other species give only approximate ages of the cycles of sedimentation and, therefore, their correlation (Figure 5) with the eustatic sea-level cycles of Vail, Mitchum, and Thompson (1977) is approximate.

The "Aymamon" foraminifers range in age from middle to late Miocene. Because of this, the time span of the "Aymamon" - Quebradillas unconformity has two possible

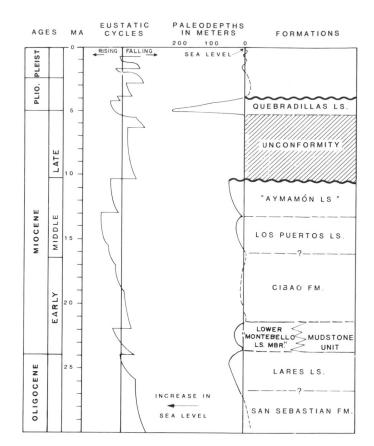

Figure 5: Comparison of Puerto Rican cycles of sedimentation with Vail, Mitchum, and Thompson (1977) cycles.

interpretations: (1) there is only one cycle of sedimentation in the "Aymamon," and accordingly the unconformity should represent a gap of about 4.5 million years (as shown in Figure 5) or, (2) there is more than one cycle, though not yet recognized, and therefore the duration of the unconformity would be only about 500,000 years.

The Quebradillas transgression is represented in outcrop and in the Islote borings by rocks indicative of paleoenvironments with water paleodepths of more than 200 m. These paleodepths are unlike those that prevailed earlier in Puerto Rico and those of Vail, Mitchum, and Thompson (1977). They suggest the possibility that tectonic activity may have lowered the sea floor, though there is no evidence to support this idea.

ECONOMIC SIGNIFICANCE OF CYCLES FOR PUERTO RICO

Sea-level changes are significant for hydrocarbon exploration. During regressions, solution of limestone by surface and underground waters increases carbonate porosity and permeability. These processes may develop not only potential hydrocarbon reservoirs, but also potential aquifers. Porosity and permeability, however, may later be modified by other diagenetic processes. The development of aquifers during the regressive episodes of the cycles is of economic significance for Puerto Rico because most of its northern aquifers are in limestones.

The "Aymamon limestone" was the formation exposed for the longest period of time. It developed a high porosity and permeability, but its exposure in a narrow belt close to the north coast of Puerto Rico limits its possibilities as an aquifer.

The "lower Montebello limestone" was close to the surface during the deposition of the Cibao Formation, which was during a time of low sea-level stand. This increased the possibilities of solution and, therefore, of porosity and permeability. This may explain why the "lower Montebello" is one of the best aquifers in northern Puerto Rico.

ACKNOWLEDGMENTS

Part of the data for this paper was collected when the authors were members of the faculty of the University of Puerto Rico at Mayaguez.

This work was supported in part by the U.S. Office of Water Resources Research Grant A-044-PR through the Institute of Water Resources Research, University of Puerto Rico at Mayaguez. We thank James Heisel of the Water Resources Division, U.S. Geological Survey, San Juan, Puerto Rico, for the thin sections of samples from the Dupont No. 5 and Union Carbide No. 1 wells. We also thank Nahum Schneidermann, Kenneth E. Scott and John Duncan of Gulf Oil Exploration and Production Company, Houston, Texas, and Eduardo Aguilar of Dames & Moore, Houston, Texas, for helpful suggestions.

REFERENCES CITED

Blow, W.H., 1969, Late middle Eocene to Recent planktonic foraminiferal biostratigraphy, in P. Bronnimann and H.H. Renz, eds., Geneva, Proceedings of the First International Conference on Planktonic Microfossils: Leiden, E.J. Brill, p. 1990, 422.

Bolli, H.M., and Bermudez, P.J., 1965, Zonation based on planktonic foraminifers of middle Miocene to Pliocene: Bol. Informativo Asoc. Venezolana Geol., Mineria y Petroleo, v. 8, n. 5, p. 121-150.

Frost, S.H., and R.L. Langenheim, 1974, Cenozoic reef bio-

facies, Tertiary larger foraminifera and scleractinian corals from Chiapas, Mexico: Dekalb, Northern Illinois University Press, 388 p.

Habib, D., 1971, Palynology of the San Sebastian coal (Oligocene) of Puerto Rico: St. Thomas, Virgin Islands, Transactions of the Fifth Caribbean Geological Conference, p. 141-142.

Hubbard, B., 1923, The geology of Lares District, Porto Rico: New York Academy of Science, Scientific Survey Porto Rico and Virgin Islands, v. 3, p. 1-115.

Meyerhoff, H.A., 1933, Geology of Puerto Rico: Puerto Rico University Monograph Series B., n. 1, 306 p.

——, 1975, Stratigraphy and petroleum possibilities of middle Tertiary rocks in Puerto Rico (discussion): AAPG Bulletin, v. 59, n. 1, p. 169-172.

Monroe, W.H., 1980, Geology of the middle Tertiary formations of Puerto Rico: U.S. Geological Survey Professional Paper 953, 93 p.

Moussa, M.T., and G.A. Seiglie, 1975, Stratigraphy and petroleum possibilities of middle Tertiary rocks in Puerto Rico (discussion): AAPG Bulletin, v. 59, p. 163-168.

Seiglie, G.A., 1968, Relationships between the distribution of Amphistegina and the submerged Pleistocene reefs off western Puerto Rico: Tulane Studies in Geology and Paleontology, v. 6, n. 4, p. 139-147.

——, 1971, Distribution of foraminifers in the Cabo Rojo platform and their paleoecological significance: Revista Espanola Micropaleontologia, v. 3, n. 1, p. 5-33.

——, 1978, Comments on the Miocene-Pliocene boundary in the Caribbean: Annales Centre Universite de Savoie, t. 3, Sciences Naturelles, p. 71-86.

——, K. Grove, and J.A. Rivera, 1977, Revision of some Caribbean Archaisinae, new genera, species and subspecies: Eclogae Geologicae Helvetiae, v. 70, n. 3, p. 855-883.

——, and M.T. Moussa, 1975, Paleoenvironments of Quebradillas Limestone (Tertiary), northern Puerto Rico, and their geologic significance: AAPG Bulletin, v. 59, p. 2314-2321.

Vail, P.R., R.M. Mitchum, and S. Thompson, 1977, Global cycles of relative changes of sea level, in C.E. Payton, ed., Seismic Stratigraphy--applications to hydrocarbon exploration: AAPG Memoir 26, p. 83-98.

Oxygen Isotope Record of Ice-Volume History: 100 Million Years of Glacio-Eustatic Sea-Level Fluctuation

R.K. Matthews
Brown University
Providence, Rhode Island

The decade of the 1970s saw great progress in documentation of Pleistocene glacio-eustatic sea-level fluctuations. An important fall-out of this work has been documentation that the deep-sea planktic foraminiferal $\delta^{18}O$ record from many oceanographically stable regions of the world is primarily a record of global ice-volume fluctuation, and thereby a record of glacio-eustatic sea-level fluctuation.

The deep-sea $\delta^{18}O$ technology and interpretation schemes developed for the late Quaternary can be applied to the Tertiary and Upper Cretaceous record as well. The data suggest an ice-free world at approximately 100 m.y. before present with sporadic increases in ice-volume throughout the Paleogene.

A working hypothesis for a 100-m.y. eustatic sea-level curve is compiled from three independent data sets according to explicit rules. First, calculations concerning change in sea level attributable to change in sea-floor spreading rate are taken to define a generally regressive sea-level curve for the last 100 m.y. Second, ice volume effects are estimated from the deep-sea, low-latitude, shallow-dwelling planktic foraminiferal $\delta^{18}O$ record. Third, in the absence of detailed information from the $\delta^{18}O$ record, the timing of Cretaceous and Paleogene major regressive events is taken from the published "relative sea-level curve" compiled from seismic stratigraphy. Finally, Neogene structure of the seismic stratigraphy curve is rescaled in accordance with experience where $\delta^{18}O$ and seismic stratigraphy records overlap.

The general pattern of eustatic regression throughout the Paleogene appears to have been punctuated by several relatively abrupt glacio-eustatic sea-level lowerings. Assuming typical values for basin subsidence, these glacio-eustatic events should be responsible for unconformities representing non-trivial amounts of geologic time.

INTRODUCTION

The principle of uniformitarianism is often stated, "The present is the key to the past." In a dynamical sense, the concept of uniformitarianism should be restated as a set of nested models, each dealing with a larger period of time. Just as the principle of uniformitarianism suggests that observations concerning the present should not be disregarded when examining ancient rocks, the dynamics of the Quaternary should not be disregarded when examining the dynamics of the Tertiary, and so on back through time.

During the past ten years, much progress has been made concerning quantification of the late Pleistocene glacio-eustatic sea-level model. Development of this model has relied heavily upon oxygen-isotope data on planktic and benthic foraminifers from deep-sea cores. The oxygen-isotope technology used in development of the late Quaternary glacio-eustatic model can also be applied to rocks of Tertiary and late Mesozoic age.

This paper presents a review of the Pleistocene dynamic model and then proceeds to apply this model to the last 100 m.y. of glacio-eustatic history.

THE PLEISTOCENE DYNAMIC MODEL

18,000-Year-Ago Glacial World

The last major Pleistocene glaciation occurred well within the range of carbon-14 dating. Carbon-14 time control can be used to demonstrate systematic relationships among: (a) the last great advance of ice sheets on North America and Europe; (b) sea-level history; (c) sea-surface temperature; and, (d) $\delta^{18}O_{water}$ variation in the world ocean. The carbon-14 time frame is indicated by cross-hatching in the left-hand portion of Figure 1.

The continental record of ice
For many years, it has been recognized that geomorphic features of central North America and northern Europe

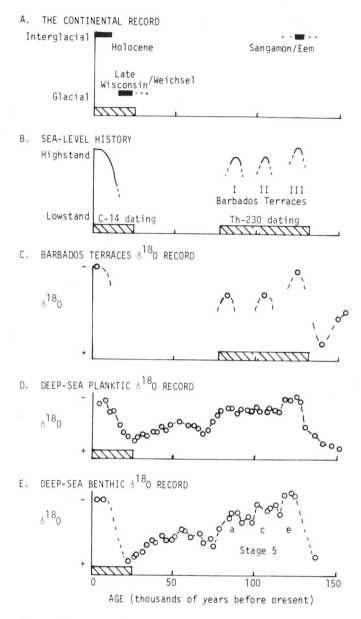

A. THE CONTINENTAL RECORD

B. SEA-LEVEL HISTORY

C. BARBADOS TERRACES δ¹⁸O RECORD

D. DEEP-SEA PLANKTIC δ¹⁸O RECORD

E. DEEP-SEA BENTHIC δ¹⁸O RECORD

Figure 1: Diagram indicating general agreement among five late Pleistocene paleoclimatic/glacio-eustatic indicators. Carbon-14 time control among various records is indicated by cross-hatched pattern at the left of the diagrams. Thorium-230 time control is indicated by cross-hatched pattern at the right of the diagrams. Record B confirms glacio-eustatic sea-level fluctuations consistent with Record A (ice-volume fluctuations on land). Records B, D, and E are directly linked to one another within the realm of carbon-14 dating. Record C provides an intermediate step between B and D/E within the realm of Thorium-230 dating.

were formed by large continental ice sheets of Pleistocene age (Bower, 1978; Bloom, 1978). Moraines and tills of the Weichsel of northern Europe and the Wisconsin of North America record the last major advances of Northern Hemisphere continental ice sheets and are extensively dated by the carbon-14 method using wood and peat associated with tills and moraines. For purposes of this discussion, the maximum advances of these two ice sheets are considered to be synchronous and to have occurred 18,000 years ago.

As a result of extensive mapping of ice-related deposits and carbon-14 dating, the periphery of global continental ice sheets at 18,000 years ago has been determined. Given the extent of the ice sheets and some general rules concerning stability of continental ice sheets, the volume of ice can be calculated. Denton and Hughes (1981, p. 274) calculate that sea level 18,000 years ago stood somewhere between 91 and 163 m (298 and 535 ft) below present sea level (depending upon assumptions concerning size of the ice sheets and rate of isostatic equilibration of the sea floor to changing load).

The last 18,000 years of sea-level history

If materials are chosen carefully, a carbon-14 dated history of sea-level rise throughout much of the last 18,000 years can be established. Favorite materials are coastal peat deposits overlying the Pleistocene subaerial exposure surfaces. Also popular are certain species of molluscs which occur only in shallow, low-salinity bays. By taking precautions to sample only good sea-level indicators and to sample them only in stratigraphic context, we can show that sea level has risen during the past 18,000 years (for example, Shepard, 1960). Though dates of greater than 12,000 years B.P. are scant, they tend to show a sea-level rise of at least 80 m (262 ft) from 17,000 to 6,000 years ago that coincides with the carbon-14-dated continental record of glacial retreat. Thus, the connection of glacio-eustatic sea-level rise and glacial retreat is made by carbon-14 dating. The absolute magnitude of the sea-level lowstand 18,000 years ago remains only loosely defined, somewhere between 80 m (262 ft) (empirical) and 163 m (535 ft) (theoretical maximum).

The 18,000-year-ago sea-surface temperature

Quantitative data concerning abundance of various taxa in core-top samples can be used to write regression equations for sea-surface temperature of the modern ocean. These regression equations may then be applied to down-core samples to estimate environmental conditions at times in the past. The basic method is discussed in detail in Imbrie and Kipp (1971) and in Kipp (1976). The CLIMAP Project used this technique to construct global maps of sea-surface temperature at the 18,000-year-ago time slice (Cline and Hays, 1976; CLIMAP, 1976, 1981). Major decreases in sea-surface temperature are noted at high latitude. Additionally, eastern equatorial regions of the Atlantic and Pacific are cooler because of increased upwelling. However, temperature estimates for large areas of the temperate and tropical ocean do not change more than a degree or so from 18,000 years ago to present.

18,000-year-ago δ¹⁸O values for benthic and tropical planktic foraminifers

The record of δ¹⁸O variation in benthic foraminifers and shallow-dwelling planktic foraminifers has been determined for numerous late Pleistocene deep-sea cores (Shackleton and Opdyke, 1973; Shackleton, 1977). A change from isotopically light values to isotopically heavy values commonly occurs within the upper meter of the core. This interval is within the carbon-14 dating time frame, and the calcium carbonate of planktic foraminifers is suitable material for carbon-14 dating. The isotopically heavy values coincide with the 18,000-year-ago glacial maximum. Total faunal analysis temperature estimates can be used to constrain the temperature effect on $\delta^{18}O$ values from shallow-dwelling planktic foraminifers. Shackleton (1967) argues that modern bottom water is sufficiently cold that the more positive $\delta^{18}O$ values given by 18,000-year-old benthic foraminifers could not have resulted from even colder conditions. Thus, both benthic foraminiferal and tropical to temperate shallow-dwelling planktic foraminiferal $\delta^{18}O$ records are taken to represent primarily fluctuations in $\delta^{18}O_{water}$ resulting from variations in the size of continental ice sheets. The average amplitude of the $\delta^{18}O$ signal in numerous high-sedimentation-rate late Quaternary cores is approximately 1.6°/oo (Shackleton, 1977).

Thus, the carbon-14 chronology confirms the general tendency to covariance among the $\delta^{18}O$ signal, sea-level fluctuations, and the size of continental glaciers. However, there remains room for improvement concerning calibration of the $\delta^{18}O$ signal to the magnitude of glacio-eustatic sea-level fluctuations. The magnitude of sea-level fluctuations is constrained only to lie somewhere between 80 and 163 m (262 and 535 ft). There is the possibility that some of the $\delta^{18}O$ signal arises from small but systematic temperature variations and/or diagenetic phenomena such as variation in dissolution of calcium carbonate at or near the sea floor. Thorium-230 dating provides additional important calibration data.

The $\delta^{18}O$ Signal as a Record of Continental Ice Volume

Carbon-14 dating of various stratigraphic records provides a picture of glaciers on continental North America and northern Europe, relatively low sea level, and relatively enriched $\delta^{18}O$ values in deep-sea foraminifers 18,000 years ago. Of these records, only the $\delta^{18}O$ signal in deep-sea cores holds any prospect of yielding an intelligible continuous record within a chronostratigraphic framework beyond the range of carbon-14 dating. Thus, it is worthwhile to develop additional documentation of the $\delta^{18}O$ signal as a reliable record of continental ice-volume fluctuations.

Chronology and $\delta^{18}O$ calibration via Thorium-230 dating of coral-reef terraces

Thorium-230 dating of coral-reef terraces provides an important chronology which includes the last interglacial. A coral-reef terrace somewhat above present sea level is well known from each of several tectonically stable regions (Neumann and Moore, 1975). This terrace consistently dates to around 125,000 years B.P., conveniently in the mid-range of applicability of the Thorium-230 dating method.

In tectonically active regions, such as Huan Peninsula of New Guinea, Ryukyu Islands of southern Japan, and the Barbados Island, Lesser Antilles, tectonic uplift allows access to several other terraces within the range of thorium-230 dating (see Bloom et al, 1974).

In the Barbados terrace sequence, Worthing (Barbados I), Ventnor (Barbados II), and Rendezvous Hill (Barbados III) terraces are extensively dated at 82,000, 105,000, and 125,000

years old The Kendal Hill terrace is only sparsely dated, but probably records a highstand of the sea in the range of 180,000 to 200,000 years ago. It has long been recognized that the Rendezvous Hill terrace is correlative with isotope stage 5e (Broecker et al, 1968; Mesolella et al, 1969) and is likewise correlative with well-dated coral-reef terraces only slightly above present sea level in tectonically stable regions. Recognizing that the Rendezvous Hill terrace equivalent is only slightly above sea level in tectonically stable regions, one can calculate an average uplift rate for the Barbados Island over the last 125,000 years. Such estimates of uplift rate vary from place to place on Barbados but do not exceed 30 cm/10 yrs. This rate of tectonic uplift is quite small compared to the rate of glacio-eustatic sea-level fluctuations. Thus, the major cause of elevation difference among the lower Barbados terraces is change in glacio-eustatic sea level. Assuming a constant uplift rate based upon the elevation of the Rendezvous Hill terrace, one can estimate that the Ventnor and Worthing terraces were deposited at sea levels 20 to 25 m (66 to 82 ft) below the sea level represented by Rendezvous Hill terrace (Matthews, 1973). Note that in the right-hand portion of Figures 1B and 1E such a relationship agrees with the relative magnitude of deep-sea oxygen-isotope stages 5e (Rendezvous Hill, "Barbados III"), 5c (Ventnor, "Barbados II"), and 5a (Worthing, "Barbados I"), thus providing additional evidence that deep-sea $\delta^{18}O$ variation reflects glacio-eustatic sea level.

Still further confirmation of the relationship between $\delta^{18}O$ and glacio-eustatic sea-level fluctuation is provided by isotopic data from Barbados molluscs (Shackleton and Matthews, 1977) and corals (Fairbanks and Matthews, 1978). As indicated in Figure 1C, the isotope data on Barbados terraces provides an undeniable link between sea-level estimates based on terrace elevation (Figure 1B) and relative peak height of deep-sea isotope stage 5e as compared to 5a and 5c (Figures 1D and E).

The isotope data on corals is especially noteworthy because the data set includes subsurface samples from below and above the subaerial exposure surface separating Kendal Hill forereef skeletal sands from skeletal sands transgressive to Rendezvous Hill highstand. Fairbanks and Matthews (1978) used these data to make a direct calibration of changing $\delta^{18}O$ values as a function of changing elevation of sea level. The relationship is estimated at -0.011°/oo m. Assuming the entire amplitude of the late Quaternary $\delta^{18}O$ signal to be caused by the ice-volume effect, this calibration would indicate full glacial lowstand of the sea some 145 m (476 ft) below present sea level. Alternatively, one could note that the amplitude of the isotope signal in the coral data approximates the amplitude in deep-sea cores whereas elevation differences are far less than 145 m (476 ft). Allowing tectonic motion to be unidirectional but discontinuous, the data can accommodate an isotope stage 6 lowstand of the sea approximately 105 m (344 ft) below present sea level. In the following discussion, the -0.011°/oo/m calibration is utilized, with the acknowledgment that as much as one-third of the signal could possibly result from global temperature variations which are synchronous with the ice-volume cycle.

Amplitude and Frequency of Pleistocene Cyclicity

Numerous curves confirm a 1.6 °/oo amplitude for the late

Quaternary global $\delta^{18}O$ signal (Shackleton, 1977). Several long records confirm that this amplitude is representative throughout the last 700,000 years. Furthermore, $\delta^{18}O$ values appear normally distributed and the amplitude of the glacial/ interglacial cycle is accurately represented as ± 2 standard deviations. The dominant periodicities within the late Quaternary $\delta^{18}O$ signal are 100,000, 40,000, and 20,000 years (Mesolella et al, 1969; Hays, Imbrie, and Shackleton, 1976).

Late Pleistocene interglacial conditions have consistently returned to a minimum ice volume approximately equal to the present situation. If all the ice on the earth today were to melt and return to the ocean, sea level would rise approximately 60 or 70 m (197 or 230 ft) and δ_{water} of the world ocean would become lighter by approximately 1 per mil (Shackleton and Kennett, 1975). Thus, the full amplitude from ice-free world to maximum late Pleistocene full-glacial conditions is approximately 220 m (722 ft) sea-level fluctuation and approximately $2.6^o/_{oo}$ $\delta^{18}O$ fluctuation.

Overview of the Pleistocene as a Dynamic Model

The major features of the late Quaternary dynamic model can be applied directly to older stratigraphic sequences. However, care must be taken in designing sampling strategy.

The deep-sea $\delta^{18}O$ signal as a monitor of glacio-eustatic sea-level fluctuations

The late Pleistocene deep-sea $\delta^{18}O$ record is highly correlated with glacio-eustatic sea-level fluctuations. Based primarily upon the direct calibration of Fairbanks and Matthews (1978), the late Quaternary 1.6 $^o/_{oo}$ amplitude is taken to represent approximately 145 m (476 m) of sea-level fluctuation. This number reasonably agrees with ice-volume estimates of Denton and Hughes (1981) (91 to 163 m, or 299 to 535 ft, sea-level equivalent). Further, the general tendency is clearly confirmed by the carbon-14 dated sea-level curve based upon submerged coastal peats and bay molluscs.

Although the general relationship of sea level, ice volume, and $\Delta\delta^{18}O$ is indisputable, precise calibration is surely subject to some error. A potential source of large error arises from questions concerning thickness of Arctic floating ice pack during glacial times (Broecker, 1975). The thickness of Arctic floating ice today is inconsequential (generally less than 7 m, or 23 ft). If floating ice attained considerable thickness, that ice could store water with a highly negative $\delta^{18}O$ value without affecting sea level. The magnitude of this effect is about 0.3 $^o/_{oo}$ uncertainty. If 0.3 $^o/_{oo}$ of the global ocean $\delta^{18}O$ signal were tied up in floating ice rather than in continental ice sheets, then the volume of water tied up in 18,000-year-ago continental ice sheets as compared to modern would be only 120 m (394 ft) sea-level equivalent.

Other uncertainties concerning $\delta^{18}O$/glacio-eustatic sea-level calibration arise from estimation of local paleotemperatures and from problems of deep-sea diagenesis. Standard error of estimate for total-faunal-analysis paleotemperatures is about 1°C. An error of 1°C in local temperature assumptions amounts to an error of about 25 m (82 ft) in sea-level

equivalent. Partial dissolution of foraminifer tests can alter their isotopic value by as much as 0.5 $^o/_{oo}$.

In attempting to apply the late Pleistocene model to older $\delta^{18}O$ data sets, problems of local temperature variation and diagenesis can be filtered out by reproducing the global $\delta^{18}O$ time-series signal at widely separated localities and in both planktic and benthic data sets. Following the lead of CLI-MAP's 18,000-year-ago temperature estimates (discussed above), strong local temperature effects can be avoided by working with tropical and low-latitude temperate sites away from regions of upwelling.

Pleistocene periodicity consistent with astronomical theory of glaciation

Though many theories of Quaternary glaciation are substantially ad hoc arguments intended only to "explain glaciation," one theory proposes an external forcing function that should be operative throughout time and predictable in its periodicity. This is the astronomical theory of climate change (Croll, 1867; Mesolella et al, 1969; Hays, Imbrie, and Shackelton, 1976).

Stated in its most general form, the astronomical theory of climate change holds that variation in the tilt of the Earth's axis, precession of the equinoxes, and variation in the eccentricity of the Earth's orbit all work together to produce systematic variation in the seasonal distribution of solar radiation. If it can be demonstrated that perturbations of the Earth's orbit have played a significant role in Pleistocene climatic change (glacial/interglacial cycles), their effects on climate may be recorded in the stratigraphic record as glacio-eustatic sea-level fluctuation or as faunal or floral alternations throughout time. In short, the perturbations of the Earth's orbit are potentially a tuning fork for geologic time.

The tilt of the Earth's axis (also referred to as the obliquity of the ecliptic) varies by about 1° with a periodicity of about 40,000 years. The higher the angle of tilt, the more pronounced the difference between summer and winter insolation in high latitudes. The two other effects--the precession of the equinoxes and variations in eccentricity of the Earth's orbit--require a more elaborate explanation than can be given here.

The eccentricity of the Earth's orbit varies with a periodicity of about 90,000 years. There are many scenarios which can be proposed by combining effects of eccentricity, precession, "critical" season, and location of regions chosen to be "climatically sensitive." One such scenario is presented here simply to show that reasonable arguments can be formulated.

At a time of high orbital eccentricity when the summer solstice occurs near the perihelion, there are exceedingly hot summers and exceedingly cold winters. One common version of the astronomical theory of climate change is that the Earth has a climatic regime that generally favors the accumulation of Northern Hemisphere ice sheets. Only during times of unusually hot summers (high orbital eccentricity and precession maximum) is there sufficient melting of continental ice sheets to cause a general retreat of the glaciers and resultant glacio-eustatic sea-level rise. Numerous other scenarios are equally plausible. Nevertheless, the principal periodicities of the late Pleistocene $\delta^{18}0$ record are precisely the periodicities of the Earth's eccentricity, tilt, and precession cycles (Hays, Imbrie, and Shackleton, et al, 1976). Thus, the Pleistocene $\delta^{18}0$ signal is empirically linked to fluctuation in continental ice volume and to chronology of calculated values of variation in the Earth's orbit.

Shifts in the glacial/interglacial $\delta^{18}0$ envelope

In the most general of terms, the envelope of the glacial/ interglacial $\delta^{18}0$ variations observed in the late Pleistocene can be taken to represent interaction of global boundary conditions with an external forcing function (the astronomical variations). As long as the boundary conditions remain more or less constant, this system oscillates between extremes, here empirically defined as $\pm 2\delta$ of the observed $\delta^{18}0$ time series.

While it is unlikely that the forcing function (astronomical perturbations) has changed substantially over the last 100 m.y., it is certain that the global boundary conditions (such as position of continents, interocean connections, etc.) have undergone major change. It is likely that changing boundary conditions cause variation in the envelope of the $\delta^{18}0$ time-series signal. Indeed, one such mode shift is well documented in several deep-sea cores somewhat below the Brunhes-Matuyama boundary (Shackleton and Opdyke, 1976; Prell, 1982). Thus, the final element of the Quaternary dynamic model is that shifts in the glacial/interglacial $\delta^{18}0$ envelope are anticipated as global boundary conditions change throughout the Tertiary. Magnitude and timing of such shifts remains a matter of empirical discovery.

The following sections propose that the deep-sea $\delta^{18}0$ record suggests the existence of very large continental ice sheets throughout the Tertiary. Presumably, such ice sheets belong in high latitudes. An argument can be made for the existence of ice sheets in either hemisphere. However, the arguments against ice sheets seem much stronger at face value for the Arctic region than for Antarctica. Therefore, this paper defers the question of Northern Hemisphere Tertiary ice sheets and concentrates on the assertion that large ice sheets have existed in Antarctica throughout the Tertiary. This is contrary to numerous literature citations that Antarctica was substantially ice-free prior to middle Miocene. It is therefore worthwhile to review briefly the facts and the arguments.

STRATIGRAPHIC RECORD OF CONTINENTAL ICE IN ANTARCTICA

Arguments Against Early Tertiary Antarctic Ice Sheets

Arguments against the existence of continental ice sheets prior to the middle Miocene fall into two categories. First, there are arguments that the absence of positive evidence for glaciation somehow indicates there was no glaciation. Secondly, there are arguments which proceed from positive evidence for existence of warm conditions within perhaps 1,000 km (621 mi) of where one might logically seek to put continental ice.

A prevalent "negative evidence" argument concerning

early Tertiary glaciation centers around the absence of ice-rafted detritus from high-latitude deep-sea cores. It is widely accepted that there is significant ice-rafted detritus in southern ocean cores from middle Miocene to Holocene. There are also reports of ice-rafted detritus in deep-sea cores tentatively ascribed to the Oligocene (Margolis and Kennett, 1971). Existing ice-rafted detritus is positive evidence for glaciation of Antarctica. It can be argued that more abundant ice-rafted detritus in younger sediments has paleoenvironmental significance. However, the absence of ice-rafted detritus in existing cores of older southern ocean sediments does not preclude continental glaciation on Antarctica. The absence of ice-rafted detritus could indicate poor sampling or that continental ice sheets, if present, melted before reaching sea level. Antarctica is a huge continent; there is abundant space for large ice sheets that need not flow as ice to sea level.

The occurrence of faunas and floras of seemingly temperate affinity can be used to argue for the presence of relatively warm conditions at high latitude in the early Tertiary. In Antarctica, such an argument was based on floral evidence from the Ross Sea region (Kemp and Barrett, 1975). Again, note that Antarctica is a huge continent. Relatively warm coastal conditions around the Ross Sea in no way negate the possibility of extremely cold conditions in the continent's interior (Oerlemans, 1982; Barron and Washington, in press).

Early Tertiary Antarctic Ice Sheets

If one wants to make a case for significant continental ice volume throughout the Tertiary, clearly Antarctica is a logical place to argue it. Field-checking the possibility of Eocene ice in Antarctica is not a simple task! The vast majority of our "field area" is presently beneath several kilometers of modern ice sheet. For this reason, global atmospheric circulation modeling experiments offer the most straightforward approach to the problem. Of secondary importance, but still very important, are some promising glimpses of early Antarctic glaciation in the Tertiary geology of the Transantarctic Mountains, the Dry Valleys, and West Antarctica.

Antarctica is beneath the South Pole today, and there is an immense ice sheet on the continent. Reconstruction of plate motion indicates Antarctica has been beneath the South Pole throughout the Tertiary. Thus, an empirical, first-principles argument would assert that a Tertiary Antarctic ice sheet is anticipated until proven otherwise. Modeling experiments lend credence to this assertion. The interior of the Antarctic continent remains quite cold regardless of specified changes in plate geometry and/or southern ocean sea-surface temperature (Barron, Thompson, and Schneider, 1981; Oerlemans, 1982; Barron and Washington, in press). While the application of global atmospheric circulation modeling experiments to geological questions is still in its infancy, results clearly indicate that this is the arena in which the argument should be settled.

Diligent field work in Antarctica has produced some promising evidence of large Antarctic ice sheets prior to the middle Miocene (Stump et al, 1980; LeMasurier and Rex, 1982; Denton et al, 1984). Denton et al (1984) present a map of ice configuration consistent with their observations and those of Stump et al (1980) and LeMasurier and Rex (1982). Such an ice sheet was considerably larger than the modern Antarctic ice sheet. Physical stratigraphy suggests this ice configuration developed several times. Biostratigraphy constrains the age of the oldest such ice sheet to be more than 9 m.y., though Denton et al argue that geomorphology suggests considerably greater antiquity. Ages of subglacially erupted volcanics range back as far as 28 m.y. (LeMasurier and Rex, 1982). Owing to geochemical and geological complications, these dates probably represent maximum age of the emplacement of the volcanic material. However, in a geological sense, the dates in no way preclude existence of ice prior to the first subglacial volcanic eruption. Thus, when viewing the antiquity of the Antarctic ice sheet, the data trail ends with a huge ice sheet sitting on the Antarctic continent. How long that ice sheet has been there and how it got there are substantially unknown.

APPLICATION OF THE LATE QUATERNARY DYNAMIC MODEL TO THE POST-EOCENE DEEP-SEA $\delta^{18}O$ RECORD

$\delta^{18}O$ Dilemma

The quantity which is measured, $\delta^{18}O_{calcite}$ for benthic or planktic foraminifers, is a function of both water temperature and the isotopic composition of the water in which the calcite grew. Regarding planktic foraminifers, local temperature can vary as a function of depth at which the foraminifer grew; likewise, $\delta^{18}O_{water}$ may vary because of local salinity effects or may reflect global $\delta^{18}O_{water}$ variation which is in turn related to global ice-volume fluctuation. Thus, random measurement of $\delta^{18}O$ on taxa of unknown habitat taken from samples of convenience could produce uninterpretable results. To estimate temperature, the $\delta^{18}O_{water}$ effect must be constrained; to estimate $\delta^{18}O_{water}$ (and thereby ice volume), temperature and local salinity effects must be constrained.

Regarding the Quaternary dynamic model, Shackleton (1967) developed the concept that $\delta^{18}O_{calcite}$ in deep-sea benthic foraminifers was primarily a measure of $\delta^{18}O_{water}$. He argued that bottom water was sufficiently cold that $\delta^{18}O$ values for 18,000-year-old benthic foraminifers could not be explained by still colder bottom-water temperatures. Therefore, the benthic isotope signal had to be substantially a $\delta^{18}O_{water}$ signal. The similarity between benthic and planktic isotope records therefore indicated that the global ice volume $\delta^{18}O_{water}$ effect was the primary signal.

In the Tertiary, the data clearly indicate a tendency to lighter benthic $\delta^{18}O$ values back through time. Thus, the arguments of Shackleton (1967) cannot be applied. There are two choices: either constrain the global $\delta^{18}O$ effect by invoking geological arguments concerning continental ice volume, or constrain tropical and low-latitude planktic taxa to have precipitated their calcium carbonate at constant temperature. If global ice volume can be constrained, then the $\delta^{18}O_{calcite}$ data record variation in temperature; if tropical and low-latitude temperate near-surface temperature can be constrained, then the $\delta^{18}O_{calcite}$ record of certain planktic foraminifers records variation in global ice volume. These two alternatives lead to profoundly different interpretations con-

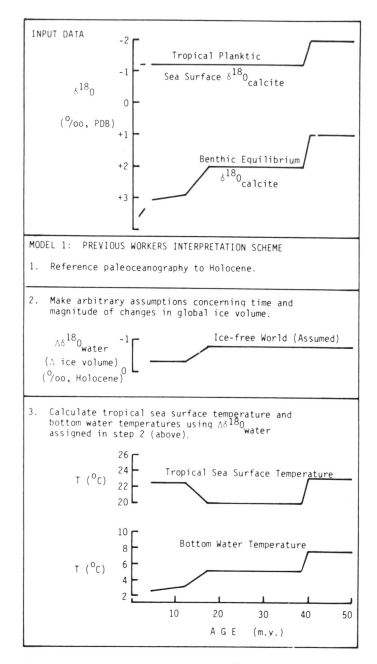

Figure 2: Diagram indicating generalized $\delta^{18}O$ record for the last 50 m.y. and the interpretation scheme applied to the data by previous workers. Age scale at bottom of figure applies to all curves. Comparison of Tertiary data to the modern ocean and assumption of pre-middle Miocene ice-free world lead to calculation of surprisingly cool tropical sea-surface temperatures. There is no published physical justification for such cool tropical sea-surface temperature in Oligocene time. Thus, it is attractive to assert that these estimates of cool temperature are an artifact of the interpretation scheme. (After Matthews and Poore, 1980)

cerning the general state of the planet throughout the Tertiary.

Benthic and Tropical Shallow-Dwelling Planktic Foraminifer $\delta^{18}O$ Record

Upper portions of Figures 2 and 3 present a generalized representation of Tertiary $\delta^{18}O$ data for tropical sea-surface

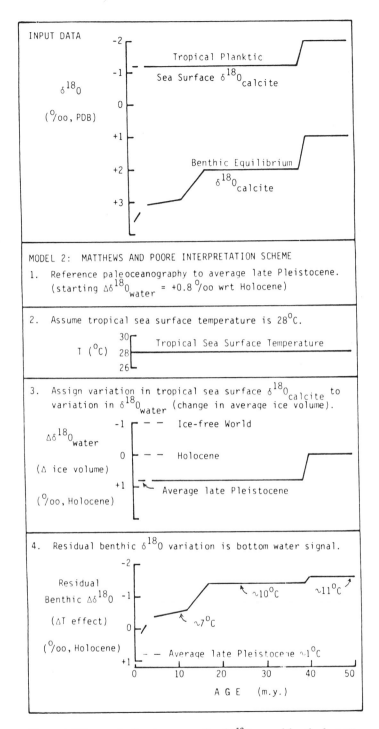

Figure 3: Diagram indicating generalized $\delta^{18}O$ record for the last 50 m.y. and the interpretation scheme proposed by Matthews and Poore (1980). Comparison of Tertiary $\delta^{18}O$ data to average late Pleistocene and assumption of constant low-latitude sea-surface temperatures suggests significant global ice-volume to 50 m.y. Ice-free world with tropical sea-surface temperature of 28°C would produce sea-surface planktic $\delta^{18}O$ values of -3. (After Matthews and Poore, 1980)

equilibrium calcite (shallow-dwelling planktic foraminifers) and for deep-sea benthic foraminiferal calcite. The curves are intended to represent average conditions. They surely smooth out real events on the scale of at least ±0.3 °/oo

(2σ), perhaps even larger. Figure 2 presents an interpretation scheme which seeks to constrain $\delta^{18}O_{water}$ by assuming the Earth was ice free prior to middle Miocene. Figure 3 presents an interpretation scheme constraining tropical sea-surface temperature at a constant value of 28°C.

Regarding interpretation of the tropical shallow-dwelling planktic foraminifer $\delta^{18}O_{calcite}$ record, the two interpretations starkly disagree. Use of the ice-free world assumption prior to the middle Miocene forces the interpretation that tropical sea-surface temperatures were quite low throughout the early Miocene and Oligocene. Taking the alternative approach and assuming constant tropical sea-surface temperature, there appears to be a major shift in global ice volume somewhere around the Eocene/Oligocene boundary, but no "ice-free world" is indicated in the last 50 m.y.

As noted above, geological arguments for an ice-free world in the early Miocene, are in my opinion, substantially without merit. The assumption of constant sea-surface temperature for large areas of tropical and low-latitude temperate ocean is consistent with the CLIMAP observation of little change between the 18,000-year-ago full glacial world and modern conditions. Further, results of atmospheric circulation modeling yield constant low-latitude temperatures even back into the Cretaceous (Barron and Washington, 1982). For these reasons, I propose that the interpretation scheme depicted in Figure 3 is the appropriate one.

Amplitude and Frequency of $\delta^{18}O$ Variation in the Post-Eocene Deep-Sea Record

By analogy with the late Quaternary dynamic model, it would be highly desirable to characterize amplitude and frequency for the Tertiary $\delta^{18}O$ signal. The requisite studies are in progress. Unfortunately, the vast majority of data collected by other workers is unsuitable to these purposes because of an excessively large sample interval and because of shortcomings of early Deep Sea Drilling Project coring methods.

Whereas the late Quaternary glacio-eustatic $\delta^{18}O$ record has been defined by sampling numerous cores at approximately 5,000-year sample intervals, the vast majority of isotopic data for the Tertiary have been gathered at something more like a 500,000-year sample interval (see Savin, 1977, summary article). Clearly, a 500,000-year sample interval will not detect high frequency cyclicity so well documented for the late Quaternary (namely, approximately 20,000, 40,000, and 100,000 year power). Nevertheless, short strings of this data can be used to estimate standard deviation of the $\delta^{18}O$ signal within the time series. Such an exercise within the Tertiary $\delta^{18}O$ data set typically results in standard deviations of about 0.25 °/oo throughout much of the post-Eocene Tertiary (Moore, Pisias, and Keigwin, 1981). By analogy with the late Quaternary model, amplitude of glacio-eustatic sea-level fluctuation may be defined as ±2 standard deviations; thus, we estimate an amplitude of about a 1 °/oo for the glacio-eustatic $\delta^{18}O$ signal for much of the post-Eocene Tertiary. This amplitude is approximately two-thirds that documented for the late Quaternary dynamic model.

Stable isotope studies at a 5,000-year sample interval are in progress for several hydraulic piston cores taken by the Deep Sea Drilling Project. Refer to the work of Prell concerning Leg 68 Pliocene samples, and to the work of Poore and Matthews concerning Leg 73 Oligocene samples.

APPLICATION OF LATE PLEISTOCENE DYNAMIC MODEL TO THE EOCENE-CRETACEOUS $\delta^{18}0$ RECORD

Relevant isotopic data from a few tropical western Pacific Deep Sea Drilling Project sites is depicted in the upper portion of Figure 4. The data indicate a continuation of the tendencies observed in the younger portions of the record (Figure 3). The $\delta^{18}0$ values of both benthic and shallow-dwelling planktic foraminifers tend to become lighter with increasing age.

Following the interpretation scheme of Matthews and Poore (1980), shallow-dwelling planktic $\delta^{18}0$ values of -3 °/oo are indicative of a tropical sea-surface temperature of 28°C with no ice sheets on continents. There are a scant four data points indicating this condition at around 100 m.y. (Albian/Aptian). Since that time, the scant data available are consistent with progressively larger amounts of continental ice volume. An especially noteworthy shift in both planktic and benthic values occurs between 80 and 65 m.y. (being approximately the transition from the Mesozoic to Cenozoic era).

COMPARING SEISMIC STRATIGRAPHY "RELATIVE SEA-LEVEL CURVE" TO THE $\delta^{18}0$ ICE-VOLUME ESTIMATES

Figure 4 compares 100 million years of deep-sea $\delta^{18}0$ record with 100 million years of "relative sea-level curve" based on seismic stratigraphy. The metric scale on the relative sea-level curve is derived from literal redrafting of Figure 6 of Vail, Mitchum, and Thompson (1977). The $\delta^{18}0$ axis is plotted to the same scale as the ΔSL axis, using the Fairbanks and Matthews (1978) calibration of $\delta^{18}0$ variation to late Pleistocene sea-level fluctuations.

Curves like those depicted in the lower portion of Figure 4 have been constructed from seismic data by estimating relative coastal onlap or by estimating sediment thickness accumulated in coastal aggradation (Vail, Mitchum, and Thompson, 1977, Figure 13, page 78). Relative coastal onlap need not necessarily be related to magnitude of sea-level fluctuations. Scaling of onlap events, from +1 for maximum Cretaceous transgression to 0 for maximum Oligocene regression, allows some 50 m.y. for epeirogenic movement to occur within any particular continental margin environment. Thus, there should be only a qualitative relationship between relative coastal onlap and glacio-eustatic sea-level fluctuations. Coastal aggradation should provide a more reliable measure of actual vertical change in relative sea level; namely, the accumulated sediment, isostatically subsided in response to accumulation of new load, should measure approximately three times the actual change in relative sea level. However, recent versions of the "relative sea-level curve" refer to it as a "relative coastal onlap curve" and do not mention using coastal aggradation in its construction. Thus, I propose that the relative coastal onlap curve captures the timing of transgression/regression events, but does not accurately measure the magnitude of relative sea-level fluctuation associated with these events. The $\delta^{18}0$ record allows calibration of glacio-eustatic sea-level fluctuations for those transgression/regression events which are attributable to fluctuations in global ice volume.

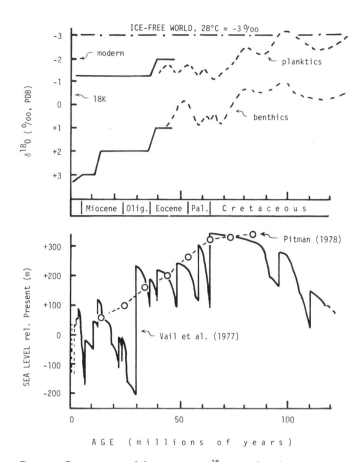

Figure 4: Comparison of the 100-m.y. $\delta^{18}0$ record to the seismic stratigraphy "relative sea-level curve." The y-axes of the two diagrams are plotted to comparable scale using the δ^{18}/Δsea-level calculation of Fairbanks and Matthews (1978). Note that both data sets are conveniently divided into three regimes, each typified by successively lower sea levels. These regimes are Late Cretaceous, early to middle Paleogene, and post-late Paleogene. However, there is severe discrepancy concerning magnitude and timing of regressive events. Isotope curves 0 to 50 m.y. from Matthews and Poore (1980); 50 to 120 m.y. from Douglas and Woodruff (1981); with modification. Relative sea-level curve replotted from Vail, Mitchum, and Thompson (1977), Figure 6, p. 91.

Note that the seismic stratigraphy relative sea-level curve includes the Pitman (1978) calculations and the generalized seismic stratigraphy relative sea-level curve in accordance with Vail, Mitchum, and Thompson (1977, page 91). The Pitman (1978) calculations of paleo sea level are based on an ice-free world. Thus, planktic isotopic values less than -3 °/oo in the upper diagram may indicate glacio-eustatic sea-level lowering below the Pitman curve.

Comparing the upper and lower portions of Figure 4, we clearly see similarities and differences among these independent records of eustatic history. The isotopic data are consistent with an ice-free world at approximately 100 m.y. ago. The isotopic data for Paleocene and Eocene generally reflect lower glacio-eustatic sea levels than for the Cretaceous. This generality is also captured in the seismic stratigraphy curve within this general time frame. Indeed, the difference between average Late Cretaceous and average lower Tertiary sea-level stands is approximately the same

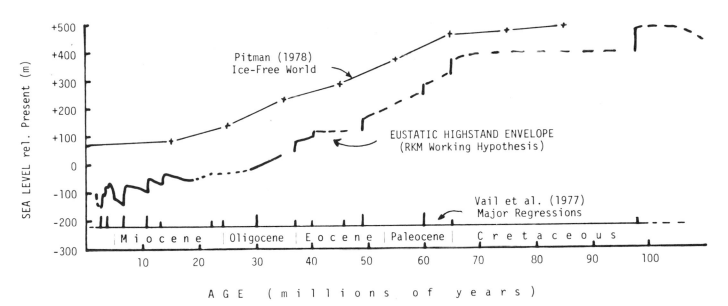

Figure 5: Eustatic sea-level curve constructed from three independent data sets according to simple rules. The construct begins with the calculated effects of changing spreading rate on sea level for an assumed ice-free world (Pitman, 1978). Next, the $\delta^{18}O$ record is used to estimate the ice-volume effect for those time frames within which there is sufficient $\delta^{18}O$ data. Next, the timing of Cretaceous and Paleogene regression events on the seismic stratigraphy relative sea-level curve (Vail, Mitchum, and Thompson, 1977) is used wherever these events are not constrained within the $\delta^{18}O$ time series data. Finally, Neogene structure of the seismic stratigraphy curve is rescaled in accordance with experience where $\delta^{18}O$ and seismic stratigraphy records overlap; namely, amplitudes indicated on the seismic stratigraphy curve (Figure 4) are cut by a factor of three.

scale as average isotopic differences.

Note, however, that major discrepancies exist regarding timing and magnitude of the major mid-Tertiary regression. Whereas the seismic stratigraphy curve suggests a 400 m (1,312 ft) sea-level fall and places it in the middle Oligocene (30 m.y. ago), the isotope data calibrate the event at around 80 to 100 m (262 to 328 ft) sea-level fall, and the deep-sea biostratigraphy places the major isotopic shift near the Oligocene/Eocene boundary (approximately 38 m.y. ago). Thus, both records recognize a major the mid-Tertiary regression, but the two records disagree as to magnitude and timing of that event. If there is a relatively distinct, single glacio-eustatic regression within the mid-Tertiary, event correlation suggests that to be the major event within the seismic stratigraphy record.

SIMPLE CONSTRUCT CONCERNING THE LAST 100 MILLION YEARS OF EUSTATIC HISTORY

Figure 5 presents a eustatic sea-level curve based upon four rules of construct. First, the calculations of Pitman (1978) are taken to represent the eustatic component ascribable to changes in the rate of sea-floor spreading from late Mesozoic to the present. Note that Figure 5 plots the value "*h*" (real, oceanic sea-level change) rather than the value "0.7 *h*" used by Vail, Mitchum, and Thompson (1977). Use of the 0.7 *h* parameter flows from the assumption that isostasy can be corrected for in a simple and uniform manner. This may be an oversimplification in the case of rapid glacio-eustatic sea-level fluctuations. Further, the error bars placed on these calculations by Kominz (this volume) indicate the choice between *h* and 0.7 *h* is trivial compared to the overall uncertainty. A more rigorous geological calibration of this important concept is probably forthcoming. In

the meantime, it is best to simply portray a straightforward representation of eustasy as the global sea-level signal and relegate questions of isostasy, epeirogeny, and calibration to each time frame and local situation.

Returning to the general rules of construct of Figure 5, the second step is estimating the glacio-eustatic component of sea level from the tropical shallow-dwelling planktic foraminifer $\delta^{18}O$ record in accordance with the calibration of Fairbanks and Matthews (1978). Where detailed time series are not available, position of the highstand glacio-eustatic envelope is taken as $(\bar{X} - 2\sigma)$ for nearby data points. Note that the $\delta^{18}O$ data set is given preference over seismic stratigraphy regarding estimation of amplitude of the glacio-eustatic signal.

Thirdly, the relative sea-level curve of seismic stratigraphy is used to locate Cretaceous and Paleogene major regressive events not sufficiently defined by the $\delta^{18}O$ record. Finally, major Neogene events in the seismic stratigraphy relative sea-level curve which are not adequately defined within the $\delta^{18}O$ record are scaled in accordance with $\delta^{18}O$ calibration provided by events which occur within both data sets. To a first approximation, this amounts to cutting the amplitude of events depicted in the lower portion of Figure 4 by approximately three.

The resultant curve is taken to represent the highstand envelope of a eustatic record which almost certainly contains events with 40,000 and 100,000-year periodicity. On the scale depicted in Figure 5, including such events with even the thinnest drafting pens available would produce a solid black bar with a width of 50 to 100 m (164 to 328 ft).

Each step in the construction of Figure 5 involves its own set of uncertainties. Thus, the question is not whether Figure 5 is an accurate representation of eustatic sea-level history; rather, the question is whether Figure 5 can be continually improved by adherence to the rules of construct. Whereas

Kominz (this volume) places large error bars on the calculated sea-level curve of Pitman (1978), the approach of Bond (1978) may develop geological arguments which better quantify the important long-term trends in sea level. Whatever becomes the ultimate ice-free eustatic curve, glacio-eustatic effects will simply be subtracted from it. The seismic stratigraphy relative sea-level curve is undoubtedly subject to further revision. Indeed, Watts (1982) questions the entire concept of eustatic origin of onlap/off-lap relationships. Regarding the glacio-eustatic component of eustatic history, Watts' (1982) arguments will simply be proved or disproved by the continuing acquisition of $\delta^{18}O$ data. The rules of construct allow for the seismic stratigraphy curve to be phased out completely as isotopic data becomes sufficiently abundant. Finally, the existence of high-frequency glacio-eustatic sea-level fluctuations (20,000-, 40,000-, and 100,000-year periodicity) throughout the Tertiary is, at this stage, little more than a hypothesis. The straightforward test of the hypothesis is continued acquisition of Tertiary $\delta^{18}O$ data at 5,000-year sample intervals.

DISCUSSION

The principal feature of the eustatic highstand envelope portrayed in Figure 5 is the general tendency to glacio-eustatic lowering of sea level at discrete intervals from 98 through 6 m.y. Note especially that the Paleogene includes glacio-eustatic regressions on the order of 50 m (164 ft). Is a 50-meter (164-ft) sea-level drop sufficient to cause a significant interregional unconformity?

To estimate the amount of missing section at an unconformity caused by a 50-meter (164-ft) glacio-eustatic sea-level fall, one must the relate the relative subsidence rate to the magnitude of the glacio-eustatic event. Of particular importance to such a calculation is the general slope of the "ice-free world" eustatic sea-level curve. As indicated in Figure 5, the general tendency throughout Paleogene time is for worldwide eustatic regression with an approximate sea-level lowering rate of 0.5 to 1.0 cm/10 yr. Thus, if subsidence were proceeding at a similar rate, any glacio-eustatic sea-level lowering exposes former shoreline, and that former shoreline will remain exposed henceforth. Alternatively, if basin subsidence exceeds the rate of the ice-free world eustatic sea-level lowering, then an unconformity surface resulting from a 50-meter (164-ft) glacio-eustatic sea-level lowering would once again become submerged. If net subsidence is the difference between regional driving subsidence and the ice-free world rate of sea-level fall, the duration of an unconformity can be conveniently calculated. If net subsidence is approximately 1 cm/1,000 yr, a 50-meter (164-ft) glacio-eustatic sea-level lowering would crease an unconformity which would be returned to its original relation to sea level only after 5 m.y. Given a regional net subsidence rate of approximately 5 cm/1,000 yr, this figure drops to 1 m.y. In either case, the actual amount of missing section would also depend on erosion.

CONCLUSIONS

1. A model for interpretation of the deep-sea $\delta^{18}O$ record as a glacio-eustatic ice-volume signal has been developed from a very large and diverse late Quaternary data set.

2. Application of the late Quaternary dynamic model to deep-sea $\delta^{18}O$ data suggests an ice-free world around 100 m.y. ago, followed by progressive ice buildup 65, 60, 50, and 38 m.y. ago. These glacio-eustatic events contributed significantly to the major interregional unconformities of the Paleogene.

3. Displacement of the glacio-eustatic highstand envelope is no more than 50 m (164 ft) at each of the major events mentioned above. The fact that these glacio-eustatic events are superimposed upon a general trend of declining sea level tends to amplify their importance in the time domain.

4. The late Quaternary dynamic model suggests giving serious consideration to the possibility of 40,000- to 100,000-year glacio-eustatic cycles throughout the Tertiary, these being major periodicities of the perturbations of the Earth's orbit. Proper evaluation of this prospect requires sampling of Tertiary sequences at approximately 5,000-year sample intervals.

REFERENCES CITED

Barron, E.J., S.L. Thompson, and S.H. Schneider, 1981, An ice-free Cretaceous? Results from climate model simulation: Science, v. 212, n. 4494, p. 501-508.
———, and W.M. Washington, 1982, The atmospheric circulation during warm geologic periods: Is the equator-to-pole surface temperature gradient the controlling factor?: Geology, v. 10, p. 633-636.
Bloom, A.L., 1978, Geomorphology: Englewood Cliffs, N.J., Prentice-Hall, Inc., 510 p.
———, et al, 1974, Quaternary sea-level fluctuations on a tectonic coast: new $^{230}Th/^{234}U$ dates from the Huon Peninsula, New Guinea: Quaternary Research, v. 4, p. 185-205.
Bond, G., 1978, Speculations on real sea-level changes and vertical motions of continents at selected times in the Cretaceous and Tertiary periods: Geology, v. 6, p. 247-250.
Bower, D.Q., 1978, Quaternary geology, a stratigraphic frame-work for multidisciplinary work: London, William Clowes & Sons, Ltd., 221 p.
Broecker, W.S., 1975, Floating glacial ice caps in the Arctic Ocean: Science, v. 188, p. 1116-1118.
———, et al, 1968, Milankovitch hypothesis supported by precise dating of coral reefs and deep-sea sediments: Science, v. 159, p. 297-300.
CLIMAP Project Members, 1976, The surface of the ice-age Earth: Science, v. 191, p. 1131-1137.
———, 1981, Seasonal reconstructions of the Earth's surface at the last glacial maximum: Geological Society of America Map and Chart Series MC-86.
Cline, R.M., and J.D. Hays, eds., 1976, Investigation of late Quaternary paleoceanography and paleoclimatology: Geological Society of America Memoir 145, 464 p.
Croll, J., 1867, On the change in the obliquity of the ecliptic, its influence on the climate of the polar regions and on the level of the sea: Phil. Magazine, v. 33, p. 426-445
Denton, G.H., and T.J. Hughes, eds., 1981, The last great ice sheets: New York, Wiley Interscience, 484 p.
———, et al, 1984, Tertiary history of the Antarctic ice sheet: evidence from the dry valleys: Geology, v. 12, p. 263-267.

Douglas, R., and F. Woodruff, 1981, Deep-sea benthic foraminifera, *in* C. Emiliani, ed., The sea: New York, Wiley
Interscience, v. 7, p. 1233-1327.

Fairbanks, R.G., and R.K. Matthews, 1978, The marine
oxygen isotope record in Pleistocene coral, Barbados,
West Indies: Quaternary Research, v. 10, p. 181-196.

Hays, J.D., J. Imbrie, and N.J. Shackleton, 1976, Variations
in the Earth's orbit: pacemaker of the ice ages: Science, v.
194, p. 1121-1132.

Imbrie, J., and N.G. Kipp, 1971, A new micropaleontological method for quantitative paleoclimatology: application to a late Pleistocene Caribbean core, *in* K.K.
Turekian, ed., The late Cenozoic glacial ages: New
Haven, Yale University Press, p. 71-181.

Kemp, E.M., and P.J. Barrett, 1975, Antarctic glaciation
and early Tertiary vegetation: Nature, v. 258, p. 507-
508.

Kipp, N.G., 1976, New transfer function for estimating past
sea-surface conditions from sea-bed distribution of
planktonic foraminiferal assemblages in the North Atlantic, *in* R.M. Cline and J.D. Hays, eds., Investigation of
late Quaternary paleoceanography and paleoclimatology: Geological Society of America Memoir 145, p. 3-41.

Kukla, G.J., 1977, Pleistocene land-sea correlations; I.
Europe: Earth-Science Review, v. 13, p. 307-374.

LeMasurier, W.E., and D.C. Rex, 1982, Volcanic record of
Cenozoic glacial history in Marie Byrd Land and western
Ellsworth Land II: revised chronology and evaluation of
tectonic factors, *in* C. Craddock, ed., Antarctic geoscience: Madison, University of Wisconsin Press, p. 725-
734.

Margolis, S.V., and J.P. Kennett, 1971, Cenozoic paleogla-
cial history of Antarctica recorded in subantarctic deep-
sea cores: American Journal of Science, v. 271, p. 1-36.

Matthews, R.K., 1973, Relative elevation of late Pleistocene
high sea-level stands: Barbados uplift rates and their
implications: Quaternary Research, v. 3, p. 147-153.

———, and R.Z. Poore, 1980, Tertiary δ¹⁸0 record and
glacio-eustatic sea-level fluctuations: Geology, v. 8, p.
501-504.

Mesolella, K.J., et al, 1969, The astronomical theory of climatic change: Barbados data: Journal of Geology, v. 77,
p. 250-274.

Moore, T.C., Jr., N.G. Pisias, and L.D. Keigwin, Jr., 1981,
Ocean basin and depth variability of oxygen isotopes in
Cenozoic benthic foraminifera: Marine Micropaleontol-
ogy, v. 6, p. 465-481.

Neumann, A.C., and W.S. Moore, 1975, Sea level events
and Pleistocene coral ages in the northern Bahamas:
Quaternary Research, v. 5, p. 215-224.

Oerlemans, J., 1982, A model of the Antarctic ice sheet:
Nature, v. 297, p. 550-553.

Pitman, W.C., III, 1978, Relationship between eustasy and
stratigraphic sequences of passive margins: Geological
Society of America Bulletin, v. 89, p. 1389-1403.

Prell, W.L., 1982, Oxygen and carbon isotope stratigraphy

for the Quaternary of Hole 502B: evidence for two modes
of isotopic variability, *in* W.L. Prell, et al, Initial Reports
of the Deep Sea Drilling Project, 68: Washington, D. C.,
U. S. Government Printing Office, p. 455-464.

Ruddiman, W.F., and A. McIntyre, 1979, Warmth of the
subpolar North Atlantic Ocean during Northern Hemisphere ice-sheet growth: Science, v. 204, p. 173-175.

———, and A. McIntyre, 1981a, The mode and mechanism
of the last deglaciation: oceanic evidence: Quaternary
Research, v. 16, p. 125-134.

———, and A. McIntyre, 1981b, Oceanic mechanisms for
amplification of the 23,000-year ice-volume cycle: Science, v. 212, p. 617-627.

Savin, S.M., 1977, The history of the Earth's surface temperature during the past 100 million years: Annual
Review of Earth and Planetary Science Letters, v. 5, p.
319-355.

Shackleton, N.J., 1967, Oxygen isotope analyses and Pleistocene temperatures reassessed: Nature, v. 215, p. 15-17.

———, 1977, The oxygen isotope stratigraphic record of
the late Pleistocene: Philosophical Transactions of the
Royal Society of London, B, v. 280, p. 169-182.

———, and J.P. Kennett, 1975, Paleotemperature history of
the Cenozoic and the initiation of Antarctic glaciation:
Oxygen and carbon isotope analyses in DSDP Sites 277,
279, and 281, *in* J.P. Kennett, R.E. Houtz, et al, eds., Initial Reports of the Deep Sea Drilling Project, 29: Washington, D.C., U.S. Government Printing Office, p.
743-755.

———, and N.D. Opdyke, 1976, Oxygen-isotope and
paleomagnetic stratigraphy of Pacific core V28-239 late
Pliocene to latest Pleistocene, *in* R.M. Cline and J.D.
Hays, eds., Investigation of late Quaternary paleoceano-
graphy and paleoclimatology: Geological Society of
America Memoir 145, p. 449-464.

———, and R.K. Matthews, 1977, Oxygen isotope stratigraphy of late Pleistocene coral terraces in Barbados: Nature, v. 268, p. 618-620.

———, and N.D. Opdyke, 1973, Oxygen isotope and
palaeo-magnetic stratigraphy of equatorial Pacific core
V28-238: Oxygen isotope temperatures and ice volumes
on a 10 and 10 year scale: Quaternary Research, v. 3, p.
39-55.

Shepard, F.P., 1960, Rise of sea level along northwest Gulf
of Mexico, *in* F.P. Shepard, F.B. Phleger, and T.H. van
Andel, eds., Recent sediments, northwest Gulf of Mexico: AAPG Special Publication, p. 338-344.

Stump, E., et al, 1980, Early Miocene subglacial basalts, the
East Antarctic ice sheet, and uplift of the Transantarctic
Mountains: Science, v. 207, p. 757-759.

Vail, P.R., R.M. Mitchum, Jr., and S. Thompson III, 1977,
Seismic stratigraphy and global changes of sea level, part
4: Global cycles of relative changes of sea level, *in* C.E.
Payton, ed., Seismic stratigraphy--applications to hydrocarbon exploration: AAPG Memoir 26, p. 83-97.

Oceanic Ridge Volumes and Sea-Level Change - An Error Analysis[1]

Michelle A. Kominz
Lamont-Doherty Geological Observatory of Columbia University
Palisades, New York

Sea-level fluctuations due to changing mid-ocean ridge volumes have been calculated for the last 80 m.y. in 5-m.y. intervals. An analysis of the errors involved in this set of calculations includes: the effect of omitting calculations for crust more than 70-m.y. older than the ridge crest at any given time; inaccurate estimates of stage poles and ridge lengths; subducted ridges for which only a remnant triple junction remains; completely subducted ridges; and uncertainty in absolute dating of magnetic anomalies. The maximum possible sea level 80 m.y. ago was 365 m (1,198 ft); the minimum was 45 m (148 ft) with a most probable height of about 230 m (755 ft) above present sea level. A decrease in spreading rates since the Late Cretaceous was the primary cause of a volume decrease in mid-ocean ridges.

INTRODUCTION

Studying Cretaceous to Recent sea-level variation has resulted in considerable controversy in the recent literature. Different methods have been used to obtain both relative and absolute sea-level history of the Cenozoic and Mesozoic, including continental-margin stratigraphy (Vail, Mitchum, and Thompson, 1977), stratigraphic data combined with hypsometry (Bond, 1979; Harrison et al, 1981), elevation of Cretaceous shorelines (Sleep, 1976; Hancock and Kauffman, 1979), analysis of subsidence of continental margins (Watts and Steckler, 1979; Wood, 1981; Brunet and Le Pichon, 1982), and analysis of the changing volumes of the ocean basins (Hays and Pitman, 1973, Pitman, 1978; Schlanger, Jenkyns, and Premoli-Silva, 1981). The resulting sea-level curves differ in the amplitude of sea-level change but all show an overall decrease from Late Cretaceous to present. It cannot be claimed that the history of sea-level change is fully understood until analyses of basin-volume change and analyses of stratigraphic data are shown to be fully compatible.

The purpose of this paper is to perform an error analysis of sea-level calculations obtained from data on the variation in the volume of oceanic spreading centers with time. Such a calculation was made by Pitman (1978) without an estimate of error. The results of an error analysis are not expected to solve the sea-level problem; detailed study of all other factors influencing ocean volume (hot spots, orogeny, trenches, sediment deposition, formation of passive margins, etc.) must eventually be made to complete the picture.

METHOD

Analysis of the volume of spreading centers as a cause of sea-level variation is a straightforward process. Sclater, Anderson, and Bell, (1971) and Parsons and Sclater (1977) showed that the depth of the ocean floor is primarily a function of its age. This is because the young, warm lithosphere is less dense than the older, cool, contracted lithosphere. As a result, fast spreading ridges occupy more volume than do slow spreading ridges. Knowing the age of the ocean floor (from magnetic anomaly data) and the relationship between age and depth of the ocean floor (from Parsons and Sclater, 1977), the volume of a mid-ocean ridge system can be calculated. Conversion to sea-level change with respect to the present air-sea interface is made according to the following equation:

$$d(t) = \frac{\sqrt{(A_o/0.7)^2 + 340 \times \Delta V} - A_o/0.7}{170} \qquad (1)$$

from Pitman (1978), where A_o is the surface area of the oceans (taken to be 360×10^6 km^2) and ΔV (in 10^6 km^3) is the change in volume of the ocean ridges from time, t, to the present. The present day average increase in ocean area as sea level rises 1 km is 170×10^6 km^2; and, although continental hypsometry may have been different in the past, the

[1]Lamont-Doherty Geological Observatory Contribution Number 3579

error resulting from using this value is less than ± 2 percent and is minor compared to the other sources of error which are considered in this paper.

The bathymetric depth of the ocean ridges follows a $t^{1/2}$ curve for 0 to 70-m.y.-old crust, whereas crust older than 70 m.y. (called the ridge flanks in this paper) deepens according to an exponential decay curve (Parsons and Sclater, 1977). In this paper, the ridge volume is calculated only for the 70 m.y. and younger portion of the ridges. This introduces a minor error which is addressed later in the text. Ridge volumes are calculated using previously published poles of opening from plate-motion studies wherever possible. Thus, calculations are obtained for each stage of opening. The age versus depth relation of Parsons and Sclater (1977) is rewritten as age versus height above the 70-m.y. contour:

$$\text{height of ridge} = 2928 - 350t^{1/2} \qquad (2)$$

where 2928 is the height, in meters, of the ridge crest above the 70-m.y. contour and time is in millions of years. The area of the ocean crust under the new ridge is:

$$\text{Area} = R^2\phi(\cos\theta_1 - \cos\theta_2) \qquad (3)$$

where R is the radius of the earth, ϕ is the angle, in radians, of rotation about a given stage pole, and θ is the colatitude of the end points of the ridge (see Figure 1). The volume of the ridge for any given stage is obtained by integrating equation (2) with respect to time, dividing by the duration of the stage and multiplying by the area, from equation (3).

SPREADING DATA

North Atlantic, South Atlantic and Indian Ocean: Gondwana Ridge Data

A compilation of ridge-length and spreading-rate data for Gondwana ridges is presented in Table 1a. Gondwana separation poles of Norton and Sclater (1979) were used to calculate ridge volumes for the South Atlantic, the Indian Ocean and spreading between Antarctica and all southern hemisphere continents. To maintain internal consistency, their maps were used to determine the ridge lengths for these rotations. Red Sea and Gulf of Aden poles were taken from Cochran (1981). Rotation poles between Africa and North America were obtained from Pitman and Talwani (1972) from the present to anomaly 34 and from Schouten and Klitgord (1977) for the M-series anomalies. Poles of Pitman and Talwani (1972) were used for North America-Europe spreading. Greenland-Europe spreading poles were obtained from Talwani and Eldholm (1977). Labrador Sea spreading poles were taken from Kristofferson and Talwani (1977). The lengths of all North Atlantic spreading centers were determined using magnetic anomaly maps which were compiled in the preparation of a new set of ocean-age maps by Larson, Golovchenko and Pitman (1982).

Pacific Ridges and Remnant Triple Junctions

Spreading parameters for the Pacific ridges were calculated using the magnetic and ocean-age maps published by Sclater, Jaupart, and Gelson (1980) and are compiled in Tables 1b and 1c. For most Pacific ridges sufficient data from fracture-zone curvature were available in the portion

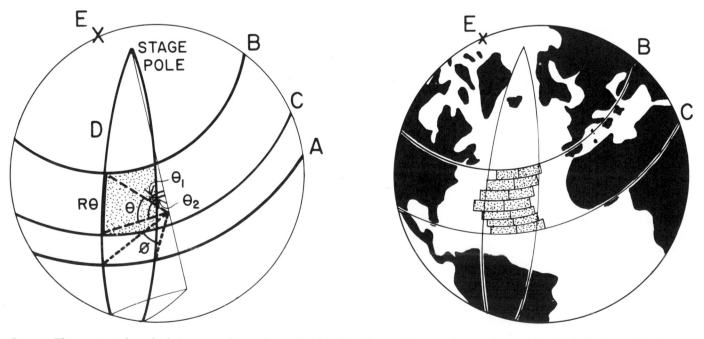

Figure 1: The construct for calculating area of spreading, ridge length and spreading rate about a given stage pole. In the figure, E is the pole of rotation, while A, B and C are lines of latitude about the stage pole and D is a line of longitude with respect to the stage pole. The angle of opening, ϕ, is measured in the equatorial plane of the stage pole (A). The colatitude of the top and bottom of the spreading, θ_1 (B) and θ_2 (C) are used to calculate the area of the ocean which has formed; equation (3) in text, and the ridge length (Rθ). The stippled pattern is the area of new ocean floor.

of the plate which is not presently subducted to define rotation poles using the method of Morgan (1968). The Pacific ocean ridge system was divided into segments with internally consistent spreading histories. These segments are the Kula-Pacific, Farallon-North Pacific, Farallon-South Pacific, Phoenix-South Pacific, and the Juan De Fuca ridges, the Galapagos spreading center, the now extinct Galapagos rise, and the Chile rise. The Farallon plate includes today's Nazca plate. The Phoenix plate is also known as the Aluk plate. Spreading of the Antarctic-Pacific plate is included with Pacific-Phoenix spreading, and spreading of the Chile rise is included under Farallon-Phoenix spreading in Tables 1b and 1c. Poles were calculated whenever there was a significant change in spreading rate or direction. Basin limits were defined using the same maps. In this study, volume calculations were made for symmetric ridge systems and corrected for the amount of ridge subducted beneath the Americas or Antarctica using the reconstructions of Figure 2.

The motion of the Pacific plate, with respect to the surrounding plates, can be obtained for the last 80 m.y. from its spreading history with respect to Antarctica (Suarez and Molnar, 1980, Molnar et al, 1975), or for the last 113 m.y. by assuming that hot spots are fixed with respect to each other and determining the relative motion between the hot spots and each individual crustal plate (Duncan, 1981; Morgan, 1981). These two sets of assumptions do not yield identical results prior to about 50 m.y. ago. The discrepancy requires one to assume either: 1) all Pacific hot spots were moving together but in a different direction than the rest of the hot spots in the world prior to 50 m.y. ago or, 2) there was motion between East and West Antarctica prior to 30 m.y. ago. In this paper, both assumptions are used to determine the position of the Pacific ridges and remnant triple junctions with respect to the surrounding continents. Reconstructions from three time periods, 113 (only possible with the hot spot model), 80, and 36 m.y. ago are shown in Figure 2 along with a map of the Pacific today showing the

ridge segments used in the reconstructions. It is important to realize that the Pacific motion is specified only with respect to Marie Byrd Land and that the position of the Antarctic Peninsula, which is drawn in its present relation to the rest of West Antarctic, is not constrained to have been in this position. Thus, any overlap with South America or the Faulkland Plateau simply requires a repositioning of the Antarctic Peninsula and does not invalidate the reconstruction. The difference between the two reconstructions (Figure 2a and b) can be seen to have very little effect on ridge-volume calculations.

The spreading rate and direction of the Kula-Farallon

Table 1a

	N. Alt.		S. Alt.		AA-AU		Afr.		SOB	
m.y.	RL	SR	RL	SR	RL	SR	RL	SR	RL	SR
0	6,240	1.3	7,660	2.0	10,390	2.6	3,680	1.0	3,680	1.8
10	6,610	1.0	7,660	2.0	10,390	2.6	3,680	1.0	3,780	1.9
20	6,610	1.0	7,660	2.0	10,390	2.6	3,680	1.0	5,330	2.5
30	6,610	1.0	7,660	2.0	10,390	2.6	3,680	1.0	3,640	1.8
40	6,970	1.1	7,560	2.0	10,080	2.2	2,480	2.6	2,070	1.4
50	7,000	1.7	7,560	2.0	10,080	2.2	2,480	2.6	2,630	1.9
60	5,510	2.4	7,340	2.6	5,060	4.8	2,320	7.0	2,230	2.8
70	3,840	1.6	7,340	2.6	4,070	5.4	1,890	4.4	1,280	3.3
80	3,980	2.4	7,030	2.4	6,080	3.6	1,730	1.3	550	1.1
90	2,880	2.3	7,030	2.4	6,080	3.6	1,730	1.3	550	1.1
100	2,880	2.3	7,030	2.4	6,080	3.0	2,580	2.0	550	1.1
110	2,880	2.3	7,030	2.4	6,080	3.0	2,580	2.0	550	1.1
120	2,880	0.7	1,360	2.9	6,100	1.5	1,540	2.5		
130	3,010	1.2			5,530	1.5	1,540	2.5		
140	3,060	1.8			4,020	1.3	970	0.9		

Table 1a: Gondwana and small ocean-basin ridge lengths (RL, in km) and spreading rates (SR, in cm/year) using the LGP time scale (Larson, Golovchenko, and Pitman, 1982). Ridges include: the North Atlantic (N. Atl.), the South Atlantic (S. Atl.), Australia and Antarctic spreading (AA-AU); Africa, India and Madagascar spreading (Afr.); and small ocean basins and marginal basins (SOB).

Table 1b

Time	Ph/Fa		Ku/Fa		Teth & Red		Pa/Fa		Pa/Ph		Pa/Ku		Pa/W. Ku	
(m.y.)	RL	SR	RL	SR	RL	SR	RL	SR	RL	SR	RL	SR	RL	SR
0	1,210	4.6			2,680	1.5	8,280	7.0	5,070	2.9				
10	1,320	4.3			2,680	1.5	8,630	4.4	5,070	2.9				
20	2,320	2.6			500	1.3	10,290	4.9	3,930	2.5	200	3.4		
30	1,810	2.6			3,870	1.3	12,300	4.9	3,930	2.5	1,500	3.4		
40	1,300	2.6	310	4.2	7,240	1.3	12,250	4.7	3,930	2.5	2,790	3.4	1,010	3.2
50	790	2.6	1,040	4.2	8,210	1.3	1,1810	4.7	3,930	2.5	2,450	3.4	3,420	3.2
60	550	4.9	1,670	4.2	9,180	1.3	11,640	4.8	2,660	4.3	2,230	3.7	4,960	3.0
70	370	4.9	2,270	4.2	10,190	3.1	11,460	4.8	2,020	5.5	1,920	3.7	6,200	3.0
80	200	9.0	2,880	7.1	11,230	2.9	10,430	7.8	1,550	10.7	1,540	7.3	6,300	6.4
90	910	9.0	3,400	7.1	11,500	2.9	9,490	7.7	2,560	10.7	1,000	7.3	6,300	6.4
100	1,610	9.0	3,930	7.1	11,780	2.9	9,490	7.5	2,780	10.7	7,320	6.7		
110	2,320	9.0	4,460	7.1	12,050	2.9	9,490	7.5	3,000	10.7	7,320	6.7		
120	2,350	2.6	4,720	3.4	12,260	5.9	9,000	4.2	3,220	6.1	7,960	5.2		
130	2,350	2.6	4,950	3.4	12,660	5.9	8,500	4.2	3,440	6.1	7,960	5.2		
140	2,350	2.6	5,190	3.4	12,660	5.9	7,600	4.2	3,660	6.1	7,960	5.2		

Table 1b. Pacific ridge lengths (RL, in 10^6 km) and spreading rates (SR, in cm/year) using the LGP time scale (Larson, Golovchenko, and Pitman, 1982). Ph/Fa is the Phoenix-Farallon ridge, at present the Chile rise. Ku/Fa is the Kula-Farallon ridge; Teth & Red includes the Tethys ridge and the Red Sea ridge; Pa/Fa includes the Pacific-Farallon ridges and the Juan de Fuca Ridge; Pa/Ph is the Pacific-Phoenix ridge; the Pa/Ku is the Pacific-Kula ridge and Pa/W.Ku is the addition to the Pacific-Kula ridge for maximum ridge length assumptions.

Table 1c

	80		70		60		50		40		30		20		10		0	
Phoenix-Farallon																		
0-10	200	9.0	370	4.9	550	4.9	790	2.6	1,300	2.6	1,810	2.6	2,320	2.6	1,320	4.3	1,210	4.6
10-20	100	9.0	170	9.0	350	4.9	500	4.9	790	2.6	1,200	2.6	1,700	2.6	1,320	2.6	1,210	4.3
20-30			50	9.0	100	9.0	200	4.9	300	4.9	450	2.6	650	2.6	1,210	2.6	1,210	2.6
30-40							100	9.0	180	4.9	270	4.9	390	2.6	650	2.6	900	2.6
40-50									180	4.9	270	4.9	390	2.6	650	2.6
50-60											180	4.9	270	4.9	390	2.6
60-70													180	4.9	270	4.9
Kula-Farallon																		
0-10	2,880	7.1	2,270	4.2	1,670	4.2	1,040	4.2	310	4.2								
10-20	2,880	7.1	2,270	7.1	1,670	4.2	1,040	4.2	310	4.2								
20-30	3,000	7.1	2,270	7.1	1,670	7.1	1,040	4.2	310	4.2								
30-40	3,000	7.1	2,390	7.1	1,670	7.1	1,040	7.1	310	4.2								
40-50	3,000	3.4	2,390	7.1	1,600	7.1	1,040	7.1	310	7.1								
50-60	3,000	3.4	2,390	3.4	1,600	7.1	500	7.1	310	7.1								
60-70	3,000	3.4	2,390	3.4	1,600	3.4	500	7.1	100	7.1								
Tethys and Red Sea																		
0-10	11,230	2.9	10,190	3.1	9,180	1.3	8,210	1.3	7,240	1.3	3,870	1.3	500	1.3	2,680	1.5	2,680	1.5
10-20	11,230	2.9	10,190	2.9	9,180	3.1	8,210	1.3	4,110	1.3	3,000	1.3					2,680	1.5
20-30	5,890	2.9	5,750	2.9	8,420	2.9	5,500	3.1	2,290	1.3	1,500	1.3						
30-40	4,820	2.9	3,110	2.9	5,750	2.9	3,500	2.9	2,000	3.1								
40-50	2,820	5.9	3,110	2.9	3,110	2.9	3,000	2.9	500	2.9								
50-60	2,820	5.9	2,820	5.9	2,800	2.9	1,500	2.9										
60-70	2,820	5.9	2,820	5.9	2,800	5.9												
Juan de Fuca																		
0-10			1,030	6.7	1,210	5.1	1,380	4.4	1,820	4.4	1,870	5.9	1,660	5.9	1,290	1.7	1,070	6.9
10-20					1,030	6.7	1,210	5.1	1,380	4.4	1,820	4.4	1,870	5.9	1,330	5.9	970	1.7
20-30							1,030	6.7	1,210	5.1	1,380	4.4	1,360	4.4	940	5.9	830	5.9
30-40									510	6.7	910	5.1	690	4.4	910	4.4	930	5.9
40-50											510	6.7	610	5.1	690	4.4	910	4.4
50-60													510	6.7	610	5.1	690	4.4
60-70															510	6.7	610	5.1
North Pacific-Farallon																		
0-10	5,590	7.1	5,590	5.2	5,590	5.2	5,590	5.2	5,590	5.2	5,590	5.2	3,790	5.2	2,500	4.6	2,370	6.4
10-20	4,120	7.1	3,940	7.1	5,590	5.2	5,590	5.2	5,030	5.2	4,480	5.2	3,910	5.2	3,200	5.2	2,500	4.6
20-30	4,120	7.1	3,940	7.1	3,730	7.1	5,590	5.2	5,030	5.2	4,480	5.2	3,910	5.2	3,080	5.2	3,010	5.2
30-40	5,840	7.1	3,940	7.1	3,730	7.1	3,610	7.1	5,030	5.2	4,480	5.2	3,910	5.2	3,080	5.2	2,800	5.2
40-50	5,840	3.4	5,710	7.1	3,730	7.1	3,610	7.1	2,410	7.1	4,480	5.2	3,910	5.2	3,080	5.2	2,800	5.2

ridge was made by reconstructing the two limbs of the triple junction which remain, following Pitman and Hayes (1968) and McKenzie and Morgan (1969). The third ridge of the triple junction was found by vector addition around a presumed ridge-ridge-ridge triple junction. The spreading rates presented in Tables 1b and c are average spreading rates for the entire ridge and not the spreading rates at the triple junctions. The location of the triple junction was taken from the plate reconstructions, while the location of the intersection of the ridge with the North American continent was constrained by geological data presented by Atwater (1970). A ridge jump was assumed to have occurred prior to anomaly 34 (80 m.y.), about 97 m.y. ago, to be consistent with the ocean-age interpretation of Sclater, Jaupart, and Gelson (1980).

The direction and rate of spreading of the Phoenix-Farallon ridge 113 m.y. ago can be constructed in the same way as for the Kula-Farallon ridge. One piece of information which is missing is the position where this ridge inter-

sects the South American plate, chosen arbitrarily to occur off the coast of Peru in Figure 2c, 113 m.y. ago. Because this ridge is almost entirely subducted by 80 m.y., the volume of this ridge has no effect upon sea-level variation from 80 m.y. to the present. The ridge separating the Antarctic (Phoenix in Table 1b and c) and Farallon plates from 80 m.y. to the present is the Chile rise, and its spreading history can be determined reasonably well from its southwest side. This reconstruction, however, requires some spreading north of the Chile rise, which can be accommodated by the Phoenix (Aluk)-Farallon spreading center of Cande, Herron, and Hall (1982) and Weissel, Hayes, and Herron (1977). We did not attempt to calculate the volume of this ridge.

Fully Subducted Ridges

There are little data available to constrain the spreading history of a Tethys ridge in the Tethys Seaway (Dewey et al,

	80		70		60		50		40		30		20		10		0	
50-60	4,300	3.4	5,710	3.4	5,220	7.1	3,610	7.1	2,410	7.1	2,050	7.1	3,910	5.2	3,080	5.2	2,800	5.2
60-70	3,300	3.4	3,690	3.4	5,220	3.4	4,300	7.1	2,410	.7.1	2,050	7.1	2,010	7.1	3,080	5.2	2,800	5.2
South Pacific-Farallon																		
0-10	4,840	8.6	4,840	4.3	4,840	4.3	4,840	4.3	4,840	4.3	4,840	4.3	4,840	4.3	4,840	4.3	4,840	8.1
10-20	4,390	8.6	4,290	8.6	4,840	4.3	4,840	4.3	4,690	4.3	4,700	4.3	4,700	4.3	4,700	4.3	4,840	4.3
20-30	2,950	8.6	3,680	8.6	4,290	8.6	4,300	4.3	4,600	4.3	4,600	4.3	4,600	4.3	4,600	4.3	4,230	4.3
30-40	1,680	8.6	1,680	8.6	2,700	8.6	2,520	8.6	4,110	4.3	4,600	4.3	4,600	4.3	4,600	4.3	4,230	4.3
40-50	1,680	4.7	1,680	8.6	1,680	8.6	2,080	8.6	2,520	8.6	4,110	4.3	4,500	4.3	4,600	4.3	4,230	4.3
50-60	1,680	4.7	1,680	4.7	1,680	8.6	1,680	8.6	2,000	8.6	1,680	8.6	4,110	4.3	4,500	4.3	4,230	4.3
60-70	1,680	4.7	1,680	4.7	1,680	4.7	1,680	8.6	1,680	8.6	1,680	8.6	1,680	8.6	4,110	4.3	4,230	4.3
Pacific-Phoenix																		
0-10	1,550	10.7	2,020	5.5	2,660	4.3	3,930	2.5	3,930	2.5	3,930	2.5	3,930	2.5	5,070	2.9	5,070	2.9
10-20	1,280	10.7	1,160	10.7	2,020	5.5	2,660	4.3	3,930	2.5	3,930	2.5	3,930	2.5	3,930	2.5	5,070	2.9
20-30	1,390	10.7	1,280	10.7	780	10.7	2,020	5.5	2,660	4.3	3,930	2.5	3,930	2.5	3,930	2.5	3,930	2.5
30-40	1,500	10.7	1,390	10.7	1,150	10.7	780	10.7	2,020	5.5	2,660	4.3	3,930	2.5	3,930	2.5	3,930	2.5
40-50	1,620	6.1	1,500	10.7	1,250	10.7	770	10.7	780	10.7	2,020	5.5	2,660	4.3	3,930	2.5	3,930	2.5
50-60	1,720	6.1	1,620	6.1	1,350	10.7	830	10.7	640	10.7	780	10.7	2,020	5.5	2,660	4.3	3,930	2.5
60-70	1,830	6.1	1,720	6.1	1,460	6.1	900	10.7	690	10.7	640	10.7	780	10.7	2,020	5.5	2,660	4.3
Pacific-Kula																		
0-10	1,540	7.3	1,920	3.7	2,230	3.7	2,450	3.4	2,790	3.4	1,500	3.4	200	3.4				
10-20	1,140	7.3	1,540	7.3	1,920	3.7	2,230	3.7	1,960	3.4	2,240	3.4	1,000	3.4				
20-30	7,050	6.7	1,140	7.3	1,540	7.3	1,920	3.7	1,110	3.7	1,220	3.4	1,400	3.4	1,000	3.4		
30-40	7,050	6.7	6,100	6.7	1,140	7.3	1,540	7.3	1,060	3.7	1,110	3.7	1,220	3.4	1,400	3.4	50	3.4
40-50	7,000	5.2	5,900	6.7	5,800	6.7	1,140	7.3	770	7.3	1,060	3.7	1,110	3.7	1,220	3.4	760	3.4
50-60	6,500	5.2	5,700	5.2	5,600	6.7	5,300	6.7	570	7.3	770	7.3	1,060	3.7	1,110	3.7	1,100	3.4
60-70	6,000	5.2	5,000	5.2	5,000	5.2	4,900	6.7	4,700	6.7	570	7.3	770	7.3	1,060	3.7	1,110	3.7
Pacific-West Kula																		
0-10	6,300	6.4	6,200	3.0	4,960	3.0	3,420	3.2	1,010	3.2								
10-20	6,300	6.4	5,500	6.4	4,500	3.0	2,500	3.0	500	3.2								
20-30			4,000	6.4	2,000	6.4												
30-40																		
40-50																		
50-60																		
60-70																		

Table 1c: Ridge lengths (in 10^6 km) and spreading rates (in cm/year) through time for Pacific ridges utilizing the LGP magnetic anomaly time scale of Larson, Golovchenko, and Pitman (1982). For a given ridge, any column gives the ridge length and spreading rates of 0- to 70-m.y.-old crust at the time indicated above the column.

1973). A maximum ridge volume was calculated by inserting a ridge along the strike of the Indo-European suture zone. The limits of the Tethys Seaway, taken from reconstructions of Smith and Briden (1977), were used to constrain both the length of the ridge and the amount of subduction of the ridge flanks. Spreading rates were chosen to be extremely high while the seaway was broad but were decreased as the seaway narrowed, because subduction of very young ocean crust should result in slow subduction rates (Molnar and Atwater, 1978). A minimum ridge-volume calculation was made by reducing the early spreading rates to half those used in maximum rate assumptions and reducing spreading ridge lengths by 15 percent throughout its spreading history. The Tethys Seaway is assumed to have closed entirely about 20 m.y. ago (Dewey et al, 1973; Smith and Briden, 1977). Maximum ridge length and spreading rates for the Tethys ridge are given in Tables 1b and c.

There is also no evidence of spreading between the Pacific and Kula plates west of 170°E. This implies that there would have been either a fracture zone separating the east and west Kula plate, (a minimum-ridge-length assumption), or that spreading continued between the west half of the Kula plate and the Pacific plate until subduction of this ridge (a maximum-ridge-length assumption). The ridge length and spreading rates that would result from the maximum assumption are presented in Table 1c as a combination of Kula-Pacific and West Kula-Pacific spreading. The minimum assumption utilizes only the Kula-Pacific ridge.

Small Ocean Basins and Marginal Basins

Calculation of ridge volumes was also carried out for small ocean basins and marginal basins which presently make up about 9 percent of the area of the ocean floor. Subduction zones outboard of marginal basins tend to consume only old ocean floor (ridge flanks; Molnar and Atwater, 1978). Thus, no correction need be made for the ocean floor

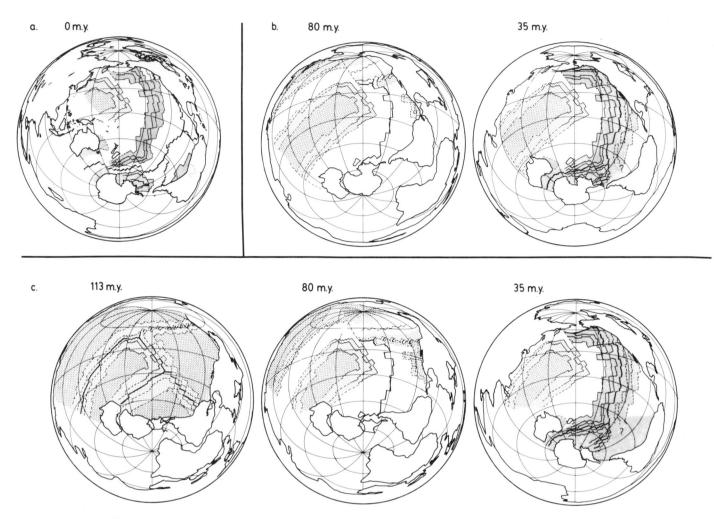

Figure 2: Pacific paleoreconstructions used in this study. Shaded area is anomaly 13 through 34 (35-80 m.y.). Dotted area is anomaly M0 through M22 (113-140 m.y.). Patterned dots represent anomaly M22 and older (140 + m.y.). a) Pacific magnetic anomaly map showing present locations of simplified magnetic anomaly patterns (modified from Sclater et al, 1980). b) Plate reconstructions holding the Pacific plate fixed and assuming that there is no motion within the Antarctic plate for 80 and 35 m.y. (rotations from Suarez and Molnar, 1980). c) Plate reconstructions holding the Pacific plate fixed and assuming a fixed hot spot frame. This requires motion between East and West Antarctica (rotations from Duncan, 1981).

displaced by opening of marginal basins. Age ranges of marginal basins were taken from Sclater, Jaupart, and Gelson (1980), from Hayes and Taylor (1978), and from Barker and Dalziel (in press). Although spreading in small ocean basins appears to be more complex than that of ocean ridges (Watanabe, Langseth, and Anderson, 1977), it is assumed, for purposes of calculation, that the age-depth relation for oceanic crust holds for small ocean basins. The maximum spreading rates were assumed to occur in the center of a given basin with rotation poles 90° away, along the strike of the ridge. For those basins in which ridge and magnetic data were not available, the ridge was assumed to be located parallel with the long axis of the basin, and the pole of opening was taken to be where the extension of the basin edges intersects.

ERROR ANALYSIS

Errors inherent in the reconstruction of ocean-ridge volumes fall under two general headings—incorrect dating and

missing information. Incorrect dating is a problem if the magnetic anomaly pattern is correlated incorrectly with the biostratigraphic time scale, or if the biostratigraphic time scale is incorrect. Missing information includes: 1) lack of information on age and depth of the ridge flanks; 2) insufficient information from which to delineate ocean-basin limits; 3) incomplete or total lack of information on a given spreading center, and; 4) contamination of the age versus depth curve for ocean-crust by hot spots.

Time Scales

To determine the age of the ocean floor, one must know the ages of all magnetic-polarity-reversal boundaries, how they fit into the biostratigraphic column, and the radiometric age of the biostratigraphic time scale. Table 2 includes the ages of critical magnetic anomalies used and their dates according to four time scales, including that of Heirtzler et al (1968) and the three time scales used in this paper.

It is of primary importance to determine if the age versus

Table 2

Anomaly No.	Heirtzler et al, (1968)	LGP*	NVH*	BIO*
3.2	4.4	4.0	4.2	4.0
4	7.5	7.0	7.3	7.0
5	9.0	9.0	8.7	9.0
5'	9.9	9.7	10.3	9.7
5A	12.2	11.4	12.0	11.4
6	21.0	20.0	20.0	20.0
6 old	25.4	23.8	23.8	25.5
7 old	28	26.2	26.8	28
11 old	34	32	31.5	32
13	38.4	35.8	35.2	36
16	41	39	38.2	37.5
21	53	50	48.7	50
22	55.5	52	50.5	52
24	61	57	55.5	57
25	63	59	58.2	59
28	68.5	64	65.3	64
29	69.5	65	66.7	65
32	74.5	70	73.1	70
34	-	80	86.1	82
M0	-	113	108	122
M1	-	116	120	126
M5	-	120	124	127
M11	-	126	127.6	130
M16	-	134	133	134
M22	-	143	140.5	151
M25	-	147	146	157

*Where LGP is the magnetic time scale of Larson, Golovchenko, and Pitman (1982); NVH is the Cenozoic magnetic time scale of Ness, Levi, and Couch (1980) combined with the magnetic anomaly time scales of Van Hinte (1976a and 1976b); and BIO is based upon the time scales of Berggren (1972) for the Paleogene, Dickinson and Rich (1972) for the Cretaceous, and Lambert (1971a and 1971b) for the Jurassic.

Table 2: Critical time-scale dates of magnetic anomalies (age in m.y.).

depth curve of Parsons and Sclater (1977), in which the magnetic reversal time scale of Hiertzler et al (1968) was used, is consistent with the time scales used in this paper. In general, time scales for magnetic anomalies 1 through 30 are reasonably consistent and agree within ±5 percent (Table 2). The major differences arise in dating of Jurassic and Cretaceous anomalies. Thus, Parsons and Sclater's (1977) equation for the youngest 70 m.y. of crust should be essentially correct for any of these time scales.

The errors caused by missing information are presented in terms of error bars in the calculation of ocean-ridge volumes. The error resulting from time-scale uncertainties cannot be presented in this manner, because ages must be assigned to ridge length and spreading angles in accordance with one specific time scale. Ridge-volume estimates are calculated for each time scale, and these estimates are compared to obtain a measure of the error introduced by uncertainties in dating.

The most recent Lamont-Doherty-compiled magnetic-anomaly time scale (Larson, Golovchenko, and Pitman, 1982; referred to as LGP) is taken as the standard against which other time scales are compared. This magnetic-anomaly time scale is revised from that of LaBrecque, Kent, and Cande (1977) and Larson and Hilde (1975), with revisions based on the Jurassic and Cretaceous biostratigraphic

time scales of Van Hinte (1976a and 1976b) and recent Deep Sea Drilling Project ocean-floor dates. The major uncertainty in dating of the magnetic time scale as it affects ridge-volume estimates is in establishing the duration of the Cretaceous magnetic quiet zone. The two time scales which are compared to the LGP time scales are designed to give maximum and minimum estimates of the duration of this period. The Ness, Levi, and Couch (1980) magnetic-anomaly time scale, which covers the period from the present to anomaly 34 (the end of the Cretaceous quiet zone) is combined with the magnetic anomaly time scales of Van Hinte (1976a and 1976b). When the Ness, Levi, and Couch (1980) magnetic anomaly pattern is calibrated to the biostratigraphic ages used in LGP, anomaly 34 is assigned an age of 86 m.y. rather than LGP's 80 m.y. date, decreasing the Cretaceous quiet zone by 6 m.y. Using the magnetic anomaly pattern proposed for the Cretaceous by Van Hinte (1976a), anomaly M0 (the beginning of the Cretaceous quiet zone) falls at 108 m.y. which shrinks the Cretaceous quiet zone by an additional 5 m.y., or a total of 11 m.y. This causes estimates of Albian to Santonian Cretaceous spreading rates to be faster than estimates made using the LGP time scale. This combination magnetic anomaly time scale is referred to as the NVH (Ness-Van Hinte) time scale. The duration of the Cretaceous quiet zone is maximized by utilizing the Cretaceous biostratigraphic ages of Dickinson and Rich (1972). By adjusting the biostratigraphic ages of the LGP magnetic anomaly pattern to conform to the biostratigraphic ages of Dickinson and Rich (1972), the duration of the quiet zone is increased by 7 m.y. Although estimates of biostratigraphic ages do not vary a great deal for the Jurassic or the Paleogene, some minor changes in calculated ridge volumes will result from these differences. Thus, in addition to the Cretaceous Dickinson and Rich (1972) time scale, the ages of the Paleogene from Berggren (1972), and the ages set forth by the Geological Society of London for the Jurassic (Lambert, 1971a and 1971b) are used to calibrate the magnetic anomaly pattern of Larson, Golovchenko, and Pitman (1982) and to create a third time scale, subsequently referred to as the BIO (biostratigraphic) time scale. A more recent biostratigraphic time scale by Harland et al (1983), if applied to the Larson, Golovchenko, and Pitman (1982) magnetic anomaly time scale, would result in a magnetic time scale intermediate between the LGP and NVH magnetic time scales used in this work.

Missing Information

1) Uncalculated Ocean (Ridge Flanks)

The ridge volumes presented in this paper are calculated for only the portion of the sea floor which subsides according to a $t^{1/2}$ curve. This is the youngest 70 m.y. of ocean crust. The ridge flanks (the portion of the ocean crust older than 70 m.y.) continue to subside, exponentially, approaching a depth of 6,400 m (21,000 ft). The volume occupied by the ridge flanks can be determined by using a modified Parsons and Sclater (1977) equation for determining the height of the ridge above 6,400 m (21,000 ft) (for crust older than 70 m.y.):

$$\text{Height} = 3200e^{-t/62.8} \qquad (4)$$

Table 3

Calculation	LGP*	NVH*	BIO*	Units
(1) Spreading rate 0-70 m.y.	3.0	2.8	2.9	cm/y
(2) Area ocean 0-70 m.y. $\{(1) \times 2 \times 70 \times RL\}$	201.6	188.2	194.9	10^6 km^2
(3) Area ocean older than 70 m.y. $\{360 - (2)\}$	158.4	171.8	165.1	10^6 km^2
(4) Spreading rate 70-125 m.y.	4.2	4.8	3.8	cm/y
(5) Time needed to make ocean floor > 70 m.y. $\{(3)/[2 \times (4) \times RL^{\#}]\}$	39.3	37.3	45.4	m.y.
(6) Time needed to make ocean floor > 70 m.y. $\{(3)/[2 \times (1) \times RL^{\#}]\}$	55.0	63.9	59.3	m.y.
(7) Volume ridge flank (5) in eq. (4)	61.5	67.7	61.4	10^6 km^3
(8) Volume ridge flank (6) in eq. (4)	55.1	56.3	55.8	10^6 km^3
(9) ΔVolume ridge flanks $\{(5) - (7)\}$	6.4	11.4	5.6	10^6 km^3
(10) Sea level overestimated at 80 m.y.	12.4	22.1	10.9	m

*Where LGP is the magnetic time scale of Larson, Golovchenko, and Pitman (1982); NVH is the Cenozoic magnetic time scale of Ness, Levi, and Couch (1980) combined with the magnetic anomaly time scales of Van Hinte (1976a and 1976b); and BIO is based upon the time scales of Berggren (1972), Dickinson and Rich (1972), and Lambert (1971a and 1971b).

RL # is the ridge length, taken to be 48,000 km (29,826 mi).

Table 3: Ridge flank corrections.

using this equation for calculating the volume of the ridge flanks in conjunction with equation (2) for calculating the volume of the ridges and knowing the age distribution of the entire ocean floor, one can obtain the exact volume taken up by the ocean floor at the present time (with respect to ridges alone). However, to calculate the volume taken up by the ocean floor 10 m.y. ago the age distribution of the ocean crust which has been subducted in the last 10 m.y. would have to be known to calculate the ridge flank volume at that time. When attempting to calculate ocean-crust volumes for earlier times, the amount of subducted ridge flank increases and an estimate of ridge-flank volume becomes increasingly uncertain, indeed is impossible to determine. Instead, only the volume of the rapidly subsiding ridge crests are calculated and ridge-flank volumes are assumed to remain constant through time.

Knowing the world average spreading rates for the past 125 m.y. (Figure 3a, calculated from spreading data, above) it is possible to estimate the error introduced by ignoring ridge-flank subsidence. In Figure 3a, it is evident that spreading rates from 80 m.y. to the present were slower than spreading rates of 80 to 125 m.y.. Because these higher spreading rates occur in the ridge-flank portion of the ocean floor at present, ignoring ridge-flank subsidence in calculations of earlier ridge volumes is equivalent to assuming that the older, now subducted ridge flanks had also been accreted at times of high spreading rates. An estimate of the error that this assumption causes if the 80-m.y. ridge flanks were created at slowly spreading ridges can be made using the available information. Calculations of this error based

on averaging the spreading rate data of Figure 3a for each time scale are tabulated in Table 3. The ridge length is taken to be 48,000 km (29,826 mi) exclusive of small ocean basins (this work), and the ocean area is taken to be 360 × 10^6 km^2. The average 0- to 70-m.y. spreading rates [(1) in Table 3] are used to calculate the area of ocean which is currently ridge flank (3). The volume of the ridge flanks is calculated assuming either high (4) or low (8) spreading rates, and the difference (9) is the overestimate that would result from assuming that the ridge-flank spreading rates 80 m.y. ago were as high as ridge-flank spreading rates today. This volume error is translated into sea level using equation (1), (10, in Tables 3). The error is largest for the NVH time scale because it shows the greatest change in spreading rates and least for the BIO time scale which requires the least variation in spreading rates.

2) Lack of Magnetic Anomaly Data

Incomplete data from ocean-floor magnetic anomalies is primarily a problem in the calculation of ocean-ridge lengths. A record of the spreading history of an entire spreading center is not required to calculate the rotation pole; if parts of the ocean basin have a complete record it is sufficient. However, without good magnetic data at the limits of a given spreading center it is difficult to determine ridge lengths. Also, if the total-rotation poles calculated for a given spreading center are incorrect, the stage poles calculated from them are skewed with respect to the true ridge, resulting in the calculation of incorrect ridge lengths. For

Figure 3: Average spreading rates (a) and total ridge lengths (b) for the world oceans, including Pacific, Gondwana and small-ocean-basin ridges. Maximum and minimum estimates of ridge lengths in the Pacific based on different Pacific reconstructions are calculated separately and plotted as dashed and solid lines respectively. LGP utilizes the time scale of Larson, Golovchenko, and Pitman (1982); NVH uses the time scales of Ness, Levi, and Couch (1980) and Van Hinte, (1976a and b); and BIO uses a time scale combining the time scales of Lambert (1971a and b), Berggren (1972), and Dickinson and Rich (1972).

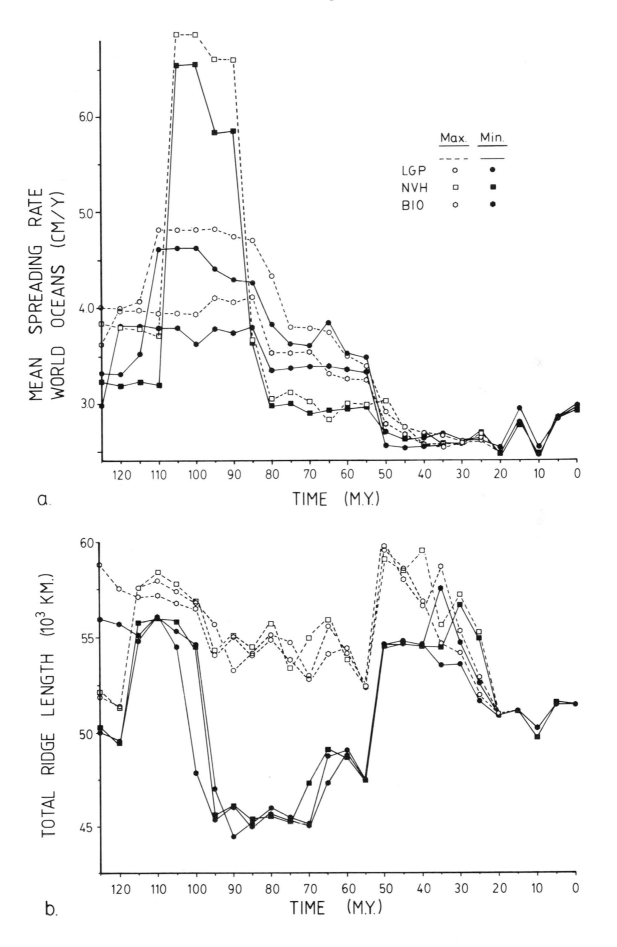

this work, the ridge lengths were determined in the following manner (see Figure 1). If a small circle (i.e., of latitude) about the rotation pole is defined for each of the two limits of the ocean basin, the angle (in radians) between these two small circles, as measured in a great circle which passes through the rotation pole, multiplied by the radius of the earth is the ridge length. In this way, ridge offsets (transforms) are ignored and only the actual spreading centers are considered.

An estimate of the error in the ridge-volume calculation due to inaccurate assessment of the ridge length of any given ocean basin was obtained through use of the Gondwana and Atlantic spreading centers. For these spreading centers, rotation poles are available in the literature and are assumed to be reasonably accurate (Ladd, 1976). Stage poles are available or can be calculated for the opening of either side of any ocean, holding the other side fixed. The limits of the entire ocean basin at any given time are positioned on the magnetic anomalies of that age on the fixed side of the ocean. By computing ridge volumes from data first on one side of any ocean, and then on the other side, the difference between the two results approximates the error due to determining ridge lengths from limited data. By adding up the total difference for all Gondwana and Atlantic spreading centers, the total error is obtained.

A plot of percent error versus age of spreading center is presented in Figure 4. Error increases steadily as the calculated ridge becomes older. This is probably due to a number of factors. The magnetic anomaly data around Antarctica is sparse; thus, ridge lengths are difficult to determine for continents moving away from Antarctica, which represents much of the early rifting of Gondwana. In most cases initial opening phases were slow, resulting in indistinct anomaly patterns in the oldest ocean floor (Tisseau and Patrait, 1981). Also, in the Indian Ocean, major plate reorganization took place about 50 m.y. ago, leaving a confused pattern for some period of time (Norton and Sclater, 1979).

3) Incomplete Ocean Basins

Ocean basins which have been partially or fully subducted introduce a serious error in ridge-volume calculations. In basins for which half or more of the sea floor remains (Pacific-Kula, Pacific-Farallon and Pacific-Phoenix spreading centers), a ridge-volume calculation can be made by assuming spreading with the Pacific as a fixed reference frame. The error in obtaining basin limits is assumed to be equivalent to the error derived from the Gondwana and Atlantic ridges in the section above.

The method used for obtaining ridge volumes of fully subducted ridges, in which only a remnant triple junction remains, is described in the methods section above. Errors due to assumed ridge length are estimated by using two different Pacific reconstructions (Figure 2) and also adding an error equivalent to that obtained from the Gondwana and Atlantic ridges for lack of magnetic-anomaly data (see previous section).

The Tethys ridge and the West Kula ridge pose a more difficult problem. The minimum and maximum spreading-rate and ridge-length assumptions were described in the spreading data section on fully-subducted ridges. For the Tethys ridge,

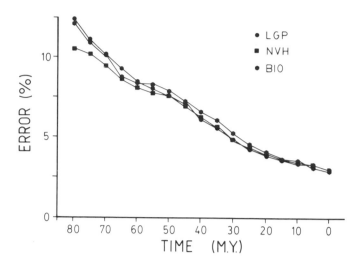

Figure 4: The error in percent of mean ridge length due to inaccuracy in defining ocean limits and stage poles of Gondwana and North Atlantic ridges. LGP is the time scale from Larson, Golovchenko, and Pitman (1982); NVH is the time scales of Ness, Levi, and Couch (1980) and Van Hinte (1976a and b); and BIO is a time scale combining the time scales of Lambert (1971a and b), Berggren (1972), and Dickinson and Rich (1972).

only the maximum assumed length and spreading rates are presented in Table 1c. The minimum ridge-length assumptions for the Kula-Pacific ridge are presented in Table 1c as Kula-Pacific spreading. The maximum assumption is the sum of this set of spreading data plus the tabulated West Kula-Pacific spreading data.

The method used for calculating small-ocean-basin ridge volumes, as described in the spreading data section, maximizes the calculated ridge volumes. In general, marginal basin sea floor averages about 500 m (1,640 ft) deeper for its age than normal ocean crust (Anderson, 1980; Watanabe, Langseth, and Anderson, 1977; Langseth, personal communication). Because 500 m (1,640 ft) is 17 percent of the height assumed for mid-ocean ridges (2,928 m, or 9,606 ft), a possible error of -17 percent is used. A source of error may arise due to small-ocean-basins which have been present within the past 150 m.y. and were subsequently subducted (e.g., Dalziel, 1981; Dewey et al, 1973). This would result in higher ridge volumes in the past than those calculated. This error was estimated by assuming that an average volume contribution for small ocean basins may be about 5.0×10^6 km^3, the mean volume calibrated for small-ocean-basins (Figure 6a). This value is taken as a maximum volume for small-ocean-basin crust prior to 35 m.y. ago, while the observed small-ocean-basin volume is taken as a minimum volume.

4) Hot Spot Contamination

An alternative equation for the relationship between the age of the sea floor and its depth is presented in a recent study by Heestand and Crough (1981). Their empirical curve is derived from Atlantic ocean floor which is at least 1,200 to 1,800 km (745 to 1,118 mi) from any hot spot trace. This equation (depth = 2,700 + 295t$^{1/2}$) has a lower slope than that of Parsons and Sclater (1977). Thus, for any given ridge length and spreading rate, the calculated ridge vol-

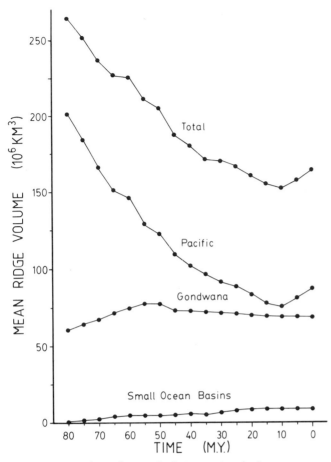

Figure 5: Mean ridge-volume calculations using the Larson, Golovchenko, and Pitman (1982) time scale of Pacific, Gondwana (and North Atlantic) and small-ocean-basin ridges. All are plotted using the same ridge-volume scale in order to emphasize the differences in importance of the three ridge categories employed in this study.

umes will be less than those calculated using the Parsons and Scalter (1977) relation. If hot spot distribution with respect to the age of the ocean crust affected has been constant through time, then it is not necessary to correct the age-depth curve for hot spot contamination or to make separate calculations of the effects of hot spot activity on ocean volume through time. This assumption is, however, unwarranted (see for example Schlanger, Jenkyns, and Premoli-Silva, 1981; Watts, Bodine, and Ribe, 1980), and a sea-level curve based on the Heestand and Crough (1981) age-depth relation may provide a more realistic value than the curves based upon the Parsons and Sclater (1977) relation for ocean-volume change due to ridges, and an ocean-volume change due to hot spots should then be superimposed on this sea-level curve.

RESULTS

The estimates of mean ridge volume (Figures 5 and 6) show that in a comparison of the ridge-volume contributions of Pacific, Gondwana, and small-ocean-basin ridges, the Pacific ridge volumes are dominant. Small-ocean-basin ridge volumes increase from past to present, but the magni-

tude of the change is insignificant when compared with the rest of the ridges. Gondwana ridge volumes also increase with decreasing time from 80 to about 50 m.y. and decrease or remain about constant from 50 m.y. to the present. Pacific ridges show a significant decrease in volume of about 50 percent from 80 to 15 m.y. This drop is the main source of the sea-level fall as calculated from ridge volumes.

A comparison of the mean ridge-volume contribution of small ocean basins, Gondwana ridges, and the Pacific ridges as indicated by the three different time scales (Figure 6) shows that variation between the different time scales is minor. The general increase in volume resulting from small ocean basins and marginal basins is at least partially due to scarcity of information about old basins which are now closed (i.e., Dalziel, 1981; Dewey et al, 1973). In the case of the Gondwana ridges (Figure 6b), volumes calculated with the LGP and BIO time scales show a similar pattern of volume changes but differ in magnitude. Because the NVH time scale has very high Late Cretaceous spreading rates and low early Cenozoic spreading rates relative to the other time scales (Figure 3a), ridge volumes calculated according to the NVH time scale show larger volumes at 80 m.y. and do not increase as rapidly as the LGP and BIO calculated volumes. This relationship can also be seen clearly by comparing the Pacific ridge-volume curves (Figure 6c) calculated according to the NVH and LGP time scales. Although all Pacific ridge-volume curves show a significant decrease, the NVH time scale gives the greatest drop, whereas the BIO time scale shows the least drop, as expected from Cretaceous spreading-rate assumptions.

It has been argued that because spreading rates are determined from absolute time scales and absolute time scales vary considerably for the Cretaceous, it may be possible to eliminate high Cretaceous spreading rates entirely (Baldwin, Coney, and Dickinson, 1974). It has also been suggested that variation in total ridge length may be the cause of sea-level variation (Hallam, 1977). It is important to understand the possible range of error involved in calculating paleo-ridge volumes in terms of both spreading rates and ridge lengths. The Dickinson and Rich (1972) Cretaceous time scale, which is incorporated in the BIO time scale, maximizes the span of time in the Late Cretaceous within the error of available absolute dates. This is the time period in which spreading rates appear to be at a maximum when one uses other time scales. In Figure 3a, it is clear that for all time scales average oceanic spreading rates have decreased from 105 to about 20 m.y. This decrease is most dramatic and most abrupt when one uses the NVH time scale and most gradual for the BIO time scale. Highest spreading rates occur at the Pacific-Farallon and Pacific-Phoenix ridges at the time of the Cretaceous quiet zone (Table 1). Figure 3b shows that by maximizing ridge lengths for the now absent Pacific ridges, total ridge length has decreased slightly through time in two pulses from 115 to 55 m.y., and from 50 to 20 m.y., thus adding to the fall of sea level from the Late Cretaceous in agreement with the suggestion of Hallam (1977). On the other hand, minimizing Pacific ridge lengths implies a slight increase in ridge length since 120 m.y., with major fluctuations at about 100 and 55 m.y. The increase in ridge length from 125 to 110 m.y. (depending on the time scale used) is due to early opening stages of the Gondwana

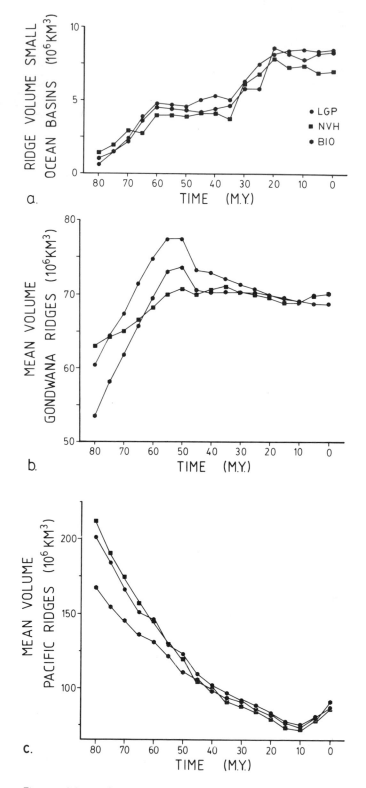

a.

b.

c.

Figure 6: Mean ridge volumes compared for the three time scales. The curve labels are LGP, which used the Larson, Golovchenko, and Pitman (1982) time scale; NVH, which uses a combined Ness, Levi, and Couch (1980) and Van Hinte (1976a and b) time scale; and BIO which uses a combination of the Lambert (1971a and b), Dickinson and Rich (1972), and Berggren (1972) time scales. a) Ridge volumes of small ocean basins and marginal basins. b) Ridge volumes of Gondwana and North Atlantic ridges. c) Pacific and Tethys ridge volumes.

Table 4

Time (m.y.)	LGP* SL	LGP* ±	NVH* SL	NVH* ±	BIO* SL	BIO* ±
0	0.0	10.9	0.0	8.5	0.0	10.7
5	2.1	12.0	−1.5	11.8	−6.0	11.2
10	6.0	12.1	−0.7	12.5	−7.1	12.3
15	32.1	13.2	15.9	13.9	14.7	13.2
20	38.0	15.0	27.4	16.4	26.6	15.8
25	48.5	18.0	36.4	19.6	29.3	18.6
30	54.7	21.9	42.6	23.1	40.3	22.5
35	57.0	26.9	45.6	26.3	42.2	28.5
40	73.4	34.4	62.0	34.7	50.3	36.1
45	87.4	40.8	60.4	41.8	65.1	41.9
50	119.8	51.4	99.5	50.1	80.4	48.2
55	131.3	56.5	116.2	57.2	99.5	55.6
60	158.1	68.6	134.5	68.4	110.9	63.0
65	160.1	79.3	159.7	80.5	112.1	70.4
70	178.7	91.7	189.2	94.5	120.8	82.6
75	205.5	100.6	214.6	104.0	130.5	89.3
80	228.4	112.4	250.6	112.6	145.4	99.1

*Where LGP is the magnetic time scale of Larson, Golovchenko, and Pitman (1982); NVH is the Cenozoic magnetic time scale of Ness, Levi, and Couch (1980), combined with the magnetic anomaly time scales of Van Hinte (1976a and 1976b) and BIO is based upon the timescales of Berggren (1972) for the Paleogene, Dickinson and Rich (1972) for the Cretaceous, and Lambert (1971) for the Jurassic.

Table 4. Sea-level variations, with errors in meters.

and Atlantic oceans. The sudden increase in ridge length at about 50 m.y. is primarily due to the splitting of Australia from Antarctica. This would shift back to 85 m.y., if the recent spreading ages of Cande and Mutter (1982) are used; however, because spreading rates are slow according to this scheme until anomaly 19, the effect of this change on ridge-volume calculations is relatively small. Apparently random changes in the last 30 m.y. are primarily due to the initiation of marginal basins. Note that neither the ridge-length plot nor the spreading-rate plot can be translated directly into a sea-level curve. Not only must both be considered, but one must integrate these effects for the preceding 70 m.y. of oceanic crustal accretion and account for crustal subduction to calculate ocean-ridge volumes. In summary, the main contribution to the decrease in ocean-ridge volumes from 125 m.y. to the present is due to a reduction in spreading rate while changes in ridge length may account for variation superimposed on the overall trend.

Relative sea-level changes calculated using equation (1) from changes in world ocean-ridge volume and ice volume are presented in Table 4 and plotted in Figure 7. Errors due to uncertainty in calculation of ridge lengths, uncertainty of ridge-flank subsidence, and lack of data for Pacific ridges are included in these plots.

The similarities in the sea-level curves are most notable. The errors involved are very large and decrease as the lengths of the unknown Pacific ridges decrease, and as magnetic anomaly data become clearer. Although the percentage of error due to the uncertainty of magnetic anomaly data is not too large (3.5 to 12.5 percent), this is the percent error in the calculation of total ridge volume, not relative ridge volume. Thus, the error is very large limits. Because each successive ridge volume is a composite of the earlier ridge volumes, coupled with a relatively small amount of newly accreted crust and some small amount of subduction. Thus, it is likely that the true trend of sea-level change does reflect that of the curves, at least to the extent that ocean-ridge volumes control sea level. The effects of orogeny, hot spot variation, formation of passive margins, evaporation of large basins, and any other processes which change oceanic volumes must be superimposed on this sea-level curve. All sea-level curves calculated here show a rapid drop of sea level from 80 m.y. (Santonian) to about 40 m.y. (late Eocene), followed by a slow but continuous drop to about 10 m.y. (late Miocene). Since this time, the calculated sea levels are approximately equal to the present value.

Comparison of the three sea-level curves of Figure 7 shows that they are not very different. The NVH time scale yields higher sea-level values than the LGP time scale for Cretaceous time, but is lower during the Paleogene and Neogene. The BIO time scale results in a sea-level curve with little or no fall from Late Cretaceous to late Paleocene. Sea level falls for the remainder of the Paleogene but shows consistently lower sea-level values as compared to results using either the LGP or NVH time scales. Utilization of the Heestand and Crough (1981) oceanic-age-versus-depth equation gives consistently lower sea-level values (by about 15 percent) than that of Parsons and Sclater's (1977) equation. Thus, if using the LGP time scale with the Heestand and Crough (1981) a mean sea-level drop from 80 m.y. to the present of about 190 m (623 ft) would be expected.

DISCUSSION

The range of uncertainty encompassed in the sea-level curves of Figure 7 are a real estimate of the range of possible sea-level variation due to changing ocean-ridge volumes. The three time scales used are each a different, soundly-based synthesis of available data. Thus, the entire range of possible sea-level change due to ridge volumes is encompassed by the different curves. Because this range of sea-level change is great, it is necessary to establish a single sea-level curve, which, with all information presently available, is the best estimate. The time scale used for the best estimate sea-level curve is that of Larson, Golovchenko, and Pitman (1982). It is a recent compilation of magnetic-anomaly data using recent biostratigraphic age data, and it does not provide the (purposely) extreme results of the NVH or the BIO times scales. With no data on ocean crust older than 150 m.y., it is only possible to obtain a rough estimate of the effect on ridge volumes owing to omitting the portion of the ocean which exceeds the age of the ridge crest by 70 m.y., — the ridge flanks. In the Atlantic and Indian oceans,

Figure 7: Sea-level change with respect to present sea level (due to ridge-volume change and post-Miocene ice-volume change) from present to 80 m.y. including errors due to inaccuracy in delineating ocean limits, lack of information about Pacific ridges, and uncertainty in ridge flank volumes. a) LGP utilizes the time scale of Larson, Golovchenko, and Pitman (1982). b) NVH utilizes the time scales of Ness, Levi, and Couch (1980) and Van Hinte (1976a and b). c) BIO utilizes the time scales of Berggren (1972), Dickinson and Rich (1972), and Lambert (1971a and b).

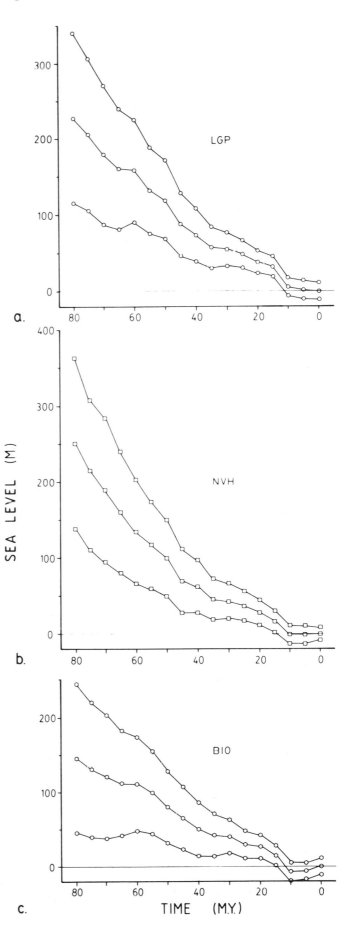

abyssal plain sediments form a horizontal surface which covers the ridge flanks causing the effect of sediment infill to compensate for the subsidence of the ridge flanks. Thus, the ridge-flank subsidence can be ignored if the effect of sediment infill is ignored as well. This sediment blanketing effect is, however, virtually absent in the Pacific Ocean where submarine trenches trap land-derived sediments along continental margins. Thus, the calculated average correction of 12.4 m (40.7 ft) less sea-level fall since 80 m.y. (see Table 3) must be added to a best estimate sea-level curve. Regarding the error in obtaining ridge lengths, the best estimate is the mean value. This curve is presented as the middle line in Figure 7a. The sea-floor age-versus-depth relationship of Heestand and Crough (1981) does not include a worldwide data base, and as such requires further investigation before being accepted as superior to that of Parsons and Sclater (1977). A more recent oceanic age-versus-depth relationship has been calculated utilizing an extensive global data base without corrections for hot spot activity, in which the slope of the curve was found to be even lower than the age-depth relation of Heestand and Crough (1981) (D. E. Hayes, in preparation). Thus, a sea-level curve based on the Heestand and Crough (1981) relation may be more correct than the LGP curve and one should not eliminate the possibility that the LGP sea-level curve is consistently at least 15 percent overestimated.

The Miocene to Holocene ice-volume change has been taken into account in the best estimate (LGP) sea-level curve. The volume of water tied up in glacial ice is sufficient to result in a 48 m (157 ft) sea-level fall from Miocene to the present (Pitman, 1978) and is added in 16 m (52 ft) increments at 5, 10, and 15 m.y. before present to all the sea-level curves based on oceanic volume calculations of Figures 7 and 8.

The best-estimate sea-level curve along with error bars is compared in Figure 8 with earlier estimates of sea-level change calculated for this time period. The curve of Pitman (1978) is the only other sea-level curve calculated from changes in ocean-ridge volumes. The form of the Pitman sea-level curve and the LGP mean curve both show a decrease in eustatic sea level from 80 m.y. to present. The magnitude of sea-level change calculated in this work is, however, consistently less than that calculated by Pitman (1978). Even with maximum error, both with respect to time scales and ridge lengths, the sea-level values calculated in this paper are consistently less than that calculated by Pitman (1978). The difference between the two curves cannot be attributed to time scale differences because although the time scale used by Pitman (1978) is essentially equal to the NVH time scale, the calculated difference in sea level is roughly the same (note Figure 7b).

The difference between the Pitman (1978) sea-level curve and the LGP curve of this work can be found in the different plate reconstructions used in the two studies. There are minor differences in the Gondwana and Atlantic plate rotations used in the two studies, but these differences are insufficient to explain the magnitude of the discrepancy. Although small-ocean-basin volume changes have been calculated for the first time in this work, their effect on sea level is too small (about 20 m, or 66 ft) to cause a significant difference between the two ridge-volume studies. The primary

cause of the discrepancy is the different assumptions that are made regarding changes in Pacific ridge lengths. Constraints on the Pacific ridges in this paper are based upon the reconstructions of Duncan (1981) and Suarez and Molnar (1980), both of which were published subsequent to Pitman (1978). Applying these reconstructions to determine ridge-length changes indicates that some ridges show a greater decrease with time, while other ridges do not decrease as rapidly with time as compared with the ridge lengths used in the reconstructions of Pitman (1978).

Ridges which decrease in volume more rapidly in this work compared with Pitman (1978) are the Kula ridges and the Tethys ridge. The spreading rates of the Kula ridges during the Cretaceous quiet zone were more than double those used by Pitman (1978). The minimum and maximum assumed ridge lengths for the Pacific-Kula ridge system encompass a large range of uncertainty between 95 and 35 m.y. While the minimum assumed Pacific-Kula ridge length is similar to that of Pitman (1978), the maximum ridge lengths are far greater and thus, the average effect is to have greater ridge lengths which decrease over the same period of time. The Kula-Farallon ridge in this work is almost twice as long as was assumed by Pitman for 140 m.y. This ridge also decreases in length through the same period of time as was assumed by Pitman until subduction at about 30 m.y. The Tethys ridge length used in this paper is also more than twice as long as the ridge length which was used by Pitman (1978). The greater length again results in a greater decrease in ridge volume from 80 m.y. until the subduction of the Tethys ridge 20 m.y. ago. By comparing the volume change of these ridges from 85 to 0 m.y. in the work of Pitman (1978) with the volume change from 80 to 0 m.y. in this work, the difference in sea-level estimates can be calculated. The volume decrease of all these ridges, when maximum ridge-length and spreading-rate assumptions are used, is greater than in the work of Pitman (1978). This implies a greater sea-level fall over this period of time of approximately 98 m (322 ft). When minimum ridge length assumptions are made this work predicts 18 m (59 ft) more sea-level fall over the past 80 m.y.

Ridges which show less volume decrease over the period studied in this work, as compared to Pitman (1978), include the Pacific-Phoenix, the Phoenix-Farallon and the Pacific-Farallon ridges. The reconstructions used in this work place the Pacific plate very close to South America at 80 m.y. (Figure 2). This results in a very small Farallon-Phoenix ridge, which grows with time until the present as do the Aluk-Farallon ridge and the Chile rise. This juxtaposition of the Pacific plate near South America also requires half of the Pacific-Phoenix ridge (the Phoenix, or Aluk, half) to have been subducted at 80 m.y. This ridge then grows by extension, with minimal subduction, from 80 m.y. to the present. This is the New Zealand-Antarctic plate of Pitman (1978). The South Pacific-Farallon ridge is also near enough to South America to require significant subduction of the Farallon portion of this ridge, in particular, that portion which is included in the Pacific-Phoenix ridge in the study of Pitman (1978). The Pacific-Phoenix ridge, as used in Pitman (1978) is a subset of both the South Pacific-Farallon ridge and the Pacific-Phoenix ridge as defined in Table 1b and c of this study. The assumptions of Pitman (1978) require that

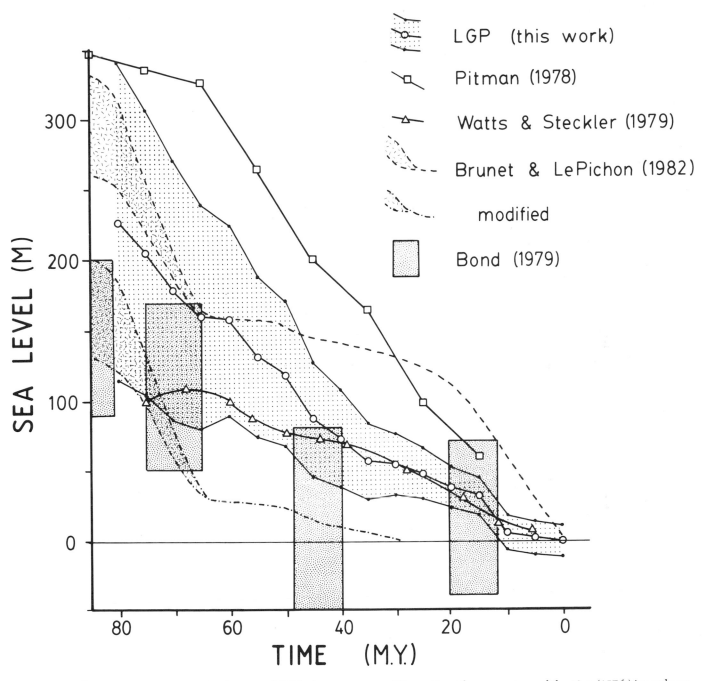

Figure 8: Late Cretaceous to present sea-level curves. LGP is the mean curve of Figure 7a with error range and the 48 m (157 ft) ice-volume correction included. This is a ridge-volume sea-level curve assuming the age-depth relationship of Parsons and Sclater (1977) and the magnetic anomaly time scale of Larson, Golovchenko, and Pitman (1982) are correct. The Pitman (1978) curve is also a ridge-volume derived sea-level curve with the same 48 m (157 ft) ice-volume correction. The Watts and Steckler (1979) sea-level curve is derived from Atlantic continental margin subsidence data. The dashed and dash-dot curves are derived from Paris basin subsidence data. The high (dashed) curve is that of Brunet and Le Pichon (1982), and assumes no post-Cretaceous tectonically driven uplift. The low (dash-dot) curve is modified from the result of Brunet and Le Pichon (1982) by assuming post-Cretaceous tectonic uplift. The hachured area is a water depth uncertainty. The Bond (1979) sea-level estimates are derived from a combination of stratigraphic data and continental hypsometry.

the Phoenix half of the Pacific-Phoenix ridge was not subducted at 80 m.y., in contrast to the assumptions of this work. Subduction of 2,000 km (1,243 mi) of this plate and the Farallon-Phoenix ridge while extension is occurring along the Antarctic-New Zealand ridge requires a reconstruction at 80 m.y. in which West Antarctica is assumed to

have undergone right lateral strike-slip motion relative to East Antarctica from 80 m.y. to the present. A rough reconstruction which attempts to recover the ridge-length and spreading-rate data presented by Pitman (1978) is presented in Figure 9. Although the reconstruction shown in Figure 9 does not quite recover all of the young oceanic crust

Figure 9: Pacific reconstruction for 80 m.y. identical to that of Figure 2b except that West Antarctica and the Pacific Ocean have been rotated counterclockwise relative to East Antarctica. This roughly illustrates a reconstruction which is consistent with the data presented by Pitman (1978).

required by Pitman (1978), there is clearly much more young ocean floor in this reconstruction than in either of the 80-m.y. reconstructions used in this study (Figure 2). The effect of the actual difference in ridge-volume calculations for the Phoenix, Farallon and Pacific ridges between this work and the reconstructions of Pitman (1978) amounts to 175 m (574 ft) less sea-level fall in this work.

In total, the ridge-volume calculations of Pitman (1978) imply a significantly greater decrease in Pacific ridge lengths through time than is implied by this work. Taking maximum ridge-length assumptions for the Tethys and Kula-Pacific ridges of this work, the sea-level difference which can be accounted for by differences in Pacific reconstructions alone is about 77 m (253 ft) while minimum Tethys and Kula-Pacific ridge-length assumptions result in about 157 m (515 ft) less sea-level fall. Additional differences are due to lesser effects, such as lower spreading rates (as a result of the time scale used) and the inclusion of small-ocean-basin calculations. It is probable that the sea-level uncertainty of 80 m (262 ft) resulting from estimating the volume of missing ridges (Tethys and West Kula ridges) in this work is too low an estimate. However, the 175 m (574 ft) of sea-level uncertainty resulting from the differences in Pacific plate reconstructions used in this work as compared to that of Pitman (1978) is, clearly, extremely significant and requires some combination of geologic and paleomagnetic studies in Antarctica for resolution.

In studying the age distribution of the ocean crust, Parsons (1982) concluded that ridge-volume changes are not likely to have caused sea-level variations in excess of 150 m (492 ft). Although Parsons (1982) did not attempt to model specific ridge volume history through time, his model is compatible with the results of this study. Utilizing average spreading rates and ridge lengths from this study to calculate the average rate of crustal generation between 125 and 80 m.y., the change in sea level predicted by the triangular age distribution model of Parsons (1982; his Figure 8) is within 30 m (98 ft) of the mean sea level predicted in this work for all three time scales used.

The problem of additional sources of ocean-volume change remains as a significant uncertainty which arises from the calculation of sea-level variation from changing ridge volumes. For example, Harrison et al (1981) calculated that the effects of increasing sedimentation rates since 100 m.y. ago could account for 77 m (253 ft) of sea-level rise. They also calculated that the collision of India with Asia adds about 18 m (59 ft) to the post-Cretaceous sea-level fall. Schlanger, Jenkyns, and Premoli-Sliva (1981) calculated that the effect of major volcanic activity between 120 and 90 m.y. might have caused a sea-level rise of as much as 80 m (262 ft) during that time. The cooling of this thermal welt would add to the sea-level fall calculated from ocean ridge data since 80 m.y. Additional factors such as continental lengthening during formation of passive margins, the growth of island arcs and accretionary prisms, crustal shortening due to orogeny, additional hot spot activity, the changing volumes of oceanic trenches, and the desiccation and filling of large basins in arid regions could have a significant effect on the shape and magnitude of sea-level change over the past 80 m.y. (Pitman, 1978; Donovan and Jones, 1979). Thus, although recent ridge-volume studies point toward a sea-level fall of about 230 m (755 ft) over the past 80 m.y., there are other factors which influence ocean volumes and should be used to modify this value.

Two of the sea-level curves presented in Figure 8 are derived from analysis of sedimentary basin subsidence. Eustatic sea level is calculated by comparing the tectonic subsidence of a basin derived from the sedimentary record to theoretical thermal subsidence models (McKenzie, 1978). The deviation between the theoretical and observed subsidence is assumed to be due to sea-level variation.

The sea-level curve of Brunet and LePichon (1982) is derived from analysis of the Paris Basin (dashed lines of Figure 8). The hachured part of the curve corresponds to paleo-water depth uncertainties in Upper Cretaceous sediments of the Paris Basin. The lower (dash-dot) curve is obtained by assuming that uplift of the basin since 30 m.y. was due to some tectonic driving force (such as a relaxation of Cretaceous alpine thrusting or uplift during the thermal heating associated with the Rhine Graben), while the higher curve is a result of the assumption that post-Oligocene uplift has been due solely to sea-level fall. The high curve is the only interpretation presented by Brunet and Le Pichon (1982). A recent study by Mackenzie and McKenzie (1983) indicates that the amount of post-Oligocene uplift assumed by Brunet and Le Pichon (1982) may be as much as 2 km too low in the eastern portion of the basin, implying that this uplift must have been tectonic in origin. The maximum sea-level curve from the Paris Basin results in a magnitude of sea-level fall since Late Cretaceous more consistent with the results of Pitman (1978). An intermediate Paris Basin curve, between the maximum and minimum, is more compatible with the LGP sea-level curve of this paper. The form of the curve is, however, very different from both ridge-volume derived sea-level curves implying that either: 1) ridge-volume change is not the primary source of ocean-volume change, and thus, sea-level variation; or, 2) the Paris Basin did not subside by simple thermal subsidence, as has been shown to be the case north of the Paris Basin, in the North Sea (Kent, 1975; Ziegler, 1977).

The sea-level curve of Watts and Steckler (1979) was derived by comparing the subsidence of the North American Atlantic margin to theoretical thermal subsidence. The resulting sea-level curve in this case is very similar to the LGP result of this study from 40 m.y. to the present and indicates less sea-level fall from 80 to 40 m.y. than that predicted by the mean LGP curve. Differences could be attributed to uncertainties in either the LGP curve or the Watts and Steckler (1979) curve. Uncertainties in the LGP curve include other factors which influence ocean volume and uncertainties in ridge lengths and spreading rates through time. Uncertainties inherent in basin-history studies include undetermined tectonic subsidence or uplift (Hardenbol and Vail, 1981; Heller, Wentworth, and Poag, 1982), uncertainty in dating of sediments (Poag 1980), and uncertainty in the basement response to the sediment load (Steckler and Watts, 1978; Sawyer et al, 1982).

An eustatic sea-level curve was calculated by Hardenbol and Vail (1981) through an analysis of the subsidence of the northwest African Margin. In this case, the deviation between the observed and theoretical subsidence curves indicated a sea-level rise in the Late Jurassic (156 m.y.) to middle Cretaceous (97 m.y.) of about 300 m (984 ft) . This is compatible with the 350 m (1,148 ft) magnitude of sea-level fall since the Late Cretaceous predicted by Pitman (1978). In addition to the errors inherent in any subsidence analysis, this study also suffers from an uncertain amount of post-Miocene uplift in Africa (Hardenbol and Vail, 1981; Bond, 1979).

Estimates of possible sea-level heights during the middle to late Miocene, middle to late Eocene, Campanian to Maestrichtian, and Turonian to Coniacian time by Bond (1979) were made by studying Mesozoic and Cenozoic stratigraphy in combination with continental hypsometry. This study allows for average tectonic uplift or subsidence of continents and removes this effect from the apparent sea-level variation to obtain an estimate of the world average sea-level change. From Figure 8 we see that the resulting sea-level estimates are slightly lower than the mean sea-level estimate of this work, prior to Miocene time, but overlap significantly the range of error due to ridge-volume calculations. The mean estimates would also fall within the range of sea-level estimates of Bond (1979) if the age-depth curve of Heestand and Crough (1981) was used instead of that of Parsons and Sclater (1977). The calculations made by Bond (1979) do not include the possibility of uniform changes in the hypsometry of all continents (Harrison, 1981) by tectonic or other processes.

In modeling sea-level fall from seismic stratigraphy, only relative sea-level changes can be obtained (Vail, Mitchum, and Thompson, 1977). A discrepancy between all the sea-level curves obtained in this study and the results of stratigraphic studies is the lack of a significant sea-level fall in the Oligocene. However, the Aluk (Phoenix)-Farallon ridge, which was not considered in this study, may cause the required sea-level fall. It is this ridge which must be postu-

lated to generate crust between 80 and 35 m.y. north and west of the Chile rise (the question marks in Figure 2b and c). This crust must then be fully subducted between 35 m.y. and the present. The maximum effect of this ridge (assuming the hot spot reconstruction) is to reduce the sea-level fall from 80 to 35 m.y. by about 36 m (118 ft), and to increase the sea-level fall subsequent to 35 m.y. by the same magnitude. If this ridge is subducted rapidly beneath South America this effect may be sufficient to satisfy the observed stratigraphic data.

CONCLUSIONS

The total volume of mid-ocean ridges has decreased since the Late Cretaceous, 80 m.y. ago. This fact is due to a decrease in overall spreading rates during that period. The slowest possible Cretaceous spreading rates allowed by magnetic-anomaly and biostratigraphic age control cannot eliminate these relatively high Cretaceous spreading rates. The longest ridges and highest spreading rates are primarily confined to the Pacific Ocean, so that changes in Pacific ridge volumes dominate world ridge-volume variations. Because less magnetic-anomaly data is available for older ocean crust, particularly in the Pacific, the range of estimates of past sea level is very large. The range of error decreases from about ± 120 m (± 394 ft) at 80 m.y. to about ± 10 m (± 33 ft) at the present. The trend of sea-level variation due to changing ridge volumes has been a general fall from 80 to 10 m.y. with little or no change from 10 m.y. to the present. A sea-level fall since the Late Cretaceous of about 230 m (± 100 m) [755 ft (± 328 ft)] is predicted by taking Pacific reconstructions based on either a fixed hot spot frame or a simple rotation of the Pacific, with respect to an undeformed antarctic continent.

It is possible to say that a magnitude of sea-level fall since the Late Cretaceous (80 m.y.) of about 180 m (± 100 m) [591 ft (± 328 ft)] is consistent with the approach used in this paper as well as that used by Watts and Steckler (1979), Parsons (1982), Bond (1979), and a modified result from Brunet and LePichon (1982). However, similar techniques also show that the magnitude of sea-level fall is more consistent with about a 350 m (1,148 ft) drop in sea level (Pitman, 1978; Hardenbol and Vail, 1981; Harrison et al, 1981). Thus, although estimates of the magnitude of eustatic sea level in the Late Cretaceous appear to be converging, ambiguities still remain. In particular, resolution of the volume change of ocean ridges requires a study of the relative motion between East and West Antarctica.

ACKNOWLEDGMENTS

I thank W.C. Pitman III, G.C. Bond, A.B. Watts, B. Parsons, S. Cande, M.S. Steckler, and J.D. Hays for critical review of the paper. I also thank W.C. Pitman and G.C. Bond for numerous discussions and recommendations (not always taken) and for moral support; and S. O'Connell for numerous suggestions. I would especially like to thank B. Parsons for removing my obstinance with respect to the Pacific Ocean. This work was supported by NSF grant OCE-79-26308.

REFERENCES

Atwater, T., 1970, Implications of plate tectonics for the

Cenozoic tectonic evolution of western North America: Geological Society of America Bulletin, v. 81, p. 3513-3536.

Anderson, R.N., 1980, Update of heat flow in the East and Southeast Asian seas, in D.E. Hayes, ed., The tectonics and geologic evolution of Southeast Asian seas and islands: Washington, D.C., American Geophysical Union, p. 319-326.

Baldwin, B., P.J. Coney, and W.R. Dickinson, 1974, Dilemma of a Cretaceous time scale and rates of sea floor spreading: Geology, v. 2, p. 267-270.

Barker, P.F., and I.W.D. Dalziel, 1983, Progress in geodynamics in the Scotia arc region, in Final report, working group II, Interunion Commission on Geodynamics: Washington, D.C., American Geophysical Union, Geodynamics Series, v. 9, p. 137-170.

Berggren, W.A., 1972, A Cenozoic time scale; some implications for regional geology and paleobiogeography: Lethaia, v. 5, p. 195-215.

Bond, G.C., 1978, Evidence for late Tertiary uplift of Africa relative to North America, South America, Australia and Europe: Journal of Geology, v. 86, p. 47-65.

——— , 1979, Evidence for some uplifts of large magnitude in continental platforms: Tectonophysics, v. 61, p. 285-305.

Brunet, M.F., and X. Le Pichon, 1982, Subsidence of the Paris Basin: Journal of Geophysical Research, v. 87, p. 8547-8560.

Cande, S.C., E.M. Herron, and B.R. Hall, 1982, The early Cenozoic history of the southeast Pacific: Earth and Planetary Science Letters, v. 57, p. 63-74.

——— , and J.C. Mutter, 1982, A revised identification of the oldest sea-floor spreading anomalies between Australia and Antarctica: Earth and Planetary Science Letters, v. 58, p. 151-160.

Cochran, J.R., 1981, The Gulf of Aden; structure and evolution of a young ocean basin and continental margin: Journal of Geophysical Research, v. 86, p. 263-287.

Dalziel, I.W.D., 1981, Back-arc extension in the southern Andes; a review and critical reappraisal: Philosopical Transactions of the Royal Society of London, v. A300, p. 319-335.

Dewey, J.F., et al, 1973, Plate tectonics and the evolution of the Alpine system: Geological Society of American Bulletin, v. 84, p. 3137-3180.

Dickinson, W.R., and E.I. Rich, 1972, Petrologic intervals and petrofacies in the Great Valley sequence, Sacramento Valley, California: Geological Society of America Bulletin, v. 83, p. 3007-3024.

Donovan, D.T., and E.J. Jones, 1979, Causes of world wide changes in sea level: Journal of the Geological Society of London, v. 136, p. 187-192.

Duncan, R.A., 1981, Hotspots in the southern oceans - an absolute frame for motion of the Gondwana continents: Tectonophysics, v. 58, p. 151-160.

Hallam, A., 1977, Secular changes in marine inundation of USSR and North America through the Phanerozoic: Nature, v. 269, p. 769-772.

Hancock, J.M., and E.G. Kauffman, 1979, The great transgression of Late Cretaceous: Journal of the Geological Society of London, v. 132, p. 175-186.

Hardenbol, J., and P. Vail, 1981, Interpreting paleoenvironments, subsidence history and sea-level changes of passive margins from seismic and biostratigraphy, in Proceedings, Continental Margin Symposium, 26th International Congress: Oceanologica Acta, v. 3, p. 33-44.

Harland, W.B., et al, 1983, A geologic time scale: Cambridge University Press, 1 p.

Harrison, C.G.A., et al, 1981, Sea level variations, global sedimentation rates, and the hypsographic curve: Earth and Planetary Science Letters, v. 54, p. 1-16.

Hayes, D.E., and B. Taylor, 1978, Tectonics map chart Series MC.-25: Boulder, Colorado, Geological Society of America.

Hays, J.D., and W.C. Pitman III, 1973, Lithospheric plate motion, sea-level changes, and climatic and ecological consequences: Nature, v. 246, p. 18-22.

Heestand, R.L., and S.T. Crough, 1981, The effect of hot spots on the oceanic age-depth relation: Journal of Geophysical Research, v. 86, p. 18-22.

Heirtzler, J.R., et al, 1968, Marine magnetic anomalies, geomagnetic field reversals and motions of the ocean floor and continents: Journal of Geophysical Research, v. 73, p. 2119-2136.

Heller, P.L., C.M. Wentworth, and C.W. Poag, 1982, Episodic post-rift subsidence of the United States Atlantic continental margin: Geological Society of America Bulletin, v. 93, p. 379-390.

Kent, P.E., 1975, Review of North Sea basin developments: Journal of Geological Society of London, v. 131, p. 435-468.

Kristoffersen, Y., and M. Talwani, 1977, Extinct triple junction south of Greenland and the Tertiary motion of Greenland relative to North America: Geological Society of American Bulletin, v. 88, p. 1037-1049.

LaBrecque, J.L., D.V. Kent, and S.C. Cande, 1977, Revised magnetic polarity time scale for Late Cretaceous and Cenozoic time: Geology, v. 5, p. 330-335.

Ladd, J.W., 1976, Relative motion of South America with respect to North America and Caribbean tectonics: Geological Society of American Bulletin, v. 87, p. 969-976.

Lambert, R., Sb.J., 1971a, The pre Pleistocene Phanerozoic time-scale; a review, in The Phanerozoic time scale - a supplement: Geological Society of London, Special Publication 5, p. 9-31.

——— , 1971b, The pre-Pleistocene time-scale; further data, in The Phanerozoic time scale - a supplement: Geological Society of London, Special Publication 5, p. 33-34.

Larson, R.L., X. Golovchenko, and W.C. Pitman III, 1982, Geomagnetic polarity time scale: AAPG Plate-Tectonic Map Circum-Pacific Region, Pacific Basin sheet, scale 1:20,000,000.

——— , and T.W.C. Hilde, 1975, A revised time scale of magnetic reversals for the Early Cretaceous and Late Jurassic: Journal of Geophysical Research, v. 80, p. 2586-2594.

Mackenzie, A.S., and D.P. McKenzie, 1983, Isomerization and aromatization of hydrocarbons in sedimentary basins formed by extension: Geological Magazine, v. 120, p. 417-528.

McKenzie, D.P., 1978, Some remarks on the development of sedimentary basins: Earth and Planetary Science Letters, v. 40, p. 25-32.

——— , and W.J. Morgan, 1969, Evolution of triple junctions: Nature, v. 224, p. 125-133.

Molnar, P., and T. Atwater, 1978, Interarc spreading and cordilleran tectonics as alternatives related to the age of subducted oceanic lithosphere: Earth and Planetary Science Letters, v. 41, p. 330-340.

———, et al, 1975, Magnetic anomalies, bathymetry, and tectonic evolution of the South Pacific since the Late Cretaceous: Geophysical Journal of the Royal Astronomical Society, v. 49, p. 383-420.

Morgan, W.J., 1968, Rises, trenches, great faults, and crustal blocks: Journal of Geophsyical Research, v. 73, p. 1959-1982.

———, 1981, Hotspot tracks and the opening of the Atlantic and Indian Oceans, in C. Emiliani, ed., The Sea, v. 7: New York, John Wiley and Sons, p. 443-487.

Ness, G., S. Levi, and R. Couch, 1980, Marine magnetic anomaly timescales for the Cenozoic and Late Cretaceous; a precise critique, and synthesis: Review of Geophysics and Space Physics, v. 18, p. 753-770.

Norton, I.O., and J.G. Sclater, 1979, A model for the evolution of the Indian Ocean and the breakup of Gondwanaland: Journal of Geophysical Research, v. 84, p. 6803-6830.

Parsons, B., 1982, Causes and consequences of the relation between area and age of the ocean floor: Journal of Geophysical Research, v. 87, p. 289-302.

———, and J.G. Sclater, 1977, An analysis of the variation of ocean floor bathymetry and heat flow with age: Journal of Geophysical Research, v. 82, p. 803-828.

Pitman, W.C., III, 1978, Relationship between eustacy and stratigraphic sequences of passive margins: Geological Society of America Bulletin, v. 89, p. 1389-1403.

———, and D.E. Hayes, 1968, Sea-floor spreading in the Gulf of Alaska: Journal of Geophysical Research, v. 73, p. 6571-6580.

———, and M. Talwani, 1972, Sea-floor spreading in the North Atlantic: Geological Society of American Bulletin, v. 83, p. 619-646.

Poag, C.W., 1980, Foraminiferal stratigraphy, paleoenvironments, and deposition cycles in the outer Baltimore Canyon Trough, in P.A. Scholle, ed., Geological studies of the COST No B-3 well, U.S. Mid-Atlantic Continental Slope area: U.S. Geological Survey Circular 833, p. 44-65.

Sawyer, D.S., et al, 1982, Extensional model for the subsidence of the northern United States Atlantic continental margin: Geology, v. 10, p. 134-140.

Schlanger, S.O., H.C. Jenkyns, and I. Premoli-Silva, 1981, Volcanism and vertical tectonics in the Pacific basin related to global Cretaceous transgressions: Earth and Planetary Science Letters, v. 52, p. 435-449.

Schouten, H., and K.D. Klitgord, 1977, U.S. Geological Survey Miscellaneous Field studies Map MF 915, scale 1:2,000,000.

Sclater, J.G., R.N. Anderson, and M.L. Bell, 1971, The elevation of ridges and the evolution of the central eastern Pacific: Journal of Geophysical Research, v. 76, p. 7888-7915.

———, C. Jaupart, and D. Gelson, 1980, The heat flow through oceanic and continental crust and the heat loss of the earth: Review of Geophysics and Space Phsyics, v. 18, p. 269-311.

Sleep, N.H., 1976, Platform subsidence mechanisms and "eustatic" sea-level changes: Tectonophysics, v. 36, p. 45-56.

Smith, A.G., and J.C. Briden, 1977, Mesozoic and Cenozoic paleocontinental maps: Cambridge University Press, 63 p.

Steckler, M.S., and A.B. Watts, 1978, Subsidence of the Atlantic continental margin off New York: Earth and Planetary Science Letters, v. 41, p. 1-13.

Suarez, G., and P. Molnar, 1980, Paleomagnetic data and pelagic sediment facies and the motion of the Pacific plate relative to the spin axis since the Late Cretaceous: Journal of Geophysical Research, v. 85, p. 5357-5280.

Talwani, M., and O. Eldholm, 1977, Evolution of the Norwegian Greenland Sea: Geological Society of America Bulletin, v. 88, p. 969-999.

Tisseau, J., and P. Patrait, 1981, Identification des anomalies magnetiques des dorsales a faible taux d'expansion; methode de taux fictifs: Earth and Planetary Science Letters, v. 52, p. 381-396.

Vail, P.R., R.M. Mitchum, Jr., and S. Thompson III, 1977, Seismic stratigraphy and global changes of sea level, part 4; global cycles of relative changes of sea level, in C.E. Payton, ed., Seismic stratigraphy - applications to hydrocarbon exploration: AAPG Memoir 26, p. 83-97.

Van Hinte, J.E., 1976a, A Cretaceous time scale: AAPG Bulletin, v. 60, p. 489-497.

———, 1976b, A Jurassic time sale: AAPG Bulletin, v. 60, p. 498-516.

Watanabe, T., M.G. Langseth, and R.N. Anderson, 1977, Heat flow in back-arc basins of the western Pacific, in M. Talwani and W.C. Pitman III, eds., Island arcs, deep sea trenches and back-arc basins: Washington, D.C., American Geophysical Union, p. 137-162.

Watts, A.B., and M.S. Steckler, 1979, Subsidence and eustacy at the continental margin of eastern North America, in M. Talwani, W. Hay, and W.B.F. Ryan, eds., Deep drilling results in the Atlantic Ocean; continental margins and paleoenvironment: Washington, D.C., American Geophysical Union, Maurice Ewing Series No. 3, p. 218-234.

———, J.H. Bodine, and N.M. Ribe, 1980, Observations of flexure and the geological evolution of the Pacific Ocean Basin: Nature, v. 283, p. 532-537.

Weissel, J.K., D.E. Hayes, and E.M. Herron, 1977, Plate tectonics synthesis; the displacements between Australia, New Zealand, and Antarctica since the Late Cretaceous: Marine Geology, v. 25, p. 231-277.

Wood, R.J., 1981, The subsidence history of Conoco well 15/30-1, central North Sea: Earth and Planetary Science Letters, v. 54, p. 306-312.

Ziegler, P.A., 1977, Geology and hydrocarbon provinces of the North Sea: Geological Journal, v. 1, p. 7-32.

Jurassic Unconformities, Chronostratigraphy, and Sea-Level Changes from Seismic Stratigraphy and Biostratigraphy

P.R. Vail
J. Hardenbol
Exxon Production Research Company
Houston, Texas

R.G. Todd
Esso Exploration and Production Norway
Stavanger, Norway

Seventeen global unconformities and their correlative conformities (sequence boundaries) subdivide the strata of the Jurassic and earliest Cretaceous into genetic depositional sequences produced by 16 eustatic cycles. These 16 cycles make up the Jurassic supercycle. Eight of the global unconformities are both subaerial and submarine (Type 1), and are believed to have been caused by rapid eustatic falls of sea level. Nine of the unconformities are subaerial only (Type 2), and are believed to be related to slow eustatic falls of sea level. In addition, 16 marine condensed sections (starved intervals) have been identified. These condensed sections are interpreted to be related to rapid eustatic rises of sea level. Unconformity recognition is locally or regionally enhanced by periodic truncation of folded and faulted strata during sea-level lowstands and onlap onto topographic highs during sea-level highstands, but we find no evidence that the tectonics caused the global unconformities. The 16 eustatic cycles that make up the Jurassic supercycle correspond to 16 global chronostratigraphic intervals that subdivide Jurassic strata into a series of genetic depositional sequences, which are recognizable from seismic, well, and outcrop data.

The Jurassic unconformities and the stratal and facies patterns between them are caused by the interaction of basement subsidence, eustatic changes of sea-level, and varying sediment supply. Detailed analyses of the sediments with seismic stratigraphy and well data permit quantification of the subsidence history and reconstruction of paleoenvironment and sea-level changes through time. The integrated use of seismic stratigraphy and biostratigraphy provides a better geologic age history than could be obtained by either method alone. Paleobathymetry, sediment facies, and relative changes of sea level can be interpreted from seismic data and confirmed, or improved on, by well control. Geohistory analysis based on geologic time-depth diagrams provides a quantitative analysis of total basin subsidence. When this subsidence is corrected for compaction and sediment loading, the tectonic subsidence and long-term eustatic changes may be determined. Short-term, rapid changes of sea level can be demonstrated from seismic, well, and outcrop data. The stratigraphic resolution of these changes rarely allows exact quantification of their magnitude, but a minimum rate of sea-level change often can be determined.

INTRODUCTION

A representative seismic line from the Inner Moray Firth in the North Sea (Figure 1) illustrates a seismic sequence interpretation of the Jurassic. Black lines indicate seismic sequence boundaries (unconformities or their equivalent conformities), arrows mark representative reflection terminations, numbers (132) indicate the age of the sequence boundaries determined where they are conformable, and symbols (J3.2) identify the sequence. Seismic stratigraphic studies tied to well control in over 60 areas located all over the globe indicate that any regional grid of reflection seismic profiles with reasonable data quality can be interpreted in a similar manner. Not only are sequence boundaries always present, but their ages dated in areas where they become conformable are the same in different basins around the world.

This paper discusses the Jurassic supercycle and shows how eustatic changes of sea level cause the global similarities of unconformity age, coastal onlap, and marine-condensed sections. It also shows how shoreline transgressions and regressions relate to these criteria and why transgressions and regressions are not necessarily globally synchronous.

The material for this paper is from a series of previously published papers: Vail et al (1977), Vail and Mitchum (1979), Vail and Hardenbol (1979), Vail et al (1980), and Vail and Todd (1981), and Hardenbol, Vail, and Ferrer (1981).

AGE, TYPES, AND CAUSES OF JURASSIC GLOBAL UNCONFORMITIES AND CONDENSED SECTIONS

Seventeen global unconformities and their correlative conformities (sequence boundaries) subdivide the strata of the Jurassic and earliest Cretaceous into genetic sequences

129

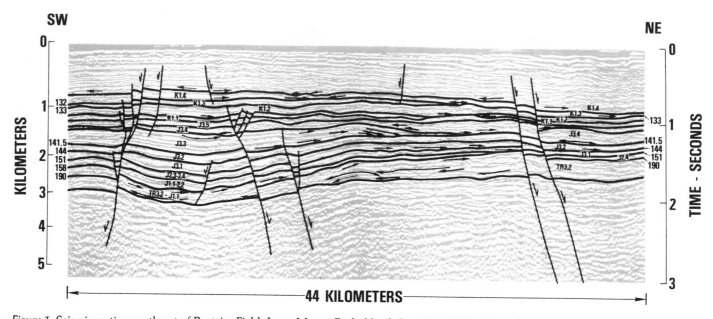

Figure 1: Seismic section northeast of Beatrice Field, Inner Moray Firth, North Sea, United Kingdom, showing seismic sequences of the Jurassic supersequence. Numbers (132) refer to the ages of the sequence boundaries where they are conformable. Symbols (J3.2) identify the individual sequences. Seismic section courtesy of Geophysical Company of Norway (United Kingdom) Ltd. (GECO).

produced by 16 eustatic cycles. These 16 cycles make up the Jurassic supercycle (Vail et al, 1977, part 8) (Table 1). Eight of the global unconformities are both subaerial and submarine (Type 1) and are believed to have been caused by rapid eustatic falls of sea level. Nine of the unconformities are subaerial only (Type 2) and are believed to be related to slow eustatic falls of sea level. In addition, 16 marine-condensed sections (starved intervals) have been identified. These condensed sections are interpreted to be related to rapid eustatic rises of sea level (Vail and Todd, 1981). Unconformity recognition is enhanced locally or regionally by periodic truncation of folded and faulted strata during sea-level lowstands and by onlap onto topographic highs during sea-level highstands, but we find no evidence that the tectonics caused the global unconformities. Table 1 is our latest summary of the ages of Type 1 and Type 2 unconformities and condensed sections. Of special significance is the fact that, by more careful age dating, we have discovered that the unconformities we previously placed at the base of the Sinemurian, Callovian, Oxfordian, Kimmeridgian, and Valanginian (Vail and Todd, 1981) are all older and occur within or at the base of the late portion of the preceding age. Figure 2 shows the relation of unconformity age and type and condensed interval age to the relative changes of coastal onlap and eustatic changes of sea level in the Jurassic supercycle.

Unconformities, marine condensed sections, coastal onlap, and marine transgressions-regressions are controlled by the interaction of these variables: subsidence, eustatics, and sediment supply. Empirical observations from seismic stratigraphic studies indicate that unconformities, condensed sections, coastal onlap, and marine transgression-regression are related as shown diagrammatically on Figure 3 and discussed below.

Unconformity is defined (Bates and Jackson, 1980, p. 675) as the structural relationship between rock strata in contact characterized by a lack of continuity in deposition, and cor-

responding to a period of nondeposition, weathering, or especially erosion (either subaerial or subaqueous) prior to the deposition of the younger beds, and often (but not always) marked by absence of parallelism between strata. We use the term unconformity for a surface representing a significant time gap with erosional truncation (subaerial or subaqueous) and /or subaerial exposure. Marine surfaces with significant hiatuses but without evidence of erosion are not unconformities, according to this usage. This definition also restricts the usage of unconformity as used in Vail et al, 1977. Angularity of the underlying strata is commonly due to erosion and truncation of tilted strata, although sedimentary bypass may produce a similar effect referred to as *toplap* (Vail et al, 1977, part 2). Discordance of the overlying rocks with an unconformity is commonly caused by terminations of overlying strata by sedimentary lapouts. These lapouts are termed onlap or downlap, depending upon whether they lap out in an updip or downdip direction, respectively, at the time of deposition (Figure 3).

Global unconformities are unconformities that are present at one place or another in all sedimentary basins with a sea-level base level at the time of deposition, and become conformable at the same geologic age. They are not necessarily unconformable surfaces everywhere throughout each basin. In other words, although a global unconformity may be present within a basin, it may be characterized by extensive conformable areas in that basin as well.

Type 1 (subaerial-submarine) unconformities are typified by downward shifts of coastal onlap (defined below), commonly below the shelf edge, typically producing subaerial exposure of the shelf, valley entrenchment, and the initiation of canyon cutting along the shelf edge (Figures 4A and 5). Fluvial sediments and lowstand deltas commonly fill the entrenched valleys, and submarine fans and slope-front fill deposits accumulate in deep water basins. Lowstand evaporite deposits and regional truncation of folded or faulted

Table 1

Type		Age	Sequence Identification	Van Hinte (1976) Time Scale in MA	Relative Magnitude		
					Maj.	Med.	Min.
Subaerial/Submarine	1	latest Berriasian	-	132	X		
Condensed Section	CS	mid Late Berriasian	K1.3	132.5			X
Subaerial	2	early Late Berriasian	-	133			X
Condensed Section	CS	basal Late Berriasian	K1.2	133.5			X
Subaerial	2	mid Early Berriasian	-	134			X
Condensed Section	CS	basal Berriasian	K1.1	135		X	
Subaerial/Submarine	1	within Tithonian	-	138		X	
Condensed Section	CS	within Tithonian	J3.5	138.5		X	
Subaerial/Submarine	1	within Tithonian	-	139		X	
Condensed Section	CS	within Tithonian	J3.4	139.5		X	
Subaerial/Submarine	1	within Tithonian	-	140	X		
Condensed Section	CS	basal Tithonian	J3.3	141		X	
Subaerial/Submarine	1	mid Late Kimmeridgian	-	141.5		X	
Condensed Section	CS	basal Kimmeridgian	J3.2	143	X		
Subaerial	2	mid Late Oxfordian	-	144		X	
Condensed Section	CS	basal Middle Oxfordian	J3.1	147	X		
Subaerial	2	basal Late Callovian	-	151		X	
Condensed Section	CS	mid Callovian	J2.4	153	X		
Subaerial/Submarine	2	mid Early Callovian	-	155			X
Condensed Section	CS	basal Callovian	J2.3	156			X
Subaerial	2	basal Late Bathonian	-	158	X		
Condensed Section	CS	basal Bathonian	J2.2	165		X	
Subaerial	2	basal Late Bajocian	-	168		X	
Condensed Section	CS	basal Bajocian	J2.1	171	X		
Subaerial/Submarine	1	mid Early Aalenian	-	173		X	
Condensed Section	CS	mid Toarcian	J1.5	175		X	
Subaerial	2	mid Early Toarcian	-	177			X
Condensed Section	CS	basal Toarcian	J1.4	178			X
Subaerial/Submarine	1	mid Late Pliensbachian	-	179		X	
Condensed Section	CS	basal Late Pliensbachian	J1.3	180		X	
Subaerial	2	basal Pliensbachian	-	184			X
Condensed Section	CS	mid Late Sinemurian	J1.2	185		X	
Subaerial/Submarine	1	latest Hettangian	-	190		X	

Table 1: Jurassic supercycle global unconformities (sequence boundaries) and condensed sections.

strata are characteristic in the proper geological setting. Type 1 unconformities are typified by both subaerial and submarine erosion and are formed when the rate of eustatic sea-level fall exceeds the rate of subsidence at the shelf edge.

Type 2 (subaerial) unconformities are characterized by downward shifts of coastal onlap to a position at or landward of the shelf edge, subaerial exposure of the landward portion of the shelf, and no evidence of canyon cutting (Figures 4B and 5). Lowstand evaporite deposits are characteristic in the proper geologic setting. Type 2 unconformities have only subaerial erosion, but their equivalent submarine conformities can commonly be recognized by submarine onlap resulting from shifts in the sites of deposition. Type 2 unconformities are formed when the rate of fall of eustatic sea-level is less than the rate of subsidence at the shelf edge, but exceeds the rate of subsidence on the inner portion of the shelf.

The chronostratigraphic significance of an unconformity is that all the rocks below the unconformity are older than the rocks above it. The ages of the strata immediately above and below the unconformity differ geographically according to the areal extent of erosion or nondeposition. The dura-

tion of the hiatus associated with an unconformity differs correspondingly, but the unconformity itself is a chronostratigraphic boundary because it separates rocks of different ages, and no chronostratigraphic surfaces cross it. Although many chronostratigraphic surfaces may merge along an unconformity, none actually cross the unconformity. For these reasons, unconformities are not diachronous but are time boundaries that may be assigned a specific geologic age dated in those areas where the hiatus is least and/ or where the rocks above and below become conformable. By careful correlation of unconformities and their correlative conformities, a sedimentary interval can be divided into genetic depositional sequences bounded by these unconformities. A depositional sequence is a chronostratigraphic interval because it contains all the rocks deposited during a given interval of geologic time limited by the ages of the sequence boundaries where they are conformities. The global Jurassic unconformities and their correlative conformities subdivide the Jurassic superrsequence (all sediments deposited within the Jurassic supercycle) (Figure 2) into a series of genetic depositional sequences with global chronostratigraphic significance. Our studies indicate that the most

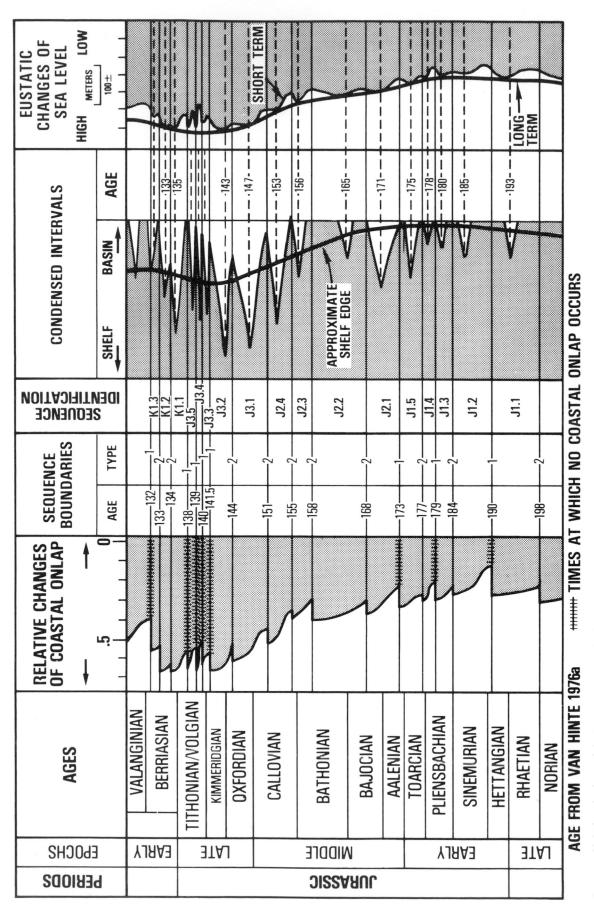

Figure 2: Global cycle chart of the Jurassic supercycle showing age and type of sequence boundaries (unconformities and their correlative conformities), sequence identification, age of condensed intervals and charts showing relative changes of coastal onlap and eustatic changes of sea level.

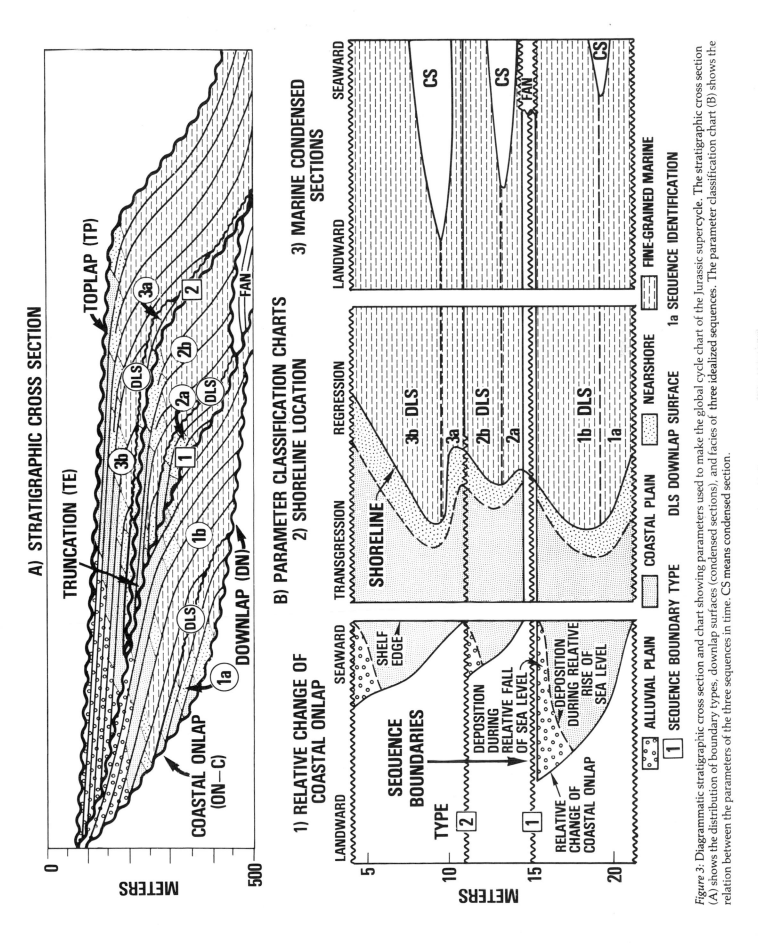

Figure 3: Diagrammatic stratigraphic cross section and chart showing parameters used to make the global cycle chart of the Jurassic supercycle. The stratigraphic cross section (A) shows the distribution of boundary types, downlap surfaces (condensed sections), and facies of three idealized sequences. The parameter classification chart (B) shows the relation between the parameters of the three sequences in time. CS means condensed section.

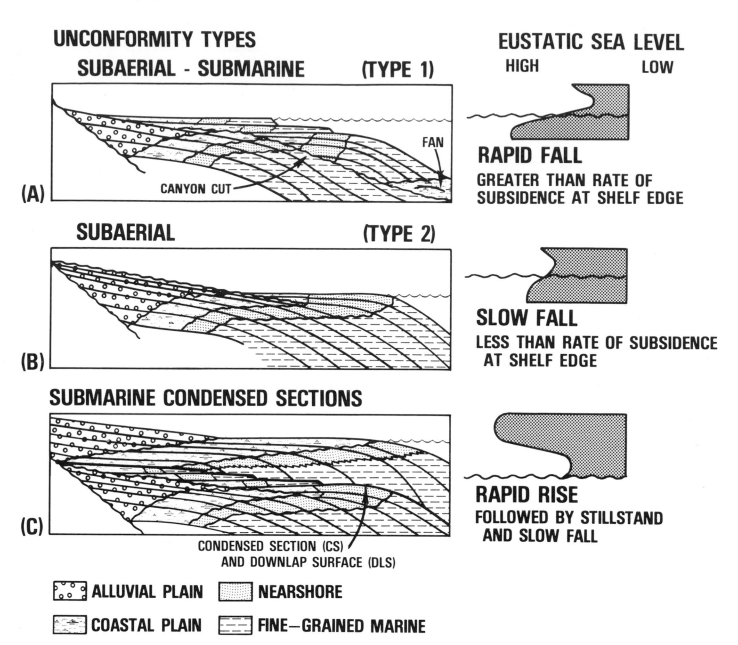

UNCONFORMITY TYPES

SUBAERIAL - SUBMARINE (TYPE 1)

EUSTATIC SEA LEVEL

HIGH   LOW

(A)   CANYON CUT   FAN

RAPID FALL
GREATER THAN RATE OF
SUBSIDENCE AT SHELF EDGE

SUBAERIAL (TYPE 2)

(B)

SLOW FALL
LESS THAN RATE OF SUBSIDENCE
AT SHELF EDGE

SUBMARINE CONDENSED SECTIONS

(C)

RAPID RISE
FOLLOWED BY STILLSTAND
AND SLOW FALL

CONDENSED SECTION (CS)
AND DOWNLAP SURFACE (DLS)

ALLUVIAL PLAIN   NEARSHORE

COASTAL PLAIN   FINE—GRAINED MARINE

Figure 4: Diagrammatic charts showing the relation between Type 1 and Type 2 unconformities and submarine condensed sections to eustatic sea-level changes.

widespread chronostratigraphic unit at the time of deposition is near the top of each sequence, as shown on Figure 5. The upper portions of clastic sequences characteristically show the greatest landward onlap of the alluvial plain, and at the same time they exhibit major seaward progradation.

In most basins of the world, Type 1 unconformities are characterized by both subaerial and submarine erosion. However, there are occasional exceptions where the shelf margin is locally subsiding so rapidly that its rate exceeds that of the fall of sea level, and the unconformity has the characteristics of a Type 2. An example is a shelf margin controlled by a rapidly subsiding growth fault, such as in the Neogene of the Gulf of Mexico. Similarly, Type 2 unconformities may locally show Type 1 characteristics in areas where structural activity causes stability or uplift so that sea level may locally fall below the shelf edge.

Condensed sections are thin marine stratigraphic intervals characterized by very slow depositional rates (less than 1 cm/1000 yrs). They are commonly associated with thin but continuous zones of burrowed or lightly lithified beds (hardgrounds, omission surfaces). They are conspicuous by the presence of scattered pebbles or high concentrations of volcanic ash, glauconite, phosphate, or radioactive minerals, and are commonly linked to marine hiatuses or faunal mixing horizons (Figures 4C and 5). In vertical deepening-upward sections, a condensed section commonly marks the greatest paleowater depth. However, in areas of rapid subsidence the greatest paleowater depths may be above the condensed section. Condensed sections should not be confused with a winnowed interval associated with high-energy removal of fine sediment.

A condensed section develops when the rate of relative

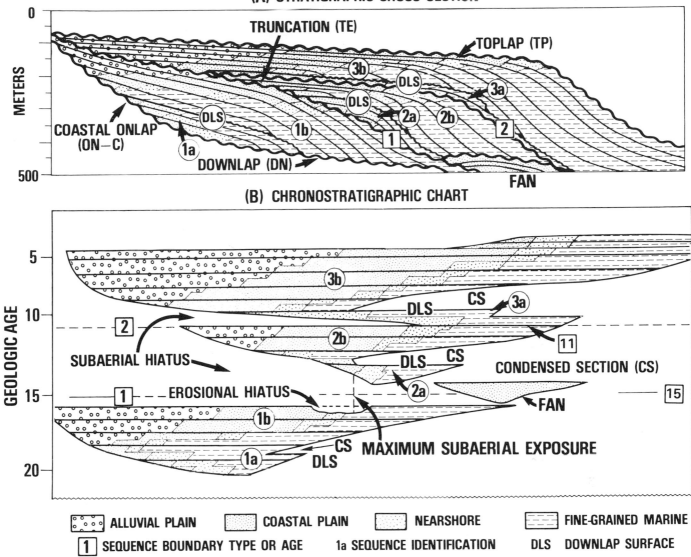

Figure 5: Diagrammatic stratigraphic cross section and chronostratigraphic chart showing the distribution of sequence boundary types, downlap surfaces (condensed sections), and facies of three idealized sequences in depth and time.

rise (eustatic rise plus subsidence) of sea level is significantly greater than the rate of accumulation, causing the depositional site to shift landward and resulting in low sedimentation rates (starved conditions) seaward of the depositional site. The marine hiatuses associated with condensed sections tend to disappear landward into essentially conformable successions developed at the site of deposition (Figures 3 and 5). The age of a condensed section within a given depositional sequence tends to be synchronous globally but may differ slightly from basin to basin with changes in rates of deposition and subsidence.

Downlap surfaces are submarine surfaces characterized seismically by a downlap over a concordant pattern and are commonly associated with a marine hiatus (Figures 4C and 5). Downlap surfaces associated with condensed sections mark the change from the end of transgression to the start of regression as the rate of eustatic rise decreases, and sediments begin to prograde out over the old starved surface.

The downlap represents deep-marine clinoform toes developed as prograding progresses seaward. Downlap surfaces may also be present above submarine fans.

Coastal onlap refers to the progressive landward encroachment of the coastal (littoral, coastal plain, or alluvial plain) deposits of a given sequence (Vail et al, 1977, part 3, Figure 3 and 5). All sequences exhibit almost continuous coastal onlap throughout the time of their deposition. Periodically, there is a major downward (basinward) shift in the coastal onlap pattern at the times of Type 1 or Type 2 unconformities. This abrupt downward shift in coastal onlap gives a chart of relative changes of coastal onlap the sawtooth character observed in Figure 2. The almost continuous presence of coastal onlap indicates that most sediments are deposited during a relative rise of sea level. Either eustatic sea level is rising, or the depositional surface is subsiding, or the two are combining to create space for the accommodation of sediments. Type 1 and Type 2 uncon-

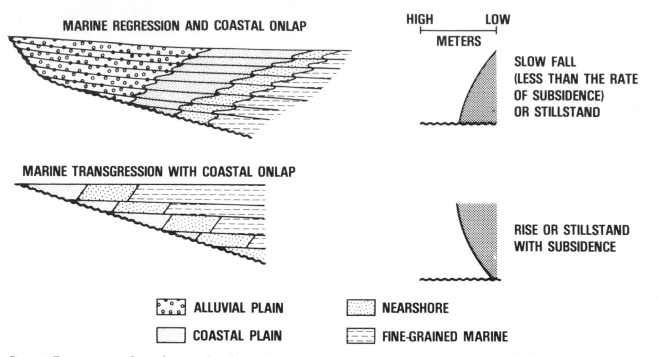

Figure 6: Diagrammatic charts showing the relation between transgression-regression and eustatic sea-level changes.

formities mark those exceptional periods of time when the rate of eustatic sea-level fall is greater than that of subsidence, producing relative falls of sea level and globally synchronous downward shifts of coastal onlap. During the periods of time involved in Type 1 relative falls, and prior to the beginning of the following relative rises, the only sediments deposited in significant volumes are submarine fans, which have no corresponding coastal onlap. Thus, the downward shift in coastal onlap appears only as a gap or discontinuity in the coastal onlap record and is instantaneous in terms of geologic time (Figures 3 and 5).

In previous papers (Vail et al, 1977; 1980), we directly equated relative changes of coastal onlap with relative changes of sea level. Pitman (1978) and M.T. Jervey (personal communication) demonstrated that changes observed in coastal onlap cannot be related directly to relative changes of sea level in the upper part of the cycle, as we have done in the past. We now refer to our sawtoothed global charts as relative changes of coastal onlap. Ideally, a chart of coastal onlap should be constructed by plotting the successive positions of the upper limit of the coastal plain (paludal and deltaic deposits). When an alluvial plain is associated with the coastal plain, the relative sea-level boundary approximates the facies change between the coastal plain and the alluvial plain (M.T. Jervey, personal communication, Figure 3). Unfortunately, seismic stratigraphic techniques do not always permit identification of this facies change between coastal and alluvial plains. Consequently, the charts developed from studies of seismic sections include data from both the coastal plain and the alluvial plain with its slightly greater alluvial dips. This greater dip is the source of the differences between the coastal onlap chart and the relative change-of-sea-level chart. We recognized this factor in Vail et al (1977, part 3), but did not take it into account when interpreting relative changes of sea level. The two charts, however, are similar. Charts of relative changes of coastal onlap versus distance typically show abrupt sawtooth shifts from widespread to restricted, whereas charts of relative changes of sea level versus distance commonly show more gradual shifts.

Transgression is defined as the landward displacement of the shoreline indicated by landward migration of the littoral facies in a given stratigraphic unit. Regression is the seaward displacement of the shoreline (Vail et al, 1977, part 3, Figures 3 and 4). Our studies indicate that transgressions and regressions are *not* necessarily globally synchronous. Other writers have reached similar conclusions (Yanshin, 1973). There is, however, a general tendency for certain periods of time to be dominated by transgressive deposits, while in others the prevalent trend is regressive. Times of maximum transgression tend to be more globally synchronous than times of regression, because of the cumulative effects of sealevel rise and subsidence. In general, transgression results when eustatic sea level rises and regression results when eustatic sea level falls at a slower rate than subsidence (Figure 6). We consider an interval transgressive if the overall tendency is to deepen upward, even though there may be a series of stacked, retrogradational regressive units.

We believe that cycles of relative changes of coastal onlap are more useful than transgressive-regressive cycles for defining stratigraphic intervals. The lows in the sawtooth pattern mark times of global unconformities that subdivide the section into depositional sequences genetically related to cycles of coastal onlap. The cycles define chronostratigraphic intervals, which include all strata deposited during the time period between the points where the unconformi-

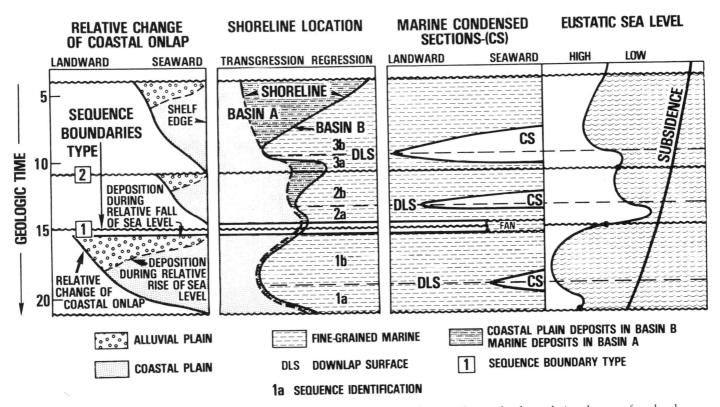

Figure 7: Summary chart showing relation of sequence-boundary types, relative changes of coastal onlap, relative changes of sea level, transgression-regression and marine condensed sections to eustatic changes of sea level and subsidence.

ties become conformable. Stratigraphic units bounded by unconformities or their equivalent conformities are defined as sequences (Vail et al, 1977, part 2). They have the advantage of being genetic units and are ideal for facies analysis and paleogeographic reconstructions. Marine condensed sections occur within sequences, and can be used to subdivide sequences.

Cycles of transgression and regression do not necessarily coincide with cycles of relative change of coastal onlap. In general, the maximum point of transgression occurs in the middle of a cycle of coastal onlap, and the maximum point of regression occurs after the downward shift of coastal onlap and thus above the associated global unconformity (Figures 3 and 5). As stated previously, this relationship does not always exist because of the variations of sediment supply. In general, transgressions are more rapid than regressions.

The stratigraphic relations among global unconformities, condensed sections, coastal onlap, relative changes of sea level, marine transgressions and regressions, subsidence, and eustatic changes of sea level are diagrammed in Figure 7. In general, as we discuss later in this report, the tectonic subsidence along most passive margins is long-term and gradually decreases in rate, because it is related to a thermal decay curve. It does not change rapidly enough to cause regional unconformities. Tectonic subsidence patterns differ from region to region, and are not globally synchronous. We therefore conclude that the many synchronous unconformities and abrupt changes in stratigraphy observed in basins globally are caused by eustatic sea-level changes, superimposed on regional tectonic and sedimentary regimes

that change at much slower rates.

Type 1 and Type 2 unconformities (Figure 7) are globally synchronous because when they are dated precisely they coincide in time with the inflection points of curves showing eustatic falls of sea level (Vail and Todd, 1981; M.T. Jervey, personal communication). An inflection point also marks the time of downward shift of coastal onlap. Global cycles of coastal onlap are chronostratigraphic intervals that can be dated precisely because they include all the rocks deposited during the interval of time between the inflection points of eustatic falls of sea level.

Truncation below unconformities caused by erosion during a relative fall of sea level has a very different pattern depending on whether or not tectonic activity (folding, faulting, uplift) is taking place during the time of the fall. If no tectonic activity other than subsidence is taking place, the truncation should be greatest at the shelf edge and should progress shelfward by headward erosion and downcutting of river valleys (Figure 8a). In general, valley and canyon truncation dies out landward away from the shelf edge, the overlying and underlying strata are generally parallel, and very detailed stratigraphic correlation is necessary to identify the truncation and onlap patterns associated with the canyon and valley fills. Farther shelfward, as truncation effects die out, the unconformities must be identified by coastal onlap and related facies patterns (Figure 8a). If folding, faulting, uplift, or large differential subsidence is taking place during the relative fall, erosion of the tilted beds produces an easily identifiable truncation unconformity (Figure 8b). In general, most global unconformities are not associated with tectonic activity.

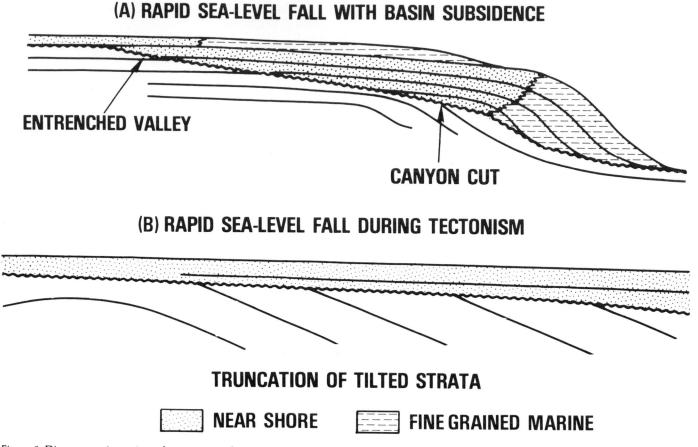

Figure 8: Diagrammatic sections showing stratal patterns associated with unconformities. Section A shows a typical headward erosion pattern associated with a Type 1 unconformity formed during basin subsidence. Section B shows the expression of a Type 1 unconformity formed during a period of tectonic uplift and folding.

The causes of global eustatic changes of sea level in the Jurassic (as shown on Figure 2) are unknown. Most workers agree, however, that the long-term changes in sea level, such as the overall rise in the Jurassic, result from changes in volume of the ocean basins due to such processes as changes in rates of sea-floor spreading (Hallam, 1963; Pitman, 1978). The causes of rapid changes of sea level (greater than 10 cm/1000 years), however, are problematical. The only known mechanism that produces rates of this magnitude is glaciation (Donovan et al, 1979), but conclusive evidence is lacking for glaciers of sufficient size to have caused the large and rapid changes of sea level we observe in the Jurassic and other Phanerozoic strata.

In summary, eustatic changes of sea level are the primary cause of major abrupt vertical changes in stratigraphy at the time of deposition and they are major factors controlling lateral changes. Eustatic changes of sea level produce inter-regional or "global" unconformities not only for continental margins, but for interior basins and deep-oceanic sites. As such, they provide a means to make exact correlations from the deep sea basins across the shelf into interior basins. Figure 2 shows our best estimate of eustatic sea-level changes during the Jurassic supercycle and how they relate to the ages of Type 1 and Type 2 regional unconformities, relative changes of coastal onlap, and age and relative magnitude of condensed sections. Transgressions and regressions are not included because of their global variability. Our studies

show that Type 1 and Type 2 regional unconformities are globally synchronous when dated precisely at the age they become conformable. The age of conformity corresponds to the time of the most rapid rate of eustatic fall of sea level (inflection point on the sea-level curve), as shown in Figures 4 and 7 (M. T. Jervey, personal communication).

Ages of condensed sections may differ slightly with changes in rate of deposition and subsidence. Downward shifts of coastal onlap, however, are synchronous globally, because they occur at inflection points on the sea-level curve at the times of highest rates of eustatic fall. The inflection points also mark the most accurate dates to establish the synchrony of the global unconformities.

When the coastal onlap patterns of restricted and widespread sequences of the same age are compared in several areas globally, the patterns tend to be similar in areas where the subsidence pattern is continuous and uninterrupted over the interval studied. If stability or uplift occurs, the onlap patterns are distorted. Such distorted patterns are useful for indicating periods of tectonism. However, the ages of the sequence boundaries remain constant in the distorted sections. Shoreline transgressions and regressions are the most variable of the parameters discussed; however, sequences tend to be mostly transgressive in the lower part of the coastal onlap cycle below the downlap surface, and mostly regressive in the upper part of the cycle above the downlap surface.

STRATIGRAPHIC NOMENCLATURE

A major operational problem is how to name and refer to unconformities (sequence boundaries) and sequences once they have been identified. Our experience is that the sequence boundaries can be marked with colored pencils to facilitate identification as a grid of seismic or well and outcrop sections is correlated. After the areal continuity of the sequence boundaries and condensed sections has been verified by seismic loop ties or well-log correlation, the geological ages of the strata are determined by means of paleontology. At this point, we believe that information is conveyed precisely by naming the unconformities after the oldest strata known to occur above the unconformity (e.g., the latest Berriasian unconformity) and by informally naming the sequences between the unconformities with symbols representative of age (e.g., Kl. 3). If precise age dating cannot be achieved because a significant hiatus is present throughout the study area or because the paleontologic data are poor or lacking, the age of the closest global unconformity is used and the error range is indicated on a summary chronostratigraphic chart. For reference purposes we designate unconformities on seismic or well-log sections by the age of their correlative conformities in million of years (m.y.). It is important to note which time scale is used, since the age in millions of years of a particular faunal zone may vary from one scale to another and may change in the future. The Jurassic global chart (Figure 2 and Table 1) shows the symbols we use for the global cycles and the ages of the global unconformities we have identified in the Jurassic supercycle. The numerical age is also useful for input into quantified stratigraphic studies, such as geohistory analysis referred to later in this paper.

Formal formation names for stratigraphic units serve a very useful descriptive purpose, but remember that formations are lithostratigraphic units and may transgress time. For proper facies and paleogeographic interpretation, the lithostratigraphy must be placed in the proper chronostratigraphic framework or serious interpretation errors will result. We also believe that naming unconformities after orogenic events is misleading. In the North Sea, for example, two major unconformities are called the Middle and Late Cimmerian. We prefer to designate the respective unconformities as latest Bathonian (158) and latest Berriasian (132), believing that more accurate stratigraphic information is thereby provided. Our studies indicate that these particular unconformities were caused by falls of eustatic sea level superimposed on longer term tectonic movements, and were not caused directly by orogenic events.

PROCEDURE FOR INTERPRETATION OF UNCONFORMITIES, CHRONOSTRATIGRAPHY, COASTAL ONLAP, TECTONIC SUBSIDENCE, AND EUSTATIC CHANGES OF SEA LEVEL

The procedure used for charting the Jurassic unconformities, chronostratigraphy, coastal onlap, tectonic subsidence, and eustatic changes of sea level consists of the following five steps:

1) Developing chronostratigraphic framework by determining sequence boundaries for a study region.

2) Dating sequence boundaries and making a representative chronostratigraphic chart for a study region.

3) Making regional chart of relative changes of coastal onlap for study regions and combining them to make a global onlap chart.

4) Determining tectonic subsidence and long-term eustatic changes of sea level with geohistory analysis corrected for loading and compaction.

5) Estimating short term eustatic changes of sea level.

Developing Chronostratigraphic Framework By Determining Sequence Boundaries for a Study Region

Sequence boundaries are interpreted most accurately by using a combination of reflection seismic data, well control, and outcrop information. Figure 5A is a diagrammatic example of a geologic cross section showing chronostratigraphic surfaces, unconformities, and facies. To interpret unconformities and sea-level changes, it is critical that we make stratal correlations in sufficient detail to recognize truncation, lapout, prograding, and other patterns associated with different depositional environments. Seismic data tied to well control or closely spaced well logs with detailed marker-bed correlations are ideal for this purpose.

Figure 4 illustrates the three types of surfaces discussed previously: Type 1 and Type 2 unconformities, and condensed sections. In general, we find that if we can trace these types of surfaces over a regional grid of seismic and well data, they will correspond to one of the global unconformities or condensed sections listed in Table I. The 15 m.y. unconformity is a Type 1; the 11 m.y. unconformity is a Type 2. Condensed sections are present at 9, 13, and 18 m.y. (Figure 5B).

The condensed sections (CS) or downlap surfaces (DLS) shown schematically within sequences 1, 2, and 3 (Figure 5) overlie transgressive deposits and thus occur within a sequence between the bounding global unconformities. In deep-water depositional sites such as ocean basins, such condensed intervals or downlap surfaces commonly directly overlie both types of unconformities in areas where lowstand deposits are absent.

We have found that unconformities and condensed sections can be recognized best on reflection seismic sections by identifying onlap, downlap, and truncation patterns and dating them with well control (Vail et al, 1977, parts 2 and 6; 1980). Because seismic reflections are produced largely by stratal surfaces and unconformities with significant velocity-density (impedance) contrasts (Vail et al, 1977, part 5), reflection patterns portray stratal configurations within the limits of seismic resolution.

Unconformities characterized by large hiatuses, such as those along basin margins, commonly extend laterally into areas where deposition has been more nearly continuous and where hiatuses gradually decrease as the unconformities approach conformity (Figure 5). A conformable surface is a chronostratigraphic horizon and must be traced with its correlative unconformity to define completely the sequence it bounds. In this way, the three-dimensional sequence framework bounded by unconformities and their correlative conformities is completely defined for subsequent seismic facies and structural analysis (Vail et al, 1977, parts 2 and 7). As shown on Figure 4, Type 1 and Type 2 unconformities

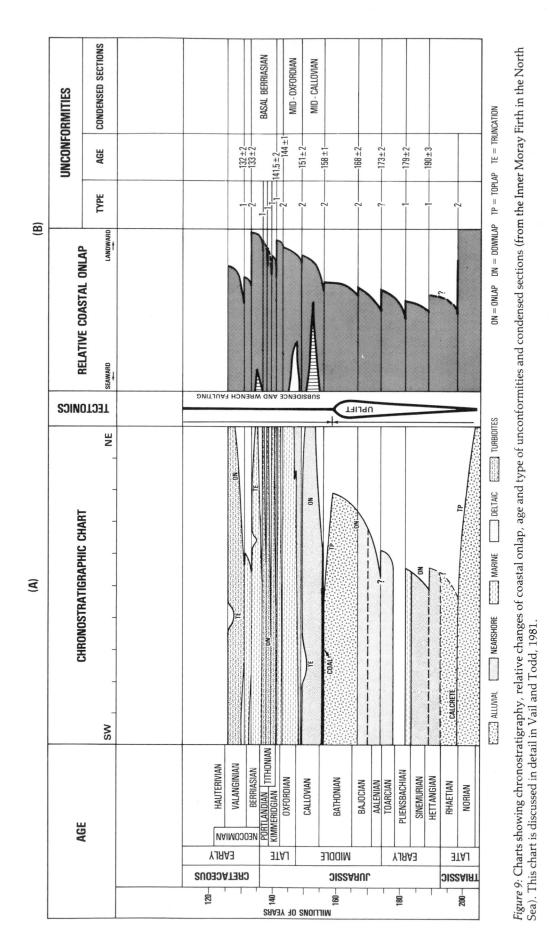

Figure 9: Charts showing chronostratigraphy, relative changes of coastal onlap, age and type of unconformities and condensed sections (from the Inner Moray Firth in the North Sea). This chart is discussed in detail in Vail and Todd, 1981.

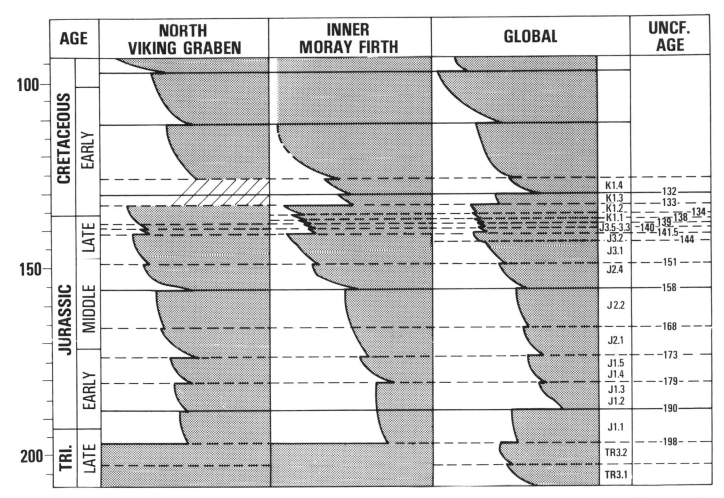

Figure 10: Charts comparing relative changes of coastal onlap from the North Viking Graben and Inner Moray Firth, in the North Sea with the global chart.

become most conformable at the point of downward shift of coastal onlap. Locally, Type 1 unconformities have the least hiatus on the upper slope landward of the submarine fan. We give the submarine fan the same age as the unconformity, since we believe it formed during the most rapid eustatic fall of sea level, but we carry the base of the fan as our principal mapping surface.

Dating Depositional Sequence Boundaries and Making a Representative Chronostratigraphic Chart for a Study Region

An unconformity is dated most accurately by establishing the ages of overlying and underlying strata at points along the unconformity where the hiatus is least. The ideal point is where the surface becomes a conformity. In many cases, however, conformity is not reached and the point of minimum hiatus is used. Dating is accomplished using biostratigraphy from wells and outcrop, and comparing with local and global onlap charts.

Figure 5B is an example of how a chronostratigraphic chart is made from a geologic cross section. The chart portrays the sequences in geological time, and thus emphasizes the genetic nature of the facies within the sequences and the abrupt discontinuities between sequences. Chronostrati-

graphic charts also emphasize the periods of time when no sediments are present because of nondeposition or erosion. Figure 9 is a chronostratigraphic chart for the Inner Moray Firth seismic section shown on Figure 1. This chart is discussed in detail by Vail and Todd (1981).

Making Regional Charts of Relative Changes of Coastal Onlap for Study Regions and Combining Them to Make a Global Onlap Chart

Charts of relative changes of coastal onlap are derived from geological cross sections or interpreted seismic sections (such as Figure 1) and from the chronostratigraphic charts (such as Figure 9). The procedure for constructing charts of relative changes of coastal onlap is discussed in Vail et al (1977, parts 3 and 4). Figure 10 shows how we combine charts of relative changes of coastal onlap to produce the global chart. In this example, unconformity age and relative patterns of coastal onlap are compared among the North Viking Graben, the Inner Moray Firth, and the global chart (see Vail and Todd, 1981). Unconformities of the same age and type are added to the global chart when they have been identified in sedimentary basins on at least three different continents. Patterns of relative changes of coastal onlap are included on the global chart only from areas undergoing

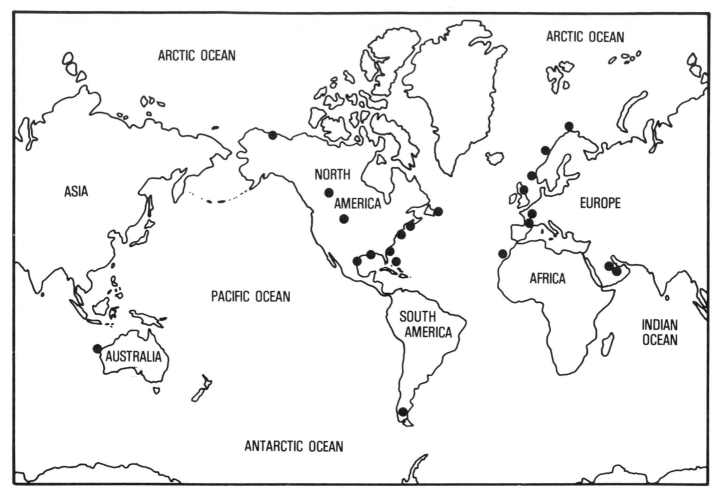

Figure 11: Map showing distribution of seismic stratigraphic and/or well studies of the Jurassic supercycle.

uniform subsidence, because the coastal onlap pattern in a given area is influenced by a variable tectonic and subsidence history. Figure 11 shows the areas in which we have made seismic stratigraphic and well-log studies used in the construction of the global Jurassic coastal onlap chart (Figure 2).

Determining Tectonic Subsidence and Long-Term Eustatic Changes of Sea Level with Geohistory Analysis (Van Hinte, 1978) Corrected for Loading and Compaction (Hardenbol, Vail, Ferrer, 1981)

Sea-floor subsidence, eustatic changes of sea level, and sediment supply are the principal interacting variables in a marine basin. This interaction is recorded in the sedimentary section in the form of sequences filling the basin. Geohistory analysis provides a technique for depicting the interaction of eustatic changes of sea level and sea-floor (tectonic) subsidence and relating these factors to the sedimentary record. This technique separates the total subsidence of an area into individual components due to the effects of sediment and water-column loading, compaction, and tectonic activity (e.g., thermal cooling, basement faulting, and tectonic loading).

Also separated are the apparent vertical movement effects of eustatic changes of sea level. By quantification of the individual effects of sediment compaction, sediment loading, and subsidence, the magnitudes of the long-term eustatic changes can be established.

An example of this geohistory analysis from offshore northwestern Africa is described in Hardenbol, Vail, and Ferrer (1981). It shows that in the early Cenomanian, sea level could have been nearly 300 m (984 ft) higher than present sea level.

The eustatic sea-level values obtained from geohistory analysis show close agreement with previous results obtained by different methods by Vail et al (1977, Figure 8, part 4), Hayes and Pitman (1973), Pitman (1978), and Kominz (this volume). A significant difference exists, however, with the much smaller values obtained by Watts and Steckler (1979), who used a similar method based on the subsidence history of the Atlantic margin of the North American east coast. Possibly their assumption of a single thermal contraction event for the thermo-tectonic subsidence of the Atlantic margin is not valid. Thermo-tectonic subsidence histories for Georges Bank and Baltimore Canyon C.O.S.T. wells suggest increased tectonic activity in the Early Cretaceous and middle Tertiary.

Estimating Short-Term Changes of Eustatic Sea Level

The pattern of short-term eustatic changes of sea level is derived from a combination of several data sets (Figure 12).

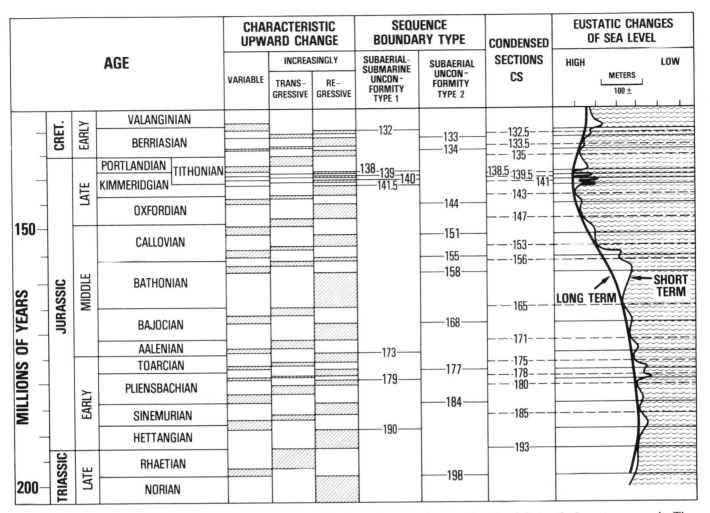

Figure 12: Diagram illustrating methodology used to construct the chart of eustatic changes of sea level during the Jurassic supercycle. The ages of Type 1 and Type 2 unconformities and condensed sections are plotted along with the intervals of time when the Jurassic sediments tend to be increasingly transgressive or regressive upward.

The ages of Type 1 and Type 2 unconformities and condensed sections are plotted, along with the intervals of time when Jurassic sediments tend to be increasingly transgressive or regressive upward. From these are derived the direction and slope of the short-term eustatic curve based on the patterns shown in Figures 4 and 6. The final eustatic sea-level curve is then derived from the configuration of the short-term changes superimposed upon the independently determined envelope of the long-term eustatic changes.

The absolute magnitudes of the rapid falls of sea level are difficult to calculate, but studies indicate that the 132-million-year and 140-million-year drops are the greatest and may approximate the magnitudes of the Pleistocene changes. The other changes appear to be less, as shown on Figures 2 and 12. These magnitudes are estimated from the downward shift of coastal onlap below the shelf edge and from the depth of subaerial erosion obtained from geohistory reconstructions.

CONCLUSIONS

Our experience indicates that analysis of stratigraphy based on depositional sequences is the most accurate approach for interpreting geologic age, depositional environment, lithofacies, and paleogeography. Not only does the approach subdivide strata into genetic depositional units, but it places them in a global context controlled by short-term variations in eustatic sea level. Sequences can be readily recognized from seismic interpretation, well-log and outcrop correlations, and facies studies. Interpretations based on lithofacies and biostratigraphy could be misleading unless they are place within a context of detailed stratal chronostratigraphic correlations. This paper illustrates how sequence concepts may be applied to the Jurassic.

ACKNOWLEDGMENT

This paper represents the joint effort of many Exxon scientists who contributed to the development, documentation, and testing of the concepts. We are especially indebted to R. M. Mitchum and M. A. Uliana, whose critical review of the manuscript provided many important improvements. In addition, we thank Exxon Production Research for permission to release this information.

REFERENCES CITED

Bates, R.E., and J.A. Jackson, 1980, eds., Glossary of geology, second edition: Falls Church, Virginia, American Geological Institute, 794 p.

Donovan, D.T., et al, 1979, Rates of marine transgressions and regressions: Journal of the Geological Society of London, v. 136, p. 187-193.

Hallam, A., 1963, Eustatic control of major cyclic changes in Jurassic sedimentation: Geological Magazine, v. 100, p. 444-450.

Hardenbol, J., P. R. Vail, and J. Ferrer, 1981, Interpretating paleoenvironments, subsidence history and sea-level changes of passive margins from seismic and biostratigraphy: Oceanologica Acta No. SP, p. 3-44.

Hayes, J.D., W.C. Pitman, III, 1973, Lithospheric plate motion, sea-level changes and climatic and ecological consequences: Nature, v. 246, p. 18-22.

Kominz, M.A., in press, Oceanic ridge volumes and sea level change - an error analysis, *in* J.S. Schlee ed., Interregional unconformities: AAPG Memoir 36, this volume.

Pitman, W.C., 1978, Relationship between sea-level change and stratigraphic sequences: Geological Society of American Bulletin, v. 89, p. 1389-1403.

Vail, P.R., and J. Hardenbol, 1979, Sea-level changes during the Tertiary: Oceanus, v. 22, p. 71-79.

——, and R.M. Mitchum, 1979, Global cycles and sea-level change and their role in exploration: Bucharest, Romania, Proceedings of the Tenth World Petroleum Congress, v. 2, Exploration Supply and Demand, p. 95-104.

——, and R.G. Todd, 1981, North Sea Jurassic unconformities, chronostratigraphy and sea-level changes from seismic stratigraphy: Proceedings of the Petroleum Geology Continental Shelf, Northwest Europe, p. 216-235.

——, et al, 1980, Unconformities of the North Atlantic: Philosophical Transactions of the Royal Society of London, A 294, p. 137-155.

——, et al, 1977, Seismic stratigraphy and global changes of sea level, *in* Seismic stratigraphy—applications to hydrocarbon exploration; AAPG Memoir 26, p. 49-212.

Van Hinte, J.E., 1976a, A Jurassic time scale: AAPG Bulletin, v. 60, p. 489-497.

——, 1976b, A Cretaceous time scale: AAPG Bulletin, v. 60, p. 498-516.

——, 1978, Geohistory analysis - application of micropaleontology in exploration geology: AAPG Bulletin, v. 62, p. 201-222.

Watts, A.B., and M.S. Steckler, 1979, Subsidence and eustacy at the continental margin of eastern North America, *in* M. Talwani, W.F. Hay, and W.B.F. Ryan, eds., Deep drilling results in the Atlantic ocean; continental margins and paleoenvironment: American Geophysical Union, Maurice Ewing Series, No. 3, p. 218-234.

Yanshin, A., 1974, The so-called global transgressions and regressions: International Geology Review, v. 16, p. 617-646 (English translation of Russian article).

Cenozoic Regional Erosion of the Abyssal Sea Floor Off South Africa

Brian E. Tucholke
Woods Hole Oceanographic Institution
Woods Hole, Massachusetts

Robert W. Embley
National Oceanic and Atmospheric Administration
Rockville, Maryland

The Cape, Agulhas, and Mozambique basins off South Africa have a well-defined and mostly continuous erosional zone along their perimeters between 4 and 5 km (2.5 and 3.1 mi) water depth. This zone lies beneath a deep boundary current of Antarctic Bottom Water. However, current speeds are generally less than 15 to 20 cm/sec (5.9 to 7.9 in/sec) and part of the erosional zone is armored by authigenic manganese deposits, so that only limited erosion is presently occurring. Erosion and corrosion of sea-floor sediments by abyssal currents probably are in dynamic equilibrium with sediment supply. The present erosional zone is largely a relict feature inherited from late Miocene time when strongly increased glaciation of West Antarctica produced large volumes of bottom water that scoured the sea floor. A deeper unconformity dating to the early Oligocene marks the onset of significant abyssal circulation in the basins, and current-controlled deposition of sediments is well defined above this unconformity. In contrast, an underlying basal Eocene unconformity shows no marked effects of control by abyssal circulation, although erosion/corrosion by weak, deep currents could have occurred. The unconformity formed primarily because of reduced sediment supply, caused by elevated sea level and probably low productivity in surface waters.

INTRODUCTION

The rock-stratigraphic and biostratigraphic records recovered by deep-sea drilling demonstrate conclusively that most areas of the ocean floor have experienced significant episodes of erosion or nondeposition. Although the resultant hiatuses thus far defined tend to be concentrated within specific intervals of geologic time (Moore et al, 1978), the nature, origin, and relative significance in time and space of the processes responsible are poorly understood. Bottom currents widely influence sedimentation patterns in the world oceans and they are the principal agents responsible for physical sea-floor erosion. In addition, deep currents probably are important in accelerating sediment removal by chemical corrosion if biogenous sediments cover the sea floor. Seismic reflection profiles provide an extensive data base for evaluating current influences on sedimentation in the ocean basins. Erosional unconformities are frequently identified by truncation of reflectors, or nondepositional zones can be detected by areas of thinned sediments. Regrettably, many disconformities formed by erosion and most formed by nondeposition or by chemical corrosion are enclosed by parallel reflectors. These can be detected only by direct sampling with piston cores or by drilling.

The southeast Atlantic Ocean and southwest Indian Ocean are unique in that, practically speaking, an uncon-

formity still is being developed or maintained over sizeable sea-floor regions in the Cape, Agulhas, and Mozambique basins (Figure 1). The unconformity results from a combination of circumstances, including attenuated supply of continental detritus, locally limited biogenic productivity in surface waters, and erosion or maintenance of nondepositional conditions by the combined effects of bottom currents and chemical corrosion by bottom waters. Physical erosion of sediments is generally recognizable by truncation of sub-bottom reflectors in both high frequency (3.5kHz) and low frequency (10 to 150Hz) seismic profiles. The observed erosion forms a band circumscribing the basins. The erosional zone is best developed between water depths of about 4 and 5 km (2.5 and 3.1 mi), and it lies beneath the principal flow pathways presently followed by Antarctic Bottom Water. This report describes the present sedimentary framework in the Cape, Agulhas, and Mozambique basins and its relation to bottom-water flow, and it then considers past effects of bottom circulation on the geologic history of the basins.

GEOLOGIC SETTING

Sediment compositions and distribution patterns across the basins and rises off South Africa are a study in contrasts. The Polar Front Zone lies at about 49°S, and rapidly accu-

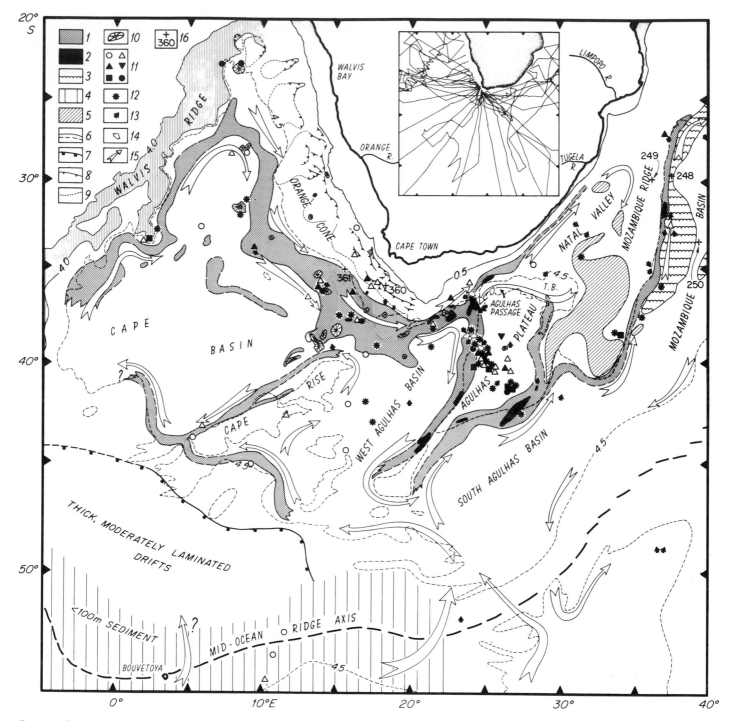

Figure 1: Generalized sedimentary framework of the basins around South Africa and its relation to bottom circulation. 1-Core of circumbasin erosional zone (significant erosion can also be found above and below this core zone); 2-basement exposed by current erosion; 3-sediment-wave field; 4-zone of thin sediment along mid-ocean ridge axis and beneath Antarctic Circumpolar Current; 5-thick sediment drifts with weak acoustic laminae; 6-generalized bathymetric contours as labelled, 4,500 m (14,764 ft) is dashed (modified from Simpson, 1974); 7-limit of thick moderately laminated drifts of diatomaceous sediment extending north of Polar Front Zone; 8-glide plane scars at head of slumps and slides on continental rise (Summerhayes, Bornhold, and Embley, 1979; Dingle, 1980; Embley and Morley, 1980); 9-approximate seaward limit of slumps and slides; 10-seamounts; 11-piston cores of pre-Quaternary sediment outcrops (from L to R, top to bottom: Pliocene, Miocene, Oligocene, Eocene, Paleocene, Cretaceous); 12-manganese nodules/pavement observed in bottom photographs; 13-current direction from bottom photographs; 14-direct current measurements in Agulhas Passage and along western Agulhas Plateau (see Figure 2); 15-flow of AABW inferred from bottom-water potential temperature (Figure 2); 16-DSDP drillsites. Compiled from unpublished data, above references, and from Bornhold and Summerhayes (1977), Emery et al (1975), Kolla et al (1976b, 1980), Ledbetter and Ciesielski (1982), Saito, Burckle, and Hays (1974), and Tucholke and Carpenter (1977). Seismic track control shown in inset and in Figure 3.

mulating diatomaceous oozes blanket the sea floor from a point at least 3° north of this oceanographic boundary (South Agulhas Basin and southern West Agulhas Basin) southward to about 60°S (Lisitzin, 1972; Defelice and Wise, 1981). Over the crest of the Atlantic-Indian Mid-Ocean Ridge, scour beneath the Antarctic circumpolar current appears to remove much of this sediment; however along the north flank of the ridge, thick (100's m) acoustically laminated oozes form drifts beneath the northern boundary of the current (Figure 1).

North of the Polar Front Zone, sediments are primarily mixtures of terrigenous detritus and biogenic carbonate; the latter is prevalent on ridges, seamounts, and plateaus that are both shallow and isolated from terrigenous sources. For example, Kolla, Be, and Biscaye (1976a) reported that calcareous oozes on the crest of the Agulhas Plateau and Mozambique Ridge contain more than 85 percent calcium carbonate. Dissolution increases and calcium carbonate content decreases with depth, so that marls (25 to 50 percent $CaCo_3$) are present in adjoining areas. In the deep West Agulhas and South Agulhas basins, dissolution has limited carbonate content to about 10 percent (Kolla, Be, and Biscaye, 1976a). Carbonate content of sediments in the Mozambique Basin also is less than 10 percent, but this is largely because of dilution by terrigenous detritus derived from the Zambesi River at 19°S and numerous smaller rivers along the southeastern African margin (Milliman and Meade, 1983; Kolla et al, 1980).

A similar pattern prevails in the Cape Basin, but it is somewhat modified by the presence of the Benguela Current along the southwest African margin. High biologic productivity results from upwelling in this area; biogenic carbonate and silica are deposited on parts of the shelf and carbonate-rich sediments are deposited farther offshore (Embley and Morley, 1980, and references therein). The carbonate content of sediments decreases with increasing water depth; chalks and oozes are present along the upper continental slope, on the Walvis Ridge and on the Cape Rise, but marls and clays occur at intermediate water depths along the continental rise. Dissolution has limited the carbonate content of clayey sediments in the deepest, central part of the Cape Basin to less than 10 percent (Biscaye, Kolla, and Turekian, 1976).

The Orange River provides almost all of the terrigenous sediment available to the Cape Basin, but the historical sediment load of the river is not much more than 35×10^6 metric tons/year (Milliman and Meade, 1983). It also appears that little of this material has reached the deep sea during Cenozoic time, as illustrated by three observations. First, the Cape Basin is the only large deep-sea basin that is adjacent to a major landmass and has no significant abyssal plain. Second, observed recent sediment accumulation rates below about 4,800 m (15,748 ft) in the Cape Basin are only about 0.1 to 0.4 mm/1000 yrs (0.004 to 0.015 in/1000 yrs) (Embley and Morley, 1980). Normal Atlantic deep-water accumulation rates are one to two orders of magnitude higher (for example, Lisitzin, 1972), so for all practical purposes the present Cape Basin sea floor is a surface of unconformity. Finally, seismic mapping of sediment thickness above Horizon D, of early Eocene age (see Figure 5), shows that there is no significant post-Paleocene depocenter on the Cape Rise seaward of the Orange River; the basic morphologic outline of the Orange Cone (Figure 1) was developed below Horizon D in the Cretaceous and Paleocene (Emery et al, 1975; Bolli and Ryan, 1978).

The foregoing summary outlines major factors presently controlling the delivery of terrigenous and biogenic sediments to the sea floor around South Africa. The remainder of this paper deals with the modern and ancient effects of the abyssal circulation, which has significantly modified sediment distribution patterns and, to some degree, sediment composition.

PRESENT DEEP CIRCULATION

Relatively little data from direct current measurements or bottom photographs is presently available to document the flow of abyssal currents off southern Africa. However, a reasonable depiction of the deep flow can be derived from these facts: 1) bottom currents will be deflected to the left in the southern hemisphere by the Coriolis force; 2) abyssal flows therefore tend to flow along basin margins and to follow bathymetric contours; and, 3) the distribution of bottom-water potential temperature can be used both as a tracer of water masses (the "core method") and as an indicator of sloping isotherms, and thus density gradients, that drive the flow.

Antarctic Bottom Water (AABW) presently reaching the basins off South America and South Africa flows principally from the Weddell Sea (Carmack and Foster, 1975; Reid, Nowlin, and Patzert, 1977) but includes an uncertain volume of water from the Enderby Land/Prydz Bay coast of Antarctica (Gordon, 1974; Jacobs and Georgi, 1977). AABW flows north into the basins around southern Africa through deep fracture-zone conduits in the Atlantic-Indian Mid-Ocean Ridge (Figures 1 and 2). Potential-temperature isotherms suggest that the principal passages are between 20° and 30°E, but some AABW may also flow through fracture-zone troughs at 5°E and at 35°E (Figure 2; Le Pichon, 1960; Wyrtki, 1971; Kolla et al, 1976b). A clockwise circulation gyre probably exists within each of the basins to the north (Cape and West Agulhas basins, Transkei Basin-Natal Valley, and South Agulhas-Mozambique Basin), and these gyres are interconnected at their eastern and/or western margins. The mean-flow arrows depicting this general circulation pattern in Figure 2 are necessarily speculative. However, they do agree with available data on distribution of potential temperatures in bottom water and with most direct indications of currents.

The basin most remote from the source of AABW is the Cape Basin. The Walvis Ridge along the northern edge of this basin has long been known as a major barrier to the northward flow of AABW (Supan, 1899; Schott, 1902; Wust, 1936). The difference in bottom-water potential temperatures (Θ) in the Cape and Angola basins on opposite sides of the Walvis Ridge is about 1.35°C (0.65°Θ vs. 2.0°Θ). Only a small gap at 36°S, 7°W, is deep enough to allow minor flow of AABW as cold as 1.0° Θ into the Angola Basin (Connary and Ewing, 1974). Thus, most AABW entering the Cape Basin is recirculated eastward along Walvis Ridge and southward along the continental

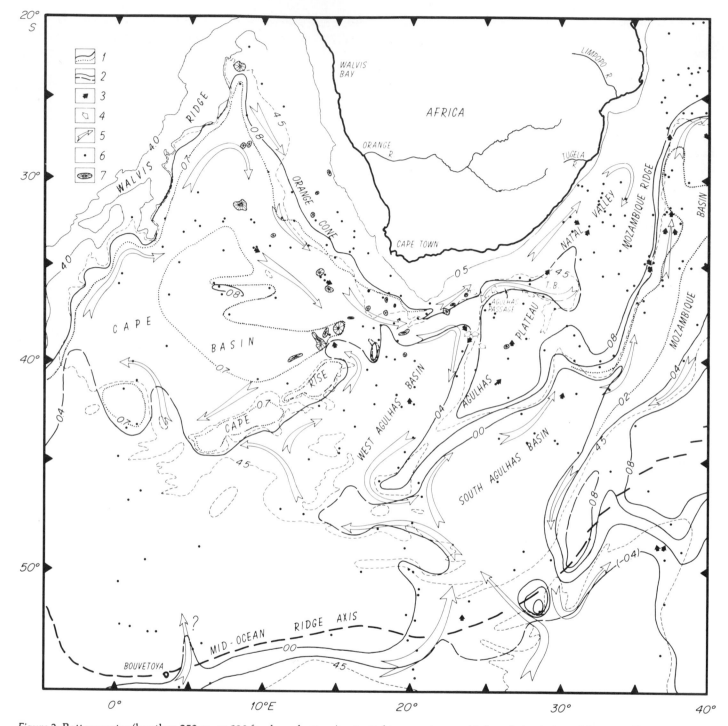

Figure 2: Bottom water (less than 250 m, or 820 ft, above bottom) potential temperature and inferred circulation of Antarctic Bottom Water. 1-potential temperature isotherms (0.4°C interval, supplementary isotherms dotted); 2-generalized bathymetric contours (4,500 m, or 14,764 ft, is dashed); 3-current directions from bottom photographs; 4-direct current measurements in Agulhas Passage (262- and 356-day records; Camden-Smith et al, 1981) and along west flank of Agulhas Plateau (24-day records; Tucholke, 1978, and unpublished); 5-flow of AABW inferred from potential temperature; 6-location of potential-temperature measurements; 7-seamounts. Compiled from Lamont-Doherty Geological Observatory thermograd traces, NODC temperature profiles, and from Fuglister (1960), Jacobs and Georgi (1977) and Kolla et al (1980).

rise into the West Agulhas Basin. The presence of this large clockwise gyre is clearly indicated by the concave-upward potential-temperature isotherms across the center of the Cape Basin (Figure 2) and by dynamical calculations (Wust, 1955).

The coldest water in the Cape Basin is about 0.65°Θ and occurs along the basin margins. At comparable latitudes and water depths in the Mozambique Basin, potential temperatures are about 0.5°C colder. The warmer water in the Cape Basin probably is the result of two factors. First, the

AABW must follow a long flow path along the periphery of deep gyres and around topographic barriers (for example, the Cape Rise) before it reaches the Cape Basin. This facilitates lateral mixing with adjacent older and warmer bottom water. Second, the North Atlantic Deep Water, with which the AABW becomes mixed at its upper boundary, is warmest over the Cape Basin, and its southeastward advection into the Indian Ocean generally opposes the northward flow of AABW (Le Pichon, 1960; Wyrtki, 1971). Vertical mixing in the shear zone between these water masses also warms the AABW that circulates through the Cape Basin.

The clockwise circulation around the West Agulhas Basin is documented only along the west flank of the Agulhas Plateau where bottom photographs and short-term (24-day) current measurements show a generally south to southwest flow between 3,900 and 4,900 m (12,795 and 16,076 ft) (Figure 2; Tucholke and Carpenter, 1977; Tucholke, 1978). In the Agulhas Passage at the northwest corner of the Agulhas Plateau, Camden-Smith et al (1981) reported long-term measurements of northeastward and westward currents on the north and south sides of the passage, respectively. These measurements agree with the concept that a clockwise gyre is present in the adjacent Transkei Basin (T.B. in Figure 2) and Natal Valley, since AABW enters and exits through Agulhas Passage. Some exchange of bottom water between the Transkei and South Agulhas basins may also occur on the east side of Agulhas Plateau.

In the South Agulhas Basin, a strong northeasterly flow across the southern margin of the Agulhas Plateau is indicated by an R/V *Conrad* hydrographic section (Jacobs and Georgi, 1977). This water is colder, fresher, and more oxygen- and nutrient-rich than the AABW in the West Agulhas Basin, indicating advection to this area of an AABW component derived directly from the source to the south (Figure 2). This flow continues north along the western margin of the Mozambique Basin, as shown both in vertical sections of potential-temperature isotherms and in bottom photographs (Kolla et al, 1976b; Embley and Tucholke, 1976; Kolla et al, 1980). Farther north the flow is poorly documented, but it probably turns east at the head of the Mozambique Basin near 25°S and recirculates southward along the basin's eastern margin. The Madagascar Ridge at 43°E longitude bounds the basin on the east, and there is no known transport of deep water from the Mozambique Basin eastward across this ridge (Le Pichon, 1960; Wyrtki, 1971; Warren, 1974; Kolla et al, 1976b). Salinity and temperature distribution across the South Agulhas Basin are known from early *Discovery* data (Deacon, 1937) and from a hydrographic section taken on R/V *Conrad* in 1974; these data indicated to Jacobs and Georgi (1977) that an elongate, clockwise gyre may be a permanent feature in the South Agulhas Basin. We suggest that this gyre also extends northward to include the Mozambique Basin (Figure 2).

Current speeds are even less well documented than the general flow directions of the bottom circulation. Bornhold and Summerhayes (1977) studied bottom photographs from the foot of Walvis Ridge at 25°S, 8°E in the Cape Basin, and concluded from the absence of significant sea-floor lineations that current speeds there were less than 10 cm/sec (3.9 in/sec). Instantaneous (photographic) measurements of current speed on the continental rise southwest of Cape Town

are generally less than 15 cm/sec (5.9 in/sec) (Connary, 1972; R.W. Embley, unpublished data). These values generally agree with speeds less than 10 cm/sec (3.9 in/sec) calculated from dynamic topography in the Cape Basin (Wust, 1955). Along the northern edge of the Agulhas Passage, Camden-Smith et al (1981) measured speeds of 2 to 49 cm/sec (.8 to 19.3 in/sec) (365-day current-meter record), and near a topographically steeper boundary along the southern edge of the passage they measured speeds of 2 to 79 cm/sec (.8 to 31.1 in/sec) (262-day record). On the deep western flank of the Agulhas Plateau, Tucholke (1978; and unpublished) measured speeds of 2 to 15 cm/sec (.8 to 5.9 in/sec) in 24-day records. Over the crest of the Agulhas Plateau bottom photographs show strongly rippled foraminiferal ooze (Tucholke and Carpenter, 1977). These ripples suggest speeds greater than 15 to 20 cm/sec (5.9 to 7.9 in/sec) because laboratory experiments show that this is the threshold of erosion for calcareous ooze (Southard, Young, and Hollister, 1971). However, the Agulhas crestal observations are quite shallow (2.8 to 3.2 km, or 1.7 to 2.0 mi), and they probably show the influence of the shallow, high-velocity Agulhas Current and its return flow (Wyrtki, 1971). The Agulhas Current appears to be largely independent of the abyssal contour-following flow. Kolla et al (1976b) also reported a set of short-term (usually less than one hour) current measurements with speeds less than 20 cm/sec (7.9 in/sec) on the Agulhas Plateau and in the Transkei and Mozambique basins. In the Mozambique Basin, Kolla et al (1980) estimated current speeds at less than 15 cm/sec (5.9 in/sec) from bottom photographs.

In general, current speeds along the margins of the circum-African basins probably have peak values of about 15 to 20 cm/sec (5.9 to 7.9 in/sec) and average values that are somewhat lower. As noted, the peak speeds are sufficient to initiate erosion in relatively low-cohesion, calcareous oozes. However, most of the basin sediments are significantly finer grained and more cohesive than are calcareous oozes (Bornhold and Summerhayes, 1977; Embley and Morley, 1980; Kolla et al, 1980). Thus, speeds greater than about 30 cm/sec (11.8 in/sec) are probably required to erode them (for example, Postma, 1967), and it is unlikely that widespread erosion presently occurs in the basins. On the other hand, Connary (1972), Bornhold and Summerhayes (1977), and Kolla et al (1976b, 1980), among others, note development of a significant bottom nepheloid layer (concentrations of 10 to 200μg/l) that requires a fairly steady sediment source. It is reasonable to assume that the nepheloid layer is maintained by erosion in two kinds of source areas: 1) in areas where sea-floor gradients are regionally steep (for example, Agulhas Passage, Cape Rise) or locally steep (as with seamounts) and currents are thereby intensified; and, 2) in broader areas that have low gradient but that may be eroded by infrequent, episodically intensified currents. We are uncertain which of these sources may be volumetrically more important over the long-term geologic history of the basins.

PRESENT CIRCUM-BASIN EROSIONAL ZONE

General Character
Long-term stability and geologic significance of the circu-

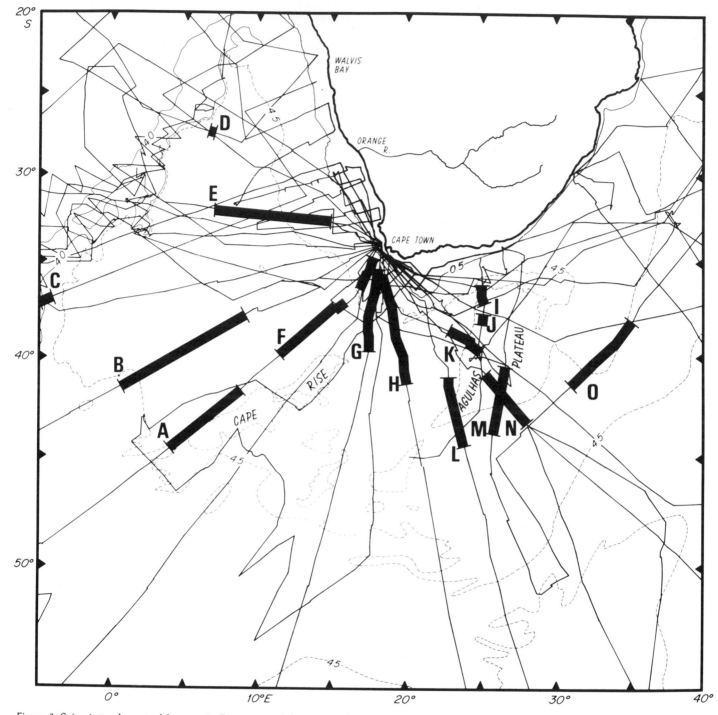

Figure 3: Seismic track control for map in Figure 1, with locations of illustrated profiles.

lation patterns just discussed are demonstrated by sedimentary patterns of erosion and deposition in the basins. Along the basin margins where abyssal currents are expected to be more intense, a zone of erosion or nondeposition can be traced from seismic profiles as a nearly continuous feature (Figure 1). The zone is identified by truncation of sedimentary reflectors at the sea floor or by anomalously thin sediments (Figures 3 through 9). The best development, or "core," of the erosional zone lies between the 4.5 and 5.0 km (2.8 and 3.1 mi) isobaths, but the erosion commonly extends to as shallow as 3.8 km (2.4 mi) or as deep as 5.2 km (3.2

mi) (Figure 1). Lateral changes in sediment sources and basin geometry cause significant variation in the surface morphology, width, and depth of erosion in the erosional zone. In some areas, the zone is identified by anomalously thin sediments with no clear truncation of reflectors (Figure 4, A and B). Such zones probably result from persistently attenuated sediment accumulation beneath the current axis, rather than from actual erosion. In other areas slightly faster currents cause some erosion, but the erosion is observed easily only in high-resolution profiles (Figures 4, profile D and 6). The most intense erosion is identified in seismic profiles by deep

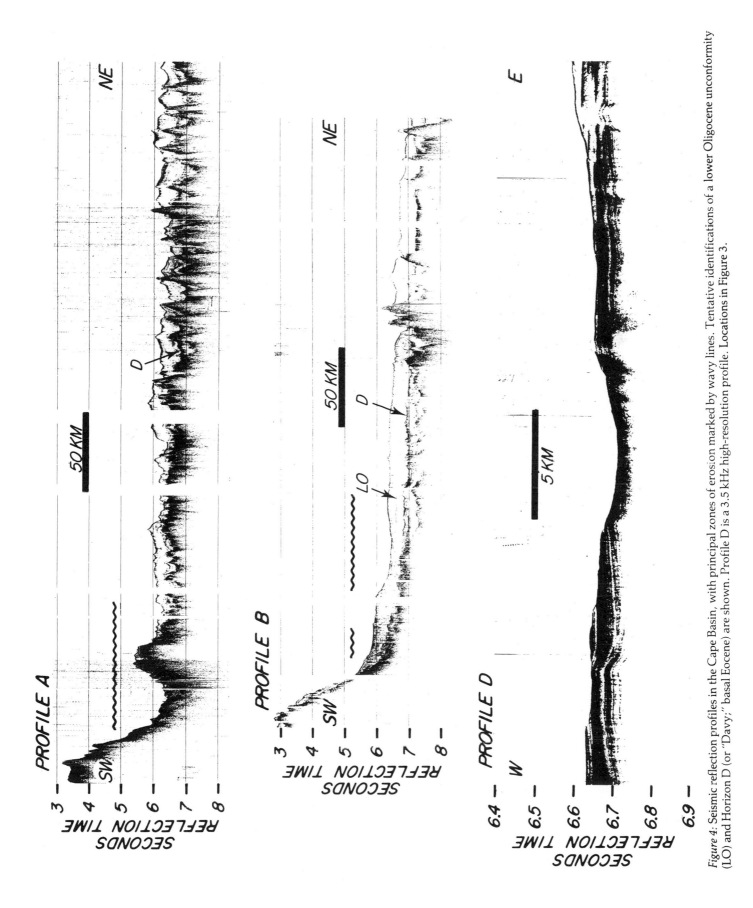

Figure 4: Seismic reflection profiles in the Cape Basin, with principal zones of erosion marked by wavy lines. Tentative identifications of a lower Oligocene unconformity (LO) and Horizon D (or "Davy;" basal Eocene) are shown. Profile D is a 3.5 kHz high-resolution profile. Locations in Figure 3.

incision of the sedimentary section, sometimes exposing basement (Figure 7).

Cape Basin

The Cape Rise forms the southern margin of the Cape Basin, and Antarctic Bottom Water enters the basin both east and west of the rise. Along the northern edge of the Cape Rise are numerous examples of attenuated accumulation, moating, and truncated reflectors beneath the AABW pathway (Figure 4, profile A and 5, profile F). At the western margin of the Cape Basin the erosional zone is rather broad and poorly defined (for example, see Figure 4; profile B). Erosion or attenuated sediment accumulation there is often concentrated around small topographic irregularities. A similar pattern is observed in the central Cape Basin. The widespread but subdued current effects along the western edge of the Cape Basin probably are caused by spreading of the AABW flow across a wide area that has a relatively low regional topographic gradient.

Against the Walvis Ridge north of about 35°S the erosional zone is locally well developed. Bornhold and Summerhayes (1977) described an erosional moat up to 80 m (262 ft) deep at the foot of Walvis Ridge just north of our profile D (Figure 4). They noted that the moat cuts into flat-lying, probably locally-ponded turbidites, and it is flanked 20 to 45 km (12.4 to 28 mi) to the east by hummocks of acoustically non-laminated sediments that may be current-deposited drifts. A comparable degree of erosion appears to persist northeastward along the foot of the Walvis Ridge (Figure 4, profile D). Deeper dissection may be limited by the occasional influx of turbidites, which originate on both the Walvis Ridge and African margin (Bornhold and Summerhayes, 1977).

Along the southwest African continental rise, identification and tracing of the erosional zone are confused by three factors: 1) There is significant along-slope and cross-slope variability in sediment supply that locally accentuates or masks the erosional zone; 2) Patterns of erosion and redeposition created by slumps, slides, and debris flows are common; and, 3) Sediment removal commonly parallels bedding planes on the gentle slope, thus making it difficult to detect the unconformity seismically.

The best-defined erosion along the continental rise generally occurs at depths between about 4.5 and 5.0 km (2.8 and 3.1 mi). However, in the area of the Orange Cone (~ 28° to 35°S), erosion normally is detected only in high-resolution 3.5 kHz records. Elsewhere, seismic profiles show gently inclined subbottom reflectors truncated at the sea floor (Figure 5, profile E). South of the latitude of Cape Town is a 350-km (217-mi) long, contour-parallel zone where the continental rise is deeply incised (Figure 5, profiles F and H; Figure 6). Near 3,800 m (12,467 ft) at the upper edge of this zone a series of sedimentary reflectors typically crop out along a gentle "scarp" 200 to 300 m (656 to 984 ft) high (Figure 6). Dingle (1980) interpreted this as the seaward edge of the "Cape Town Slump." Deep-sea drilling at Site 361 in the denuded zone below this "scarp" showed that little Neogene to Quaternary sediment cover is present (Bolli and Ryan, 1978), and this is confirmed by recovery of Paleogene sediment in several piston cores in this area (Figure 1). Very high

shear strength in the near-surface sediments at Site 361 demonstrates that significant sedimentary overburden has been removed and that this sea-floor zone is in fact erosional, rather than nondepositional (Bolli and Ryan, 1978). It is possible that the denuded zone itself is an exposed glide plane and that the "scarp" is a glide-plane scar that is strongly modified by current erosion. However, the apparent lateral continuity of the zone for a distance of more than 300 km (186.4 mi) suggests that the erosion may be primarily due to incision by abyssal currents. In either case, it is clear that bottom currents have had strong geologic effects to depths less than 4 km (2.5 mi) along this part of the margin. The numerous documented mass movements of sediment along the continental rise have undoubtedly been a significant agent of sea-floor erosion by themselves, and have probably generated sea-floor roughness that greatly facilitates erosion by abyssal currents.

Manganese nodules and pavements are commonly observed in bottom photographs in and around the erosional zone, especially in the southeastern Cape Basin (Figure 1). Accretion of these nodules is a common phenomenon in areas of erosion or low-sediment accumulation (see Kennett and Watkins, 1975), and manganese "hardgrounds" often occur at unconformities in cored sediments (see, Tucholke and Carpenter, 1977). The observation, therefore, is in accord with sedimentation rates of less than 0.4 mm/1000 yrs (0.015 in/yrs) determined by Embley and Morley (1980) from piston cores throughout the central Cape Basin and along its southeastern margin. Average sedimentation rates in the erosional zone along and north of the Orange Cone are higher (5 to 223 mm/1000 years, or 0.2 to 8.8 in/1000 yrs) and manganese nodules are observed less frequently there.

Agulhas Region

Perhaps the most complete development of the erosional zone occurs in the West and South Agulhas basins along the flanks of the Agulhas Plateau. Here, the only significant sediment supply is the suspended load carried in abyssal currents or plankton settling from surface waters. At the southern edge of Africa, the wide shelf (Agulhas Bank) and the trough of the Agulhas Fracture Zone limit most seaward transport of terrigenous detritus. On the crest and upper flanks of the Agulhas Plateau, calcareous oozes have accumulated above the carbonate compensation depth (CCD), while biosiliceous oozes are common in the deeper basins, especially south of 45°S near the Polar Front Zone. Thus, both physical erosion and chemical corrosion of biogenous tests are probably important factors in development and maintenance of the erosional zone in this area.

The erosional zone circumscribes the Agulhas Plateau. It is well developed up to depths of 4 km (2.5 mi) or less, possibly because this region is near the AABW source area and the AABW flow reaches shallower depths. In several places abyssal currents have removed all sediments above basement (Figures 1, 7, and 8). Above the erosional zone at the south edge of the Agulhas Plateau, a sedimentary ridge has been deposited and sculpted in the slack water zone peripheral to the main current (at profile crossing in Figure 8).

North of Agulhas Plateau, a narrow but persistent ero-

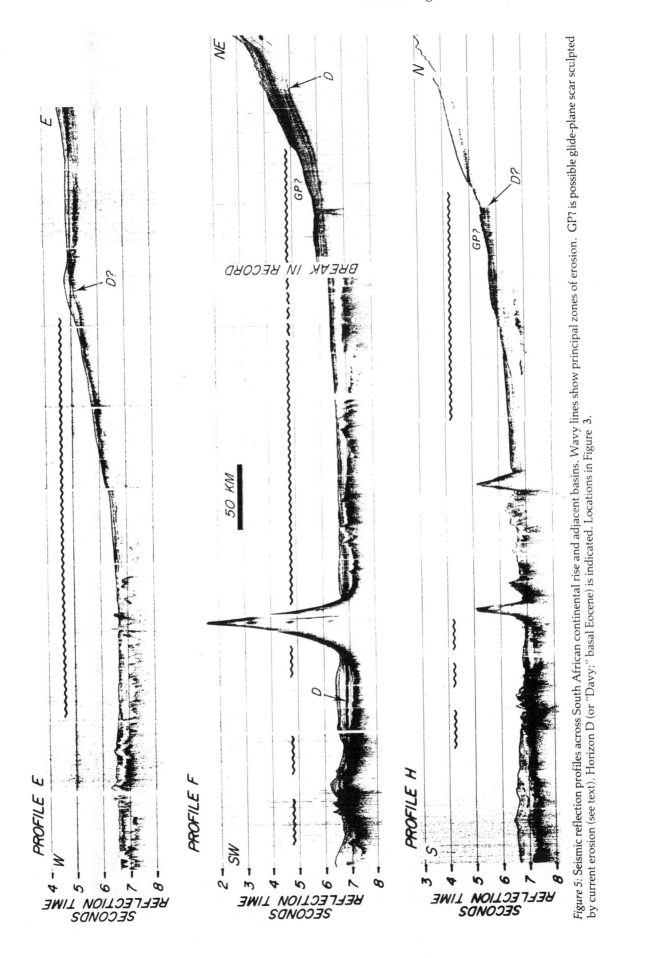

Figure 5: Seismic reflection profiles across South African continental rise and adjacent basins. Wavy lines show principal zones of erosion. GP? is possible glide-plane scar sculpted by current erosion (see text). Horizon D (or "Davy;" basal Eocene) is indicated. Locations in Figure 3.

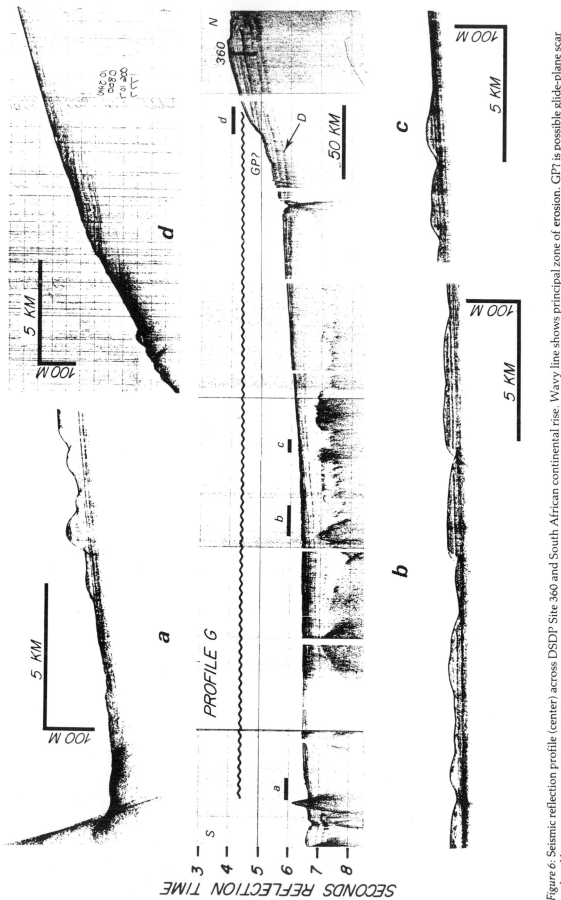

Figure 6: Seismic reflection profile (center) across DSDP Site 360 and South African continental rise. Wavy line shows principal zone of erosion. GP? is possible glide-plane scar sculpted by current erosion (see text). Horizon D (or "Davy;" basal Eocene) is indicated. Above and below are 3.5 kHz high-resolution profiles (located in center profile) that more clearly illustrate the sea-floor erosion. Location in Figure 3.

sional zone extends northeast along the base of the African margin, from Agulhas Passage into Transkei Basin (T.B. in Figure 1) and Natal Valley. There are also intermittent indications of erosion at the northern edge of Agulhas Plateau, beneath the presumed return flow exiting westward from Transkei Basin. Abyssal current effects on the sedimentary record are observed to depths less than 4 km (2.5 mi) in Transkei Basin and Natal Valley. Similarly, a zone of thinned sediments along the east flank of Agulhas Plateau suggests a geologically persistent northward current there at depths above 4.5 km (2.8 mi).

East of the Agulhas Plateau, thick (less than 1.5 km; 0.93 mi) sediments exhibiting weak seismic lamination blanket the sea floor south of the Mozambique Ridge (Figures 1 and 9). There are local indications of erosion here, but the area primarily is a locus of deposition. The sediments are marls (25 to 50% $CaCO_3$) and clays with a high quartz content (20%) that suggest a terrigenous source area probably in southeastern Africa (Kolla et al, 1980). The only reasonable way to account for accumulation of these thick sediments is by deposition from bottom currents. The sediments may have accumulated in the relatively slack water between the southward return flow from Natal Valley and the northeast flow into the Mozambique Basin (Figure 2). The Natal Valley return flow probably transports the greater, largely terrigenous, sediment load.

Mozambique Basin

The west side of the Mozambique Basin is bounded by the steep flank of the Mozambique Ridge. Although the erosional zone occurs at the foot of this topographic barrier where bottom currents are most intensified, the zone generally is not well developed (Figure 9). The erosional zone disappears at the northern end of the basin, and it has not been detected at the foot of the Madagascar Ridge bounding the basin on the east. The generally reduced erosion in the Mozambique Basin presumably indicates lower current speeds. The slower circulation might result from the fact that the basin forms a cul de sac for abyssal currents. Flanking the erosional zone along the west side of the basin is a clearly depositional zone where sediment waves are well developed (Ewing, Aitken, and Eittreim, 1968; Embley and Tucholke, 1976; Kolla et al, 1980). Kolla et al (1980) developed a model of trapped internal waves to explain the northward migration of these mud waves beneath slow (less than 10 cm/sec, or 3.9 in/sec) northward-flowing currents.

Summary

The foregoing discussions show that abyssal currents, principally concentrated along basin margins, are an important factor in controlling sedimentary patterns in the basins around South Africa. The Cape Basin is a starved basin, in part because of relatively low input and dissolution of biogenic detritus and limited riverine input, but also because abyssal currents have scoured the sea floor. Practically speaking, most of the sea floor below about 4 km (2.5 mi) in the Cape Basin is a surface of unconformity.

The Agulhas Basin and Agulhas Plateau receive significant input of hemipelagic and pelagic biogenic sediments, but in the erosional zones around these features, sediment corrosion and erosion yield net accumulation rates that are near zero. Only in and around the Transkei Basin, Natal Valley, and Mozambique Basin is regional net sediment accumulation significantly greater than zero. Here, the Zambesi, Limpopo, and numerous smaller rivers on the adjacent African margin provide a relatively high sediment input (Milliman and Meade, 1983). Clearly constructional sediment accumulations occur at the southern end of the Mozambique Ridge and in the central Mozambique Basin, and the marginal erosional zones are not so well developed as they are to the south and west.

FORMATION OF UNCONFORMITIES

The exact meaning of any unconformity or current-bedded depositional sequence in terms of bottom-current intensity is a complex and poorly-understood issue. A variety of variables determine whether the sea floor at a given location will have zero or net positive accumulation or suffer net sediment loss. The principal variables of importance include bottom-water current speed, bottom-water chemistry, sediment composition, and sediment cohesion (the last variable is used in preference to water content or grain size; although sediment "erodability" varies with changes in these factors, it is ultimately the sediment cohesive strength that determines whether erosion occurs.) Note that the variables do not include concentration of suspended load in bottom currents or rate of input to the flow from, say, continental detrital sources. These variables are excluded because sediment-transport models indicate that erosional/depositional thresholds are concentration-independent below about 300 mg/liter (Krone, 1962, McCave and Swift, 1976), and suspended-matter concentrations in the modern abyss are two to four orders of magnitude lower than this value. Thus, under most deep-sea conditions, an abyssal current flowing fast enough to erode a substrate or prevent net deposition on that substrate will not respond to increases in suspended load by depositing sediment.

Sea-floor erosion occurs if the current speed exceeds a critical erosion threshold, and the rate of erosion increases with current speed above this threshold (Partheniades, 1965). The threshold is lower for silty, sandy abyssal sediments (for example, biogenic oozes) because they are less cohesive. For example, laboratory experiments show that a typical calcareous ooze can be eroded at about 15 cm/sec (5.9 in/sec) (Southard, Young, and Hollister, 1971). Erosional thresholds several times higher apply to more cohesive sediments that are generally finer grained and have lower water content (for example, Postma, 1967).

Corrosion of sea-floor sediment is related to water chemistry. "Old" bottom waters generally are higher in CO_2 (hence lower pH) and higher in dissolved silica. These waters are more corrosive to calcareous tests (Berger, 1970; 1973), but have relatively little effect on siliceous tests (Heath, 1974). In contrast, "young" well-ventilated bottom waters tend to be silica poor. They are more corrosive to siliceous tests but less so for calcareous sediments (Moore et al, 1978). Furthermore, excepting local topographic effects, it is probably true that younger abyssal waters in a given

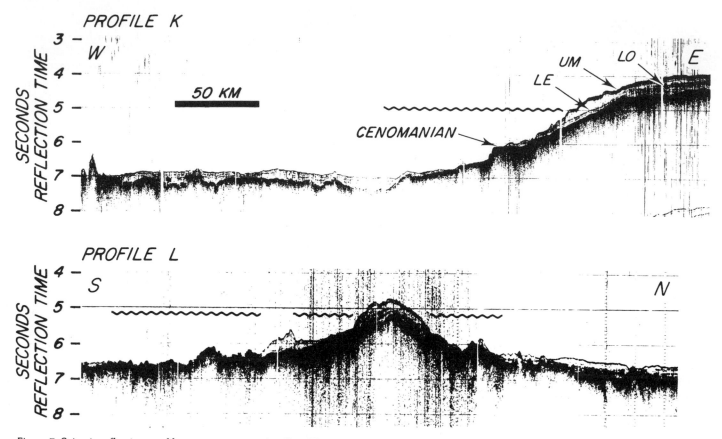

Figure 7: Seismic reflection profiles across western Agulhas Plateau. Principal zones of erosion are shown by wavy lines. Reflector identifications are from seismic mapping and correlation with piston cores: LE-basal Eocene unconformity; LO-lower Oligocene unconformity; UM-upper Miocene unconformity, essentially at sea floor. A piston core recovered Cenomanian clays from acoustic basement. Locations in Figure 3.

area circulate more vigorously than do older bottom waters. Hence, with increasing current speed we might expect increased silica dissolution in relation to carbonate corrosion.

The foregoing generalizations are especially important for a region such as that off South Africa where sediment compositions and textures vary so widely both latitudinally and from one physiographic province to another. With these considerations and the modern depositional framework in mind, we can make a preliminary assessment of the significance of Cenozoic hiatuses off South Africa.

RECORD OF CENOZOIC HIATUSES

Despite the recovery of numerous pre-Quaternary piston cores in the basins seaward of southern Africa, data are relatively limited for determining the stratigraphic level of unconformities or the duration of hiatuses in the sedimentary record. In Figure 10 we summarize hiatuses (shaded) determined from piston cores and JOIDES boreholes in the South Atlantic and the southwest Indian Ocean. We emphasize that the exact age limits of any given hiatus usually are not well constrained because of intermittent coring and often poor biostratigraphic resolution. Furthermore, the regional significance of a hiatus or trend in sediment accumulation rate determined at any one borehole is uncertain,

since each borehole is only a spot sample and may be atypical. At our present level of understanding, it is the interregionally correlated hiatuses that are most significant. The hiatuses defined in piston cores from the Falkland and Agulhas plateaus may be more reliable; the statistical sample is much larger, the cores contain rich microfossil assemblages, and ages are well-resolved by biostratigraphic and/or magnetostratigraphic studies. The factor common to all the localities in Figure 10 is that they presently are influenced by circulation of deep and bottom waters formed along the perimeter of Antarctica (including but not limited to AABW). Our implicit assumption in the ensuing discussion is that this circulation has had coherent regional effects on the sedimentary record.

The oldest intra-Cenozoic hiatus of regional significance occurs near the Paleocene/Eocene boundary. In the southwest Atlantic it probably is present in the deeper basins (Argentine, Malvinas), but its occurrence on the shallower Falkland Plateau is uncertain (Figure 10). In the Cape Basin, calculated accumulation rates at Site 361 suggest a hiatus or extremely slow sedimentation at the Paleocene/Eocene boundary. The apparent unconformity correlates with Horizon D (= Horizon Davy) which is widespread throughout the basin (Figures 4, 5, and 6). There is no clear truncation of underlying reflectors at Horizon D, but in profiles of Emery et al (1975) the overlying strata locally appear to onlap the reflector in a seaward direction. The hiatus also is

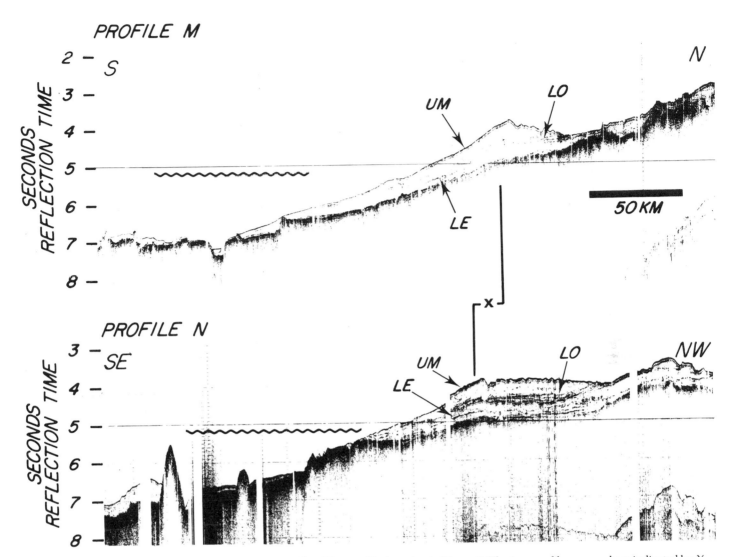

Figure 8: Seismic reflection profiles across southern Agulhas Plateau. Explanation in Figure 7. The two profiles cross where indicated by X.

present on the Agulhas Plateau (Tucholke and Carpenter, 1977) where a well-defined reflector appears at the corresponding stratigraphic level. Beds both above and below this reflector are mostly conformable (Figures 11, profile J and 12). It is uncertain whether the hiatus is present in the Mozambique Basin. Site 248 near the western margin of the basin has lower Eocene sediments present, and the hiatus at Site 249 in the center of the basin may be a result of later erosion (Figure 10).

The hiatus at the Paleocene/Eocene boundary has notable correlations with global phenomena. Moore et al (1978) noted that the Paleocene/Eocene boundary is characterized by frequent occurrences of hiatuses, both globally and in the South Atlantic and western Indian oceans. This correlates with a relative highstand of sea level (Figure 10; Vail et al, 1977) and with a period of low sedimentation rates (Davies et al, 1977; Whitman and Davies, 1979). The sea-level highstand probably restricted detrital sediment input to the basins and reduced surface-water biologic productivity by creating a more maritime climate with reduced upwelling. Both the global sea-level highstand and reduced sedimentation rates persisted into the early Eocene, yet Moore et al's

(1978) compilation shows that the early Eocene had an unusually low occurrence of hiatuses. Thus, the hiatus at the Paleocene/Eocene boundary may not be due solely to eustatic sea-level effects. It is possible that weakly circulating deep currents of "old" bottom water corroded calcareous sediments on the sea floor at this time, or even removed some strata in bedding-parallel fashion over such areas as the Agulhas Plateau. However, the general absence of seismic truncation of deeper beds at this stratigraphic level suggests that such possible current effects were slight (Figures 5, 8, 11, and 12).

Following a major lowering of sea level in the latest early Eocene, sediment accumulation rates were relatively high in the Argentine and Cape basins and probably on the Agulhas Plateau (Figure 10). A subtle seaward onlap of strata onto Horizon D occurs in the Cape Basin. This onlap is best observed in the high-vertical-exaggeration profiles of Emery et al (1975). It probably represents increased detrital input from the exposed borderland of southwest Africa during the lowstand. Whether significant thicknesses of sediments also accumulated on the Falkland Plateau and in the Mozambique Basin and were later eroded is unknown.

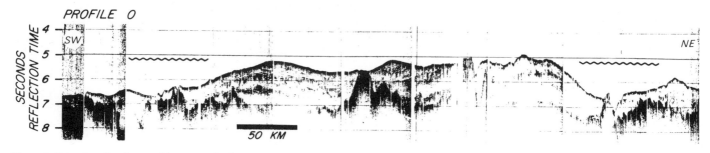

Figure 9: Seismic reflection profile across thick, weakly-laminated sediments and erosional zone bordering Mozambique Basin. Wavy lines locate zones of attenuated sedimentation or erosion. Location in Figure 3.

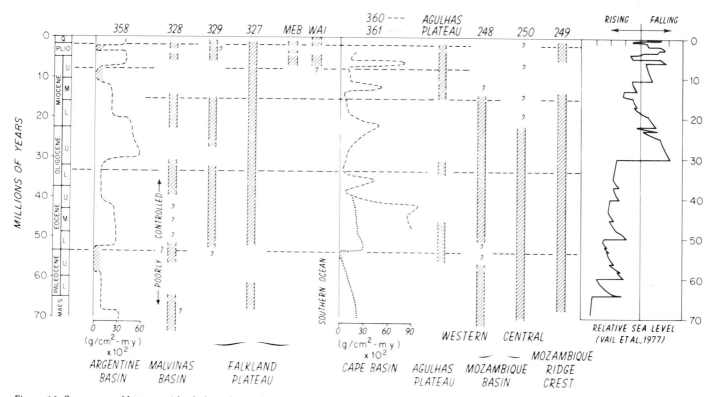

Figure 10: Summary of hiatuses (shaded, and correlated with dashed lines) and/or sediment accumulation rates at DSDP drillsites and from piston-core studies: MEB - Maurice Ewing Bank (Falkland Plateau; Ciesielski, Ledbetter, and Ellwood, 1982); WAI - Weddell Basin/ Atlantic-Indian Ridge (Ledbetter and Ciesielski, 1982); Agulhas Plateau (Tucholke and Carpenter, 1977). Drillsite information derived from Bolli and Ryan (1978), Barker and Dalziel (1976), Davies and Luyendyk (1974), Simpson and Schlich (1974), and Supko and Perch-Nielsen (1977).

A second regionally important hiatus occurs in the early to middle Oligocene (Figure 10). As during formation of the basal Eocene unconformity, sea level was also elevated during the early Oligocene, and sedimentation rates were minimal (Davies et al, 1977). However, the hiatus noted in Figure 10 slightly postdates maximum development of hiatuses on a global scale. This global maximum is centered in the late Eocene, although on a regional scale it extends into the early Oligocene in the Southern Ocean and the Atlantic (Moore et al, 1978). Our designation of early to middle Oligocene for this event is determined only by piston-core data from the Agulhas Plateau (Tucholke and Carpenter, 1977). Therefore, it is possible that the hiatus extends back into the late Eocene or that a separate late Eocene hiatus is also present.

It is clear from depositional patterns observed in seismic reflection records that abyssal currents became an important geologic agent by early Oligocene time. An unconformity truncating deeper bedding planes is often well developed at this stratigraphic level, and bed forms deposited under the influence of bottom currents appear immediately above the unconformity (Figures 11 and 12). The superposition of this erosive/corrosive agent on a regional regime of high sea level, low detrital influx, and presumed low surface productivity probably accounts for the intense and widespread development of hiatuses at this time. By the same token, it is not necessary to postulate that there was a sudden increase in intensity of abyssal circulation, although this is a possibility. The same geologic effects could be generated by gradually increasing abyssal circulation over a period of several

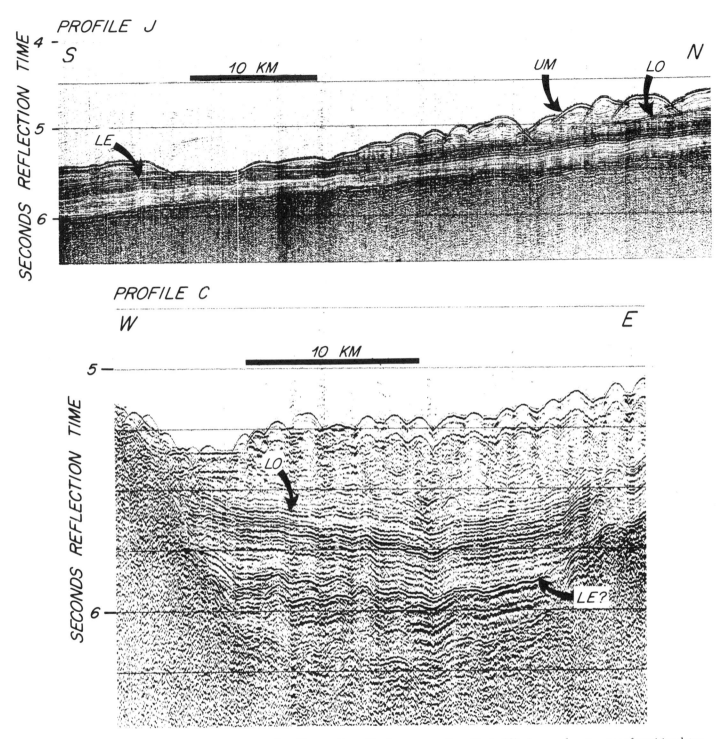

Figure 11: Seismic reflection profiles across the Agulhas Plateau (J) and in the western Cape Basin (C). Arrows locate unconformities determined from seismic mapping and correlation to piston cores. Explanation in Figure 7. In profile C, the proposed lower Oligocene (LO) unconformity correlates with reflector E of Connary (1972) that was thought to be early Oligocene to Eocene in age. Note that contorted bedding and possible sediment waves indicating current-controlled deposition first appear above the lower Oligocene unconformity. Profiles located in Figure 3.

million years.

To what factor(s) can we ascribe the increased intensity of deep circulation? An obvious candidate is the production and northward spreading of cooler bottom water in circum-Antarctic areas. Kennett and Shackleton (1976), Kennett (1977), and Keigwin (1980), among others, summarized

oxygen isotopic measurements on benthonic foraminifera that show a significant enrichment in ^{18}O beginning in the earliest Oligocene. This enrichment was interpreted as resulting from a drop in deep-water temperatures, superimposed on a long-term trend to cooler temperatures. This is thought to represent the first global development of cold

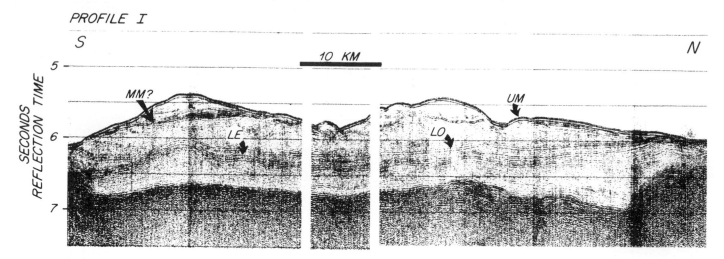

Figure 12: Seismic reflection profile across northern Agulhas Plateau showing unconformities as in Figure 7. MM? is possible middle Miocene unconformity (see text), or it may be the upper Miocene unconformity capped by current-deposited Plio-Pleistocene sediments that are unusually thick.

bottom waters (psychrosphere) in the ocean basins. However, the stimulus for this bottom-water formation is uncertain. One possibility is an abyssal "teleconnection" (Johnson, 1982) between deep waters of the Southern Ocean and deep waters of the North Atlantic. A major erosional unconformity of probably early Oligocene age is present in the northern and western North Atlantic, and it is thought to have been eroded by cool, dense Arctic waters that flowed southward into the basins as Greenland separated from Spitzbergen (Miller and Tucholke, 1983). If these bottom waters reached circum-Antarctic areas they may have provided a reinforcing teleconnection that stimulated formation of deep and bottom water currents around Antarctica (Tucholke and Miller, 1983). Another possible explanation for the enhanced bottom-water production is that Antarctic circumpolar flow first crossed shallow topographic barriers at the Tasman Rise and in the Drake Passage at this time; this could have increased the thermal isolation of Antarctica and stimulated deep-water and bottom-water production (for example, Loutit and Kennett, 1981). The time at which the Antarctic Circumpolar Current (ACC) actually began is poorly constrained. However, the Circumpolar Deep Water of the ACC is known to have had dramatic effects on sedimentation on the Falkland Plateau since the Neogene (Ciesielski, Ledbetter, and Ellwood, 1982), and the ACC may also be responsible for eroding the Paleogene unconformities there (Figure 10).

During the late Oligocene to early Miocene there are no unconformities developed other than those which may be related to later events (Figure 10). This agrees with observation of globally decreased frequency of hiatuses (Moore et al, 1978). Accumulation rates were relatively high in both the Cape and Argentine basins, corresponding with lower sea levels and with globally increased sedimentation rates (Davies et al, 1977). Abyssal currents were active in the basins but they principally transported sediment and controlled deposition in an environment of net positive sediment accumulation. Currents probably were only locally or episodically intensified to the degree where they eroded the sea floor.

In the middle Miocene, there is a strong indication of a regional hiatus (and/or attenuated sediment accumulation) as compiled in the sedimentary record at various drill sites (Figure 10). This was again a time of elevated sea level so that it is likely that reduced sediment accumulation in the basins was more a result of restricted detrital and biogenic input than a result of physical erosion by bottom currents. However, there is some evidence for erosion and redistribution of sediments by bottom currents about this time, as for example on the Agulhas Plateau (Figure 12).

Perhaps the most important regional hiatus occurred in the latest Miocene to early Pliocene, partly correlating in time with a global lowstand in sea level (Figure 10). Increased supply of detrital sediments during this lowstand may account for possibly limited development or non-development of the hiatus at locations in three basins (Argentine, Cape, Mozambique) that have access to terrigenous sediments. However, common exposure of pre-Pliocene sediments along the margins of the Cape and Mozambique basins suggests that the sea floor was strongly eroded in and around the present erosional zone. Where supplies of terrigenous sediment were more limited (Malvinas Basin, Falkland Plateau, Agulhas Plateau) the unconformity is very well developed.

On the Falkland Plateau, the unconformity has been extensively piston cored and identified by biostratigraphic and magnetostratigraphic studies (Ciesielski, Ledbetter, and Ellwood, 1982). Here the unconformity has been attributed to erosion by Circumpolar Deep Water within the Antarctic Circumpolar Current during the period 7.2 to 4.7 m.y. ago. The erosion sculpted the Maurice Ewing Bank to its present configuration and exposed sediments as old as Cretaceous in some localities (Ciesielski, Ledbetter, and Ellwood, 1982).

A similar timing for development of the unconformity is indicated by piston cores from the Agulhas Plateau (Tucholke and Carpenter, 1977). The youngest sediments cored below the unconformity date to the middle Miocene and strata as old as Cenomanian were excavated (Figure 7). Over most of the Plateau the unconformity is covered by a veneer of Pleistocene (locally Pliocene) nannofossil-

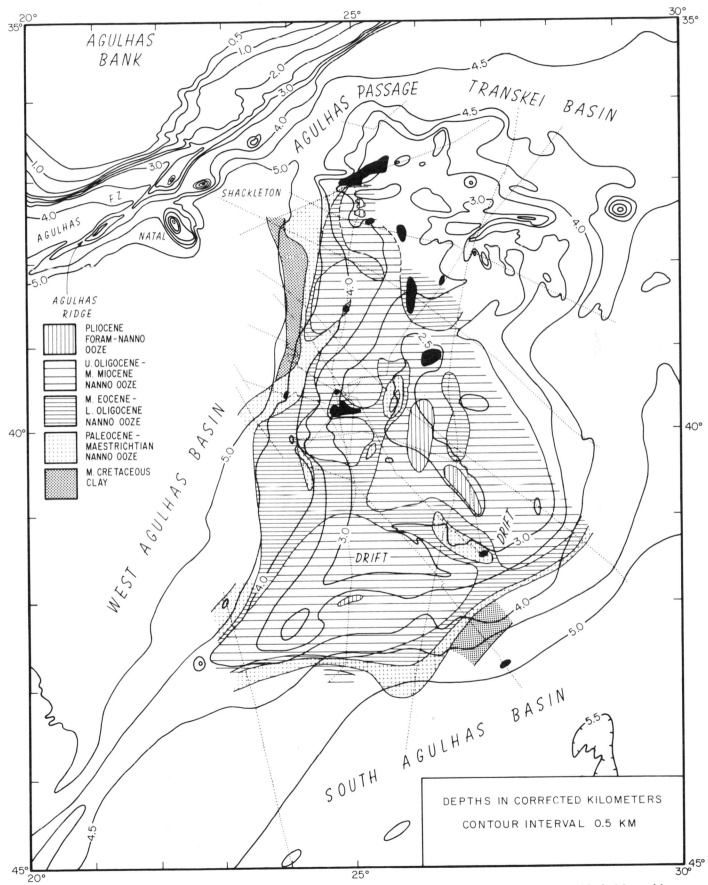

Figure 13: Preliminary subcrop map of the Agulhas Plateau beneath Pleistocene sediments. Basement outcrops are black. Mapped from seismic reflection profiles and correlations to piston cores reported by Tucholke and Carpenter (1977). Dotted lines give track control.

foraminiferal ooze only a few meters thick. Hence, in seismic reflection records the unconformity appears to coincide with the sea floor (Figures 11, profile J and 12). As indicated in the subcrop map of the Agulhas Plateau (Figure 13) and in seismic profiles (Figures 7, 8, 11 and 12), the erosion re-excavated older unconformities and exposed broad outcrops of Cretaceous and Paleogene strata.

The unconformity also appears to be well developed, at least locally, in the Weddell Basin and on the Atlantic-Indian Ridge, although only a few piston cores there have reached this stratigraphic level (Ledbetter and Ciesielski, 1982).

The strong increase in bottom circulation responsible for the erosion may relate to formation of the West Antarctic ice sheet. The development of this ice sheet is thought to have occurred in late Miocene time, stimulating the first major production of "true" Antarctic Bottom Water in the Weddell Sea and permanently altering global abyssal circulation (Ciesielski, Ledbetter, and Ellwood, 1982). Ciesielski, Ledbetter, and Ellwood (1982) attribute the extensive erosion on the Falkland Plateau to a correlative increase in the intensity of the Antarctic Circumpolar Current.

Ciesielski, Ledbetter, and Ellwood (1982) and Ledbetter and Ciesielski (1982) have also identified a younger, regionally important hiatus in the late Pliocene to early Pleistocene. Piston cores show that the unconformity is present on the Falkland Plateau, in the Weddell Basin, and on the Atlantic-Indian Ridge, but it is not as strongly developed as the upper Miocene unconformity. The unconformity does not occur in piston cores from the central part of the West Agulhas Basin. In the Cape Basin, on the Agulhas Plateau, and in the Mozambique Basin data do not exist to differentiate this event from the late Miocene episode of erosion. Most likely, the late Pliocene currents maintained or re-excavated the circum-basinal erosional zones. It is important to note that manganese pavements or nodules occur at the unconformity between middle Miocene and Pleistocene sediments in numerous piston cores from the Agulhas Plateau and adjacent areas (for example, Tucholke and Carpenter, 1977), and large areas of the sea floor in the present erosional zone have a manganese-nodule cover. These deposits probably armored the sea floor within the erosional zone against significant post-Miocene erosion.

CONCLUSIONS

A widespread zone of erosion is developed along the margins of the Cape Basin, Agulhas Plateau, and Mozambique Basin off South Africa, and it marks the principal pathways of abyssal flow of Antarctic Bottom Water through the region. However, Antarctic Bottom Water is not presently excavating the sea floor in any significant way because: 1) currents in most areas have speeds too low to erode available sediment; and, 2) a significant part of the erosional zone is armored by manganese nodules and crusts.

The erosional zone is largely an artifact of intense late Miocene erosion by a form of Antarctic Bottom Water. This erosional episode may have been caused by rapid production of "true" AABW in the Weddell Sea when West Antarctica became fully glaciated (Ciesielski, Ledbetter, and

Ellwood, 1982). Subsequent Pliocene to early Pleistocene abyssal circulation events re-excavated the erosional zone or allowed only minimal deposition, but they probably did not significantly increase the depth or lateral extent of the erosion accomplished in the late Miocene.

The oldest Cenozoic hiatus of regional significance was developed in the latest Paleocene to the early Eocene. It coincided with a global highstand in sea level and a concomitantly limited supply of terrigenous and biogenous detritus to the ocean basins. Abyssal currents, except where they were topographically intensified, probably were not significant in eroding sediments but they may have been important in accelerating corrosion of biogenous detritus exposed on the sea floor.

The first unambiguous evidence for significant abyssal circulation occurs in the lower Oligocene sedimentary record. An unconformity truncating deeper bedding planes commonly is developed at this stratigraphic level and subsequent deposition clearly was subject to control by abyssal currents. The exact tectonic and/or climatic controls that stimulated this initial production of strongly circulating bottom water are as yet unclear, but the timing agrees with observed global development of the psychrosphere beginning in the earliest Oligocene.

ACKNOWLEDGMENTS

This paper is an outgrowth of research on abyssal currents off South Africa that began in the mid-1970s. We thank John Schlee for the opportunity to present this paper at the AAPG Symposium in Woods Hole in 1980 and for the stimulus finally to put pen to paper for this volume. We also thank numerous colleagues at Lamont-Doherty Geological Observatory (particularly W.B.F. Ryan and J. Morley) for helpful discussions, and for their role in acquiring the regional geological and geophysical data that made this study possible. Age dates on piston cores were provided by L. Burkle, H. Okada, J. Morley, and P. Thompson. National Science Foundation Grant OCE81-22083 supports the L-DGO core facility. Our research was supported over the years by the Office of Naval Research, Contracts N00014-75-C-0210 to Lamont-Doherty Geological Observatory and N00014-79-C-0071 to Woods Hole Oceanographic Institution, and by National Science Foundation Grant OCE76-21782. We are indebted to K. Klitgord, K.G. Miller, J. Schlee, and E. Uchupi for constructive criticism on a late draft of the manuscript.

REFERENCES CITED

Barker, P.F., and I.W.D. Dalziel, eds., 1976, Initial reports of the deep sea drilling project, v. 36: Washington, D.C., U.S. Government Printing Office, 1079 p.

Berger, W.H., 1970, Biogenous deep-sea sediments; fractionation by deep-sea circulation: Geological Society of America Bulletin, v. 81, p. 1385-1402.

——— , 1973, Deep-sea carbonates; evidence for a coccolith lysocline: Deep-Sea Research, v. 20, p. 917-921.

Biscaye, P.E., V. Kolla, and K.K. Turekian, 1976, Distribution of calcium carbonate in surface sediments of the Atlantic Ocean: Journal of Geophysical Research, v. 81, p. 2595-2603.

Bolli, H.M., and W.B.F. Ryan, eds., 1978, Initial reports of the deep sea drilling project, v. 40: Washington, D.C., U.S. Government Printing Office, 1079 p.

Bornhold, B.D., and C.P. Summerhayes, 1977, Scour and deposition at the foot of the Walvis Ridge in the northernmost Cape Basin, South Atlantic: Deep-Sea Research, v. 24, p. 743-752.

Camden-Smith, F., et al, 1981, A preliminary report on long-term bottom-current measurements and sediment transport/erosion in the Agulhas Passage, southwest Indian Ocean: Marine Geology, v. 39, p. M81-M88.

Carmack, E.C., and T.D. Foster, 1975, On the flow of water out of the Weddell Sea: Deep-Sea Research, v. 22, p. 711-724.

Ciesielski, P.F., M.T. Ledbetter, and B.B. Ellwood, 1982, The development of Antarctic glaciation and the Neogene paleoenvironment of the Maurice Ewing Bank: Marine Geology, v. 46, p. 1-51.

Connary, S.D., 1972, Investigations of the Walvis Ridge and environs: Columbia University, unpublished Ph.D. Thesis, 228 p.

——, and M. Ewing, 1974, Penetration of Antarctic Bottom Water from the Cape Basin into the Angola Basin: Journal of Geophysical Research, v. 79, p. 463-469.

Davies, T.A., et al, 1977, Estimates of Cenozoic oceanic sedimentation rates: Science, v. 197, p. 53-55.

——, and B.P. Luyendyk, eds., 1974, Initial reports of the deep sea drilling project, v. 26: Washington, D.C., U.S. Government Printing Office, 1129 p.

Deacon, G.E.R., 1937, The hydrology of the Southern Ocean: Discovery Reports, v. 15, p. 1-124.

Defelice, D.R., and S.W. Wise, Jr., 1981, Surface lithofacies, biofacies, and diatom diversity patterns as models for delineation of climatic change in the southeast Atlantic Ocean: Marine Micropaleotology, v. 6, p. 29-70.

Dingle, R.V., 1980, Large allochthonous sediment masses and their role in the construction of the continental slope and rise off southwestern Africa: Marine Geology, v. 37, p. 333-354.

Embley, R.W., and J.J. Morley, 1980, Quaternary sedimentation and paleo-environmental studies off Namibia (South-West Africa): Marine Geology, v. 36, p. 183-204.

——, and B.E. Tucholke, 1976, A continuous erosional zone in the Cape, Agulhas, and Mozambique basins: Geological Society of America Abstracts with Programs, v. 8, p. 854.

Emery, K.O., et al, 1975, Continental margin off western Africa: Cape St. Francis (South Africa) to Walvis Ridge (South-West Africa): AAPG Bulletin, v. 59, p. 3-59.

Ewing, M., T. Aitken, and S. Eittreim, 1968, Giant ripples in the Madagascar Basin: Transactions of the American Geophysical Union, v. 49, p. 218.

Fuglister, F.C., 1960, Atlantic Ocean atlas of temperature and salinity profiles and data from the international geophysical year of 1957-1958: Woods Hole Oceanographic Institution, Atlas Series, 209 p.

Gordon, A.L., 1974, Varieties and variability of Antarctic Bottom Water: Colloques Internationaux C.N.R.S., no.

215, Processus de Formation des Eaux oceaniques profondes, p. 33-47.

Heath, G.R., 1974, Dissolved silica and deep-sea sediments, in W.W. Hay, ed., Studies in paleo-oceanography: Society of Economic Paleontologists and Mineralogists, Special Publication 20, p. 77-93.

Jacobs, S.S., and D.T. Georgi, 1977, Observations on the southwest Indian/Antarctic Ocean, in M. Angel, ed., A voyage of discovery: New York, Pergamon Press, p. 43-84.

Johnson, D.A., 1982, Abyssal teleconnections; interactive dynamics of the deep ocean circulation: Palaeogeography, Palaeoclimatology, Palaeoecology, v. 38, p. 93-128.

Keigwin, L.D., Jr., 1980, Palaeoceanographic change in the Pacific at the Eocene-Oligocene boundary: Nature, v. 287, p. 722-725.

Kennett, J.P., 1977, Cenozoic evolution of Antarctic glaciation, the circum-Antarctic Ocean and their impact on global paleoceanography: Journal of Geophysical Research, v. 82, 3843-3860.

——, and N.J. Shackleton, 1976, Oxygen isotopic evidence for the development of the psychrosphere 38 m.y. ago: Nature, v. 260, p. 513-515.

——, and N.D. Watkins, 1975, Deep-sea erosion and manganese nodule development in the southeast Indian Ocean: Science, v. 188, p. 1011-1013.

Kolla, V., A.W.H. Be, and P.E. Biscaye, 1976a, Calcium carbonate distribution in the surface sediments of the Indian Ocean: Journal of Geophysical Research, v. 81, p. 2605-2615.

——, et al, 1976b, Spreading of Antarctic Bottom Water and its effects on the floor of the Indian Ocean inferred from bottom-water potential temperature, turbidity, and seafloor photography: Marine Geology, v. 21, p. 171-189.

——, et al, 1980, Current-controlled, abyssal microtopography and sedimentation in Mozambique Basin, southwest Indian Ocean: Marine Geology, v. 34, p. 171-206.

Krone, R., 1962, Flume studies of the transport of sediment in estuarial shoaling processes: Berkeley, California University Hydraulic Engineering Laboratory, 110 p.

Ledbetter, M.T., and P.F. Ciesielski, 1982, Bottom-current erosion along a traverse in the South Atlantic sector of the Southern Ocean: Marine Geology, v. 46, p. 329-341.

LePichon, X., 1960, The deep water circulation in the southwest Indian Ocean: Journal of Geophysical Research, v. 65, p. 4061-4074.

Lisitzin, A.P., 1972, Sedimentation in the world ocean: Society of Economic Paleontologists and Mineralogists, Special Publication 17, 218 p.

Loutit, T.S., and J.P. Kennett, 1981, Australasian Cenozoic sedimentary cycles, global sea level changes and the deep sea sedimentary record: Oceanologica Acta, v. 4 supplement, p. 45-63.

McCave, I.N., and S. Swift, 1976, A physical model for the rate of deposition of fine-grained sediments in the deep sea: Geological Society of America Bulletin, v. 87, p. 541-546.

Miller, K.G., and B.E. Tucholke, 1983, Development of Cenozoic abyssal circulation south of the Greenland-

Scotland Ridge, *in* M. Bott et al, eds., Structure and development of the Greenland-Scotland Ridge: New York, Plenum Press, p. 549-589.

Milliman, J.D., and R.H. Meade, 1983, Worldwide delivery of river sediment to the oceans: Journal of Geology, v. 91, p. 1-21.

Moore, T.C., Jr., et al, 1978, Cenozoic hiatuses in pelagic sediments: Micropaleontology, v. 24, p. 113-138.

Partheniades, E., 1965, Erosion and deposition of cohesive soils: American Society of Civil Engineers Proceedings, Journal of the Hydraulics Division, v. 91, p. 105-139.

Postma, H., 1967, Sediment transport and sedimentation in the estuarine environment, *in* G.H. Lauff, ed., Estuaries: Washington, D.C. American Association for Advancement of Science, Publication 83, p. 158-179.

Reid, J.L., W.D. Nowlin, and W.C. Patzert, 1977, On the characteristics and circulation of the southwestern Atlantic Ocean: Journal of Physical Oceanography, v. 7, p. 62-91.

Saito, T., L.H. Burckle, and J.D. Hays, 1974, Implications of some Pre-Quaternary sediment cores and dredgings, *in* W.W. Hay, ed., Studies in paleo-oceanography: Society of Economic Paleontologists and Mineralogists, Special Publication 20, p. 6-36.

Schott, G., 1902, Wissenschaftliche Ergebnisse der deutschen tiefsee-Expedition *Valdivia*, 1898-1899: Oceanographie, v.1, p. 1-248.

Simpson, E.S.W., 1974, Bathymetry of the southeast Atlantic and southwest Indian oceans (map, 1st edition): Stellenbosch, South Africa, National Research Institute for Oceanology, scale 1:10 million, 1 sheet.

———, and R. Schlich, eds., 1974, Initial reports of the deep sea drilling project, v. 25: Washington, D.C., U.S. Government Printing Office, 884 p.

Southard, J.B., R.A. Young, and C.D. Hollister, 1971, Experimental erosion of calcareous ooze: Journal of Geophysical Research, v. 76, p. 5903-5909.

Summerhayes, C.P., B.D. Bornhold, and R.W. Embley, 1979, Surficial slides and slumps on the continental slope and rise of South-West Africa; a reconnaissance study: Marine Geology, v. 31 p. 265-277.

Supan, A., 1899, Die bodenformen des weltmeeres: Petermanns Geographische Mitteilungen, v. 45, p. 117-188.

Supko, P.R., and K. Perch-Nielson, eds., 1977, Initial reports of the deep sea drilling project, v. 39: Washington, D.C., U.S. Government Printing Office, 1139 p.

Tucholke, B.E., 1978, Geologic significance of abyssal currents on the Agulhas Plateau, southwestern Indian Ocean: Geological Society of America, Abstracts with Programs, v. 10, p. 507.

———, and G.B. Carpenter, 1977, Sediment distribution and Cenozoic sedimentation patterns on the Agulhas Plateau: Geological Society of America Bulletin, v. 88, p. 1337-1346.

———, and K.G. Miller, 1983, Late Paleogene abyssal circulation in North Atlantic: AAPG Bulletin, v. 67, p. 559.

Vail, P.R., et al, 1977, Seismic stratigraphy and global changes of sea level, *in* C.E., Payton, ed., Seismic stratigraphy–applications to hydrocarbon exploration: AAPG Memoir 26, p. 49-212.

Warren, B.A., 1974, Deep flow in the Madagascar and Mascarene basins: Deep-Sea Research, v. 21, p. 1-21.

Whitman, J.L., and T.A. Davies, 1979, Cenozoic oceanic sedimentation rates; how good are the data?: Marine Geology, v. 30, p. 269-284.

Wüst, G., 1936, Das Bodenwasser und die Gliederung der Atlantischen Tiefsee: Wiss. Ergeb. Deut. Atl. Exped. METEOR, 1925-1927, v. 6, n. 1, p. 3-107.

———, 1955, Stromgeschwindigkeiten im Tiefen und Bodenwasser des Atlantischen Ozeans auf Grund dynamischer Berechnung der Meteor-Profile der Deutschen Atlantischen Expedition 1925/27: Deep-Sea Research, v. 3 supplement, p. 373-397.

Wyrtki, K., 1971, Oceanographic atlas of the international Indian Ocean expedition: Washington, D.C., U.S. Government Printing Office, 513 p.

Depositional Sequences and Stratigraphic Gaps on Submerged United States Atlantic Margin

C. Wylie Poag
John S. Schlee
U.S. Geological Survey
Woods Hole, Massachusetts

Seismic reflection profiles correlated with deep stratigraphic test wells reveal a series of depositional sequences that can be traced through the three major sedimentary basins of the submerged U.S. Atlantic margin. These sequences are bounded in large part by unconformities whose stratigraphic positions are similar to those predicted by Vail's model of relative coastal onlap and eustatic sea-level change. The Mesozoic depositional sequences nearly match the supercycles of Vail's model, but the Cenozoic record indicates marked variability among the basins. Little doubt exists that these depositional sequences are the product of sea-level fluctuations modulated by changing depositional rates and variable rates of basin subsidence.

INTRODUCTION

Recent publications of Vail and his Exxon colleagues (Vail, Mitchum, and Thompson, 1977; Vail and Hardenbol, 1979; Vail and Mitchum, 1979; Vail and Todd, 1981; Hardenbol, Vail, and Ferrer, 1981) stimulated a renewed effort to understand how depositional sequences, as interpreted from seismic-reflection profiles, relate to depositional sequences sampled by boreholes. Thousands of kilometers of multichannel seismic reflection profiles are now available across the U.S. Atlantic margin and five deep continental offshore stratigraphic test wells (COST B-2, B-3, G-1, G-2, GE-1) and two commercial wells (Shell 272-1 and 273-1) have contributed geologic data from this region to the public domain (Scholle, 1977, 1979, 1980; Libby-French, 1981; Poag, 1982b; Scholle and Wenkam, 1982). These data are supplemented by thousands of kilometers of high-resolution single-channel seismic-reflection profiles, 40 shallow penetration coreholes (about 100 m, or 328 ft), and several hundred grab samples and oceanographic cores (Poag, 1978; Hathaway et al, 1979). These data constitute a research base now broad enough for us to attempt a preliminary synthesis of the depositional framework of the east coast offshore area. The synthesis is preliminary because seven deep wells, no matter how thoroughly analyzed, are obviously insufficient to fully interpret this 1 million-sq-km (about 386,000-sq-mi) area (Figure 1).

This investigation was carried out to see how well the unconformities seen on seismic profiles match those detected in nearby wells. We focus especially on the unconformities in the wells to see which are evident along the entire U.S. Atlantic continental margin, and how their ages compare with seaward shifts of coastal onlap shown by the Vail curve (Vail, Mitchum, and Thompson, 1977). Lastly, through charting of the relative paleodepths as indicated by the microfossils obtained from the offshore drill holes, we wish to evaluate the relative sea-level changes. When did sea level reach a maximum? Can we see the second order shifts of coastal onlap (supercycles) detected by Vail, Mitchum, and Thompson (1977) elsewhere in the world? What led to these shifts, and what processes formed the unconformities associated with them?

METHODS

The chief biostratigraphic data are derived from rotary cuttings and sidewall cores collected from the COST GE-1, B-2, B-3, G-1, and G-2 stratigraphic test wells (Poag, 1977, 1980, 1982a, 1982b; Poag and Hall, 1979). Foraminifera provide the primary age and paleoenvironmental data, but these are supplemented by data on other microfossil groups: calcareous nannoplankton (Valentine, 1977, 1979, 1980, 1982), ostracods, (Poag, 1980, 1982a), radiolarians (Poag, 1980, in press), diatoms (Abbott, 1978, 1980), calpionellids (Poag, 1982a, 1982b), dinoflagellates (Steinkraus, 1978, 1979, 1980; Bebout, 1980), and spores and pollen (Steinkraus 1978, 1979, 1980; Bebout, 1980); lithologic descriptions (Rhodehamel, 1979; Pollack, 1980a; Lachance, 1980; Simonis, 1980); and petrographic analyses (Halley, 1979; Pollack, 1980b).

Calculations of sediment accumulation rates, subsidence rates, and the duration of hiatuses are patterned after what van Hinte (1978) termed "geohistory analysis." This approach stresses using numerical or quantitative data (that is, paleodepths, sea-level changes, thicknesses of uncompacted sediment columns, and radiometric time scales) and techniques in stratigraphic interpretations.

The seismostratigraphic interpretations follow the techniques outlined by Vail, Mitchum, and Thompson (1977), Vail and Mitchum (1979), and Vail and Todd (1981), in which seismic reflections are treated as stratal boundaries, allowing recognition of depositional sequences and unconformities.

The two profiles shown and discussed herein are part of a broad regional grid of multichannel seismic profiles that has been collected since 1973 under contract by the United States Geological Survey (USGS) and by the Bundesanstalt für Geowissenschaften und Rohstoffe (BGR) (Schlee, Dillon, and Grow, 1979; Grow, Mattick, and Schlee, 1979; Poag, 1980, 1982a, 1982b; Grow and Sheridan, 1981; Schlee, 1981; Schlee and Fritsch, 1982; Klitgord, Schlee, and Hinz, 1982). Several of the published studies attempted to delineate major seismostratigraphic units, to map them throughout a sedimentary basin, and to analyze the character of reflections in order to infer the paleoenvironments in which the units formed. Where possible, these seismic units are tied to COST wells to allow calibration of age, lithology, and paleoenvironment (Poag, 1980, 1982a, 1982b). For both the Georges Bank basin and the Baltimore Canyon trough, Mesozoic seismic units are indicated. In the Baltimore Canyon trough, additional units of Cenozoic age can be delineated toward the outer edge of shelf because of the periodic outbuilding of the broad margin there.

For purposes of this paper, we used a single well from each of four sedimentary basins to document the depositional sequences and biostratigraphic gaps. For the U.S. basins, we used the wells located farthest offshore, which contain the richest and deepest-water microfossil assemblages. For comparison, we include the Shell Mohawk B-93 well on the Scotian shelf because it is the closest Canadian well to the Georges Bank wells (Figure 1).

LIMITATIONS OF DATA

Before proceeding with our data presentation and interpretation, we must discuss briefly the limitations of the techniques we used. Using so-called quantitative techniques in geohistory analysis (van Hinte, 1978) can result in misleading precision if taken too literally. We strongly emphasize that most of the numerical data used in this method are rather gross estimates that often are arbitrarily delimited.

For example, among the properties indirectly derived from biostratigraphic data are the "absolute" ages of various horizons and the durations of hiatuses. Under the most ideal conditions of continuous coring in tropical deep-sea locations, the accuracy of Cenozoic biozonation can vary by as much as a million years, and older biozones are much less accurate. Moreover, no concensus exists as to which published biozonation is most accurate. Several different systems are in use for each group of planktonic microfossils and nannofossils (see, Bolli, 1966; Hay et al, 1967; Blow, 1969, 1979; Bukry 1971, 1973; Martini, 1971; Stainforth et al, 1975; Thierstein, 1976). Choosing which one to use depends largely on the investigator's individual experience.

In our study, which is based on chiefly rotary cuttings samples, we are limited to using "last appearance datums" or "tops" of species ranges. This limitation dilutes the accuracy of our biozone calibration considerably. Biozone recognition also may be impaired by the presence in some intervals of shallow-water facies that prevent recognition of true "tops" of species ranges. As a result of these limitations, we can't tell precisely how much of any biozone is present or missing unless it is totally missing.

The time scale against which we calibrated the biochronology is an arbitrarily chosen composite of variably accurate segments. Several different time scales are available, and they contain significant differences (Harland, Smith, and Wilcock, 1964; Berggren and van Couvering, 1974; Obradovich and Cobban, 1975; Hardenbol and Berggren, 1978; van Hinte, 1976a, 1976b; Ness, Levi, and Couch,1980). We have used Berggren and van Couvering (1974) for the Neogene time scale; Hardenbol and Berggren (1978) for the Paleogene; van Hinte (1976a; 1976b) for the Cretaceous and Jurassic.

Determining paleodepth is a highly complex "art" that contains a certain amount of circular reasoning and even guessing (van Hinte, 1978). Such an admission is frustrating to geologists and geophysicists who need quantitative data for their models of sea-level change and basin subsidence. However, it remains a severe problem that cannot be ignored or glossed over. The basic weakness in paleodepth analysis is that most estimates are strongly keyed to the depth distribution of benthic foraminifera in *modern* oceans. Although a correlation with depth is clear (Phleger and Parker, 1951; Sliter and Baker, 1972; Murray, 1973; Poag, 1981), workers have not demonstrated convincingly that water depth is really the controlling factor. On the other hand, many recent studies have clearly shown the importance of such properties as substrate type, water-mass composition, food supply, sunlight, and environmental stability, and these properties have been drastically different in past oceans (Phleger, 1960, Lohmann, 1978; Douglas, 1979; Poag, 1981; Douglas and Woodruff, 1981; Haq, 1981). Add to this the knowledge that all biotic communities evolve in response (in part, at least) to environmental changes (Valentine, 1973), and that some species and genera are known to have changed environmental preference through time (Douglas, 1979), and the imprecision inherent in paleodepth estimates can be more fully appreciated . We have limited ourselves in this paper to the use of broad paleodepth categories, recognizing that their chief value is in delineating *relative* water depths.

The ancient level of the sea surface relative to its present position is an important element in calculating geohistory and basin subsidence (van Hinte, 1978; Watts and Steckler, 1979). Various methods have been used to calculate eustatic sea-level changes and several widely differing sea-level curves have been published (Hallam, 1963; Hays and Pitman, 1973; Vail, Mitchum, and Thompson, 1977; Pitman, 1978; Vail and Hardenbol, 1979; Hancock and Kauffman, 1979; Harrison et al, 1981; Hardenbol, Vail, and Ferrer, 1981). Compilation of each curve involves a complex series of estimates, extrapolations, and averages.

Interpretation of seismic reflection profiles is also subjective, chiefly an art, wherein the interpreter's experience and imagination are constrained by the appearance of the

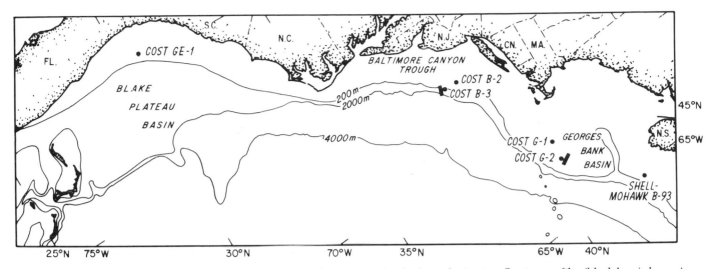

Figure 1: Index map to show the locations of offshore holes and segments of multichannel seismic-reflection profiles (black bars) shown in Figures 4 to 7.

profile—a display subject to basic geophysical principles, to the nature of the data collection system, and to the subsequent computer processing (Sheriff, 1977). In final display of the multichannel seismic-reflection profiles, we attempted to be certain that the gain was sufficiently high so that even weak reflections can be seen. Further, a comparison of reflections is made on intersecting profiles to see that key reflectors carry through and that the secondary reflections are broadly similar in appearance.

We recognize that there are different approaches to mapping key "bounding" seismic markers. In initial studies of multichannel profiles, Schlee et al (1976) used key horizons to separate seismic sequences having distinctive reflection characteristics. The difficulty with this approach is that the key reflectors can become indistinct laterally, and thus can become difficult to use as boundaries; likewise, the sequences also can change laterally in their pattern of reflections, making it difficult to separate them. We use the approach of mapping sequences based on unconformities (Vail, Mitchum, and Thompson, 1977) because of their potential as interregional boundaries. Further, the submerged Atlantic margin is an area where the sedimentary section is thicker than that of the Atlantic Coastal Plain and the deep North Atlantic basin (Tucholke, Houtz, and Ludwig, 1982). Hence, the stratigraphic column on the offshore margin has the potential to preserve the most complete sedimentary record (including unconformities) and to record the longer term trends in sea-level change.

STRATIGRAPHIC GAPS

Missing biozones were used to identify biostratigraphic gaps in the COST wells and the Shell Mohawk B-93 well on the Scotian shelf (Figure 2), and we estimated the time represented by the gaps (Poag and Hall, 1979; Poag, 1980, 1982a, 1982b). More gaps appear in the Upper Cretaceous-Tertiary section than in older rocks. This difference exists because the most refined biochronology is based on planktonic foraminifera, which are virtually absent in rocks older than Albian.

Southeast U.S. margin

In the COST GE-1 well, the following eight biostratigraphic gaps can be recognized (Figure 2): Cenomanian; late Turonian; late Maestrichtian-early Paleocene; late Paleocene-early Eocene; late Eocene-early Oligocene; early Miocene; middle Miocene-early Pliocene; and late Pliocene-early Pleistocene (Poag and Hall, 1979). Seismostratigraphy in the vicinity of the COST GE-1 well was described most recently by Paull and Dillon (1980); unconformities they identified are compared with the biostratigraphic gaps in Figure 3. The data are based on single-channel seismic-reflection profiles spaced 15 to 30 km (9.3 to 18.6 mi) apart and a wider grid of multichannel profiles (Dillon et al, 1979). The comparison shows that some of the unconformities of Late Cretaceous and Cenozoic age seen in the COST GE-1 well are represented by major reflectors on the profiles. Exceptions are between the Santonian and Campanian, between Maestrichtian and Paleocene, and between the late Eocene and the early Oligocene, where one or the other type (bio-seismo) of gap does not appear to be present. Additional biostratigraphic gaps exist in the late Tertiary and Quaternary, but these are difficult to see on reflection profiles in an area where the entire Cenozoic section is 300 to 1000 m (984 to 3,281 ft) thick (Paull and Dillon, 1980, Figure 9a).

Shipley, Buffler, and Watkins (1978) distinguished seven seismic units on a single multichannel profile across the Florida outer shelf, the Florida-Hatteras slope, and the Blake Plateau. Their subdivisions are broader than those of Paull and Dillon (1980), though they do note that additional unconformities are present in their youngest unit (BP-1). The unconformable boundaries of their seismic units are keyed to the Vail curve of coastal onlap and are placed within the Oligocene, at the base of the Tertiary, within the Santonian, within the Cenomanian, within the Albian, and within the Aptian. The ages of unconformities thought to be middle Aptian(?) and Cenomanian(?) are based on matching short-term shifts of sea level on the Vail curve (Vail, Mitchum, and Thompson, 1977) with unconformities seen on the seismic profile. Younger boundaries are tied to drilling results and to shifts in the curve. Between them, both seis-

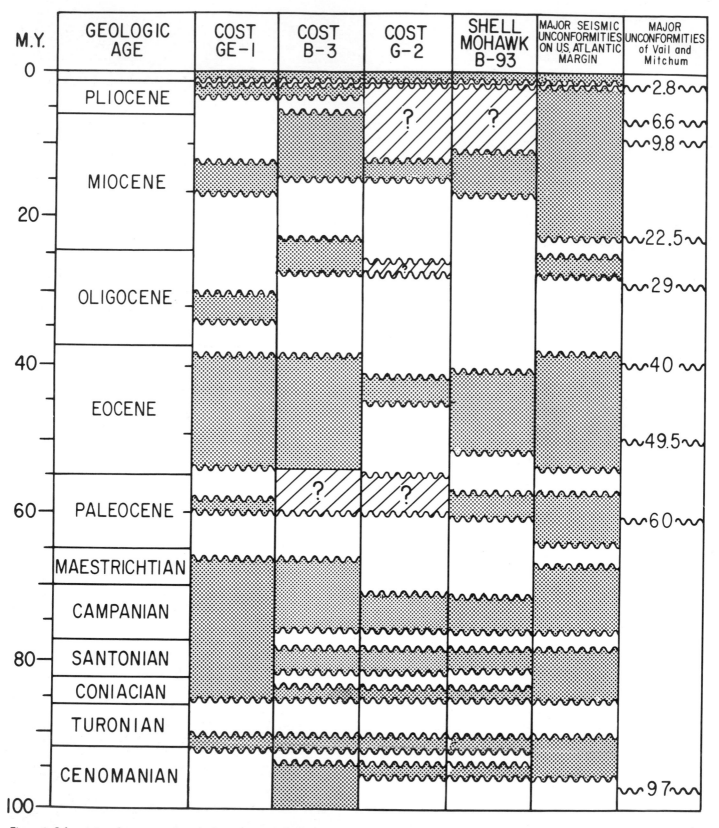

Figure 2: Schematic columnar sections (columns 1 to 4) illustrating stratigraphic positions of major unconformities documented by biostratigraphic gaps (shown as blank segments) in wells from the four Atlantic margin basins studied. Column 5 shows stratigraphic positions of major seismic unconformities identified by us in the three U.S. offshore basins. Column 6 shows position of major global unconformities of Vail and Mitchum (1979). Uncertainty as to presence of strata of a given age is indicated by diagonal hachures. Biostratigraphic resolution is too poor below the Cenomanian to justify including older strata in this figure. Sources of time scale given in "Limitations of Data."

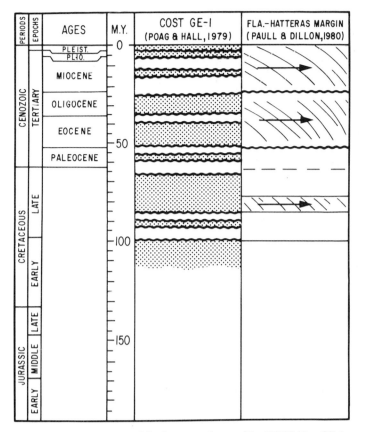

Figure 3: A comparison of the hiatuses detected in COST No. GE-1 well (white areas between wavey lines) and the seismic gaps detected by Paull and Dillon (1980). Outbuilding of the margin as progradational wedges is indicated by the arcuate lines with an arrow through them. Sources of time scale given in "Limitations of Data."

mic studies (Shipley, Buffler, and Watkins, 1978; Paull and Dillon, 1980) detect most of the gaps found in the COST GE-1 hole; an exception is the late Tertiary-Quaternary gap.

New Jersey margin

In the COST B-3 well, seven biostratigraphic gaps are present (Figure 2): late Cenomanian-early Turonian; later Turonian-early Coniacian; late Coniacian-early Santonian; late Santonian-early Campanian; late Maestrichtian-early Paleocene; late Eocene-late Oligocene; and early Miocene-middle Miocene. Two others, late Miocene-early Pliocene and late Pliocene-early Pleistocene, are postulated based on seismic evidence and nearby shallow coreholes (Poag, 1980).

A profile (part of USGS line 25) over the outer shelf and continental slope reveals several unconformities and a complex association of broad sedimentary wedges (Figure 4). The arrows indicate reflector termination through downlap, toplap, or truncation at the base or top of depositional sequences. The COST B-3 site has been projected to the line of the section from a position 10 km (6.2 mi) to the northeast (Figure 5).

The section reveals two contrasting patterns of shelf-slope buildup. Units of Cretaceous age dip gently seaward and are

fairly uniform in thickness. However, the units of Cenozoic age occur in seaward- or landward-thickening progradational wedges; the units of Miocene and younger age thin toward the present slope or are cut out by it. The units of early Tertiary age appear to be more uniform in thickness, though some appear to pinch out toward the present shelf. The change in the pattern of shelf-slope deposition from units of Cretaceous to late Cenozoic age probably results from a change in sea level. Published curves (Pitman, 1978; Vail, Mitchum, and Thompson, 1977; Kominz, this volume) show sea level reaching a maximum at the end of the Cretaceous, and dropping thereafter. Our data suggest that deepwater conditions prevailed into the Eocene (Poag, 1980) at the COST B-3 and G-2 sites. Subsequently, sea level did fall in an irregularly periodic manner. The result appears to have been a broad interval of slope erosion punctuated by buildout of progradational wedges in the Neogene. As can be seen in Figure 5, the presumed interaction of falling sea level and an active supply of sediment in the Neogene prograded the shelf and developed a well-defined, narrow, shelf-slope-rise transition. The numerous truncating unconformities shown in Figure 5, in rocks of Tertiary age, strongly indicate periodic submarine erosion of the area. The gaps are more numerous than gaps detected in the B-3 well (see Figures 2 and 5). Such obvious gaps in the seismic section inferred to be younger than middle Miocene are probably present in the wells, but were not detected because samples were not collected above the middle Miocene section. Some of the biostratigraphic gaps in the Cretaceous section show up seismically as strong reflections, but with little indication of reflector termination by truncation, downlap, or toplap.

Georges Bank Basin

In the COST G-2 well, eight biostratigraphic gaps are present (Figure 2): late Aptian-middle Cenomanian; late Cenomanian; late Cenomanian-early Turonian; late Turonian-early Coniacian; late Coniacian-early Santonian; late Santonian-early Campanian; late Campanian-middle Eocene; middle Eocene-middle(?) Miocene; late Pliocene-early Pleistocene (Poag, 1982a, 1982b; reexamination of sidewall cores subsequent to Poag, 1982a and 1982b, reveals Campanian strata in the G-2 well). The pattern of seismostratigraphic gaps on profiles across Georges Bank (Figures 6 and 7) is different from the pattern in the other two areas because the two COST holes (Figure 1) on Georges Bank were drilled well back from the shelf edge, and they penetrated a thick carbonate-clastic section of the oldest sedimentary rocks yet drilled on the U.S. Atlantic margin. Unlike the COST holes in the Baltimore Canyon trough area, the Georges Bank COST wells penetrated a relatively thin clastic sequence of Cenozoic rocks and a thick one of Mesozoic strata (Figure 7). As seen from a seismic profile (USGS Line 19, 12 km [7.5 mi] to the northeast), near the COST G-2 well site, reflections are generally parallel except below 3 secs, where a broad arcuate pattern of reflections is inferred to represent carbonate reefs(?) (Poag 1982a, 1982b; Schlee and Fritsch, 1982). The parallel reflections between 3.2 and 2 secs are moderately continuous, have low to high amplitude, and represent the carbonate section (Figure 6)

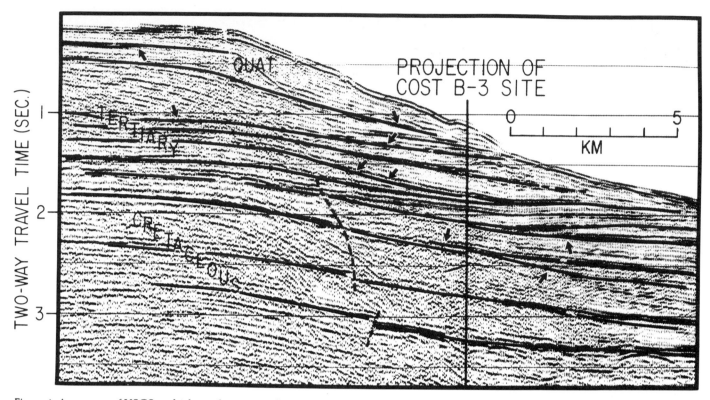

Figure 4: A segment of USGS multichannel seismic-reflection profile 25 across the outer shelf and slope off New Jersey. Inferred unconformities are marked with a heavy line, and the arrows indicate reflection terminations by truncation, downlap, or onlap. The COST No. B-3 well was drilled approximately 10 km (6.2 mi) to the northeast.

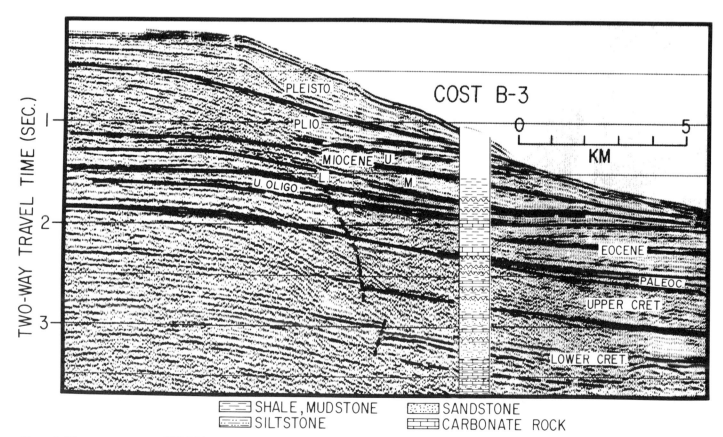

Figure 5: The same segment of USGS line 25 as shown in Figure 4. The ages of the depositional sequences are indicated and also shown is simplified lithologic log of the COST No. B-3 hole to a depth of 4,822 m (15,820 ft), unconformities are indicated by wavy lines in the log.

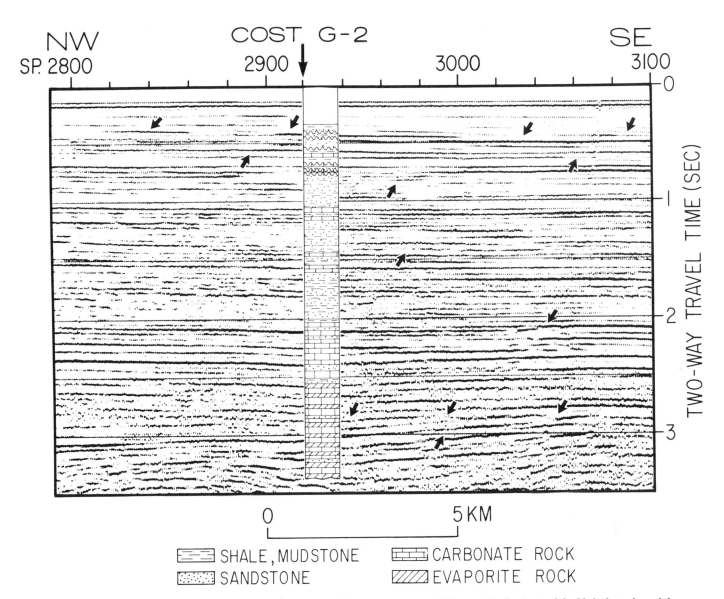

NW   COST G-2   SE
SP. 2800   2900   3000   3100

TWO-WAY TRAVEL TIME (SEC)

0   5 KM

SHALE, MUDSTONE   CARBONATE ROCK
SANDSTONE   EVAPORITE ROCK

Figure 6: A segment of USGS multichannel seismic-reflection profile 19, across part of Georges Bank. A simplified lithologic log of the COST G-2 well to a depth of 6,667 m (21,873 ft) has been projected to the line of section from 12 km (7.5 mi) to the southwest. The arrows outline horizons along which reflections are terminated by truncation, downlap, or onlap; the wavy lines indicate unconformities in hole.

drilled in the COST G-2 well. Alternating zones of weak discontinuous reflections and strong continuous ones, are evident between 2 and 1.25 secs (Figure 6); in the COST G-2 well, the corresponding section is interbedded marine shelf limestone and transitional-coastal sandstone, siltstone, and shale. In the upper 1.25 secs of the profile (Figure 6), high-amplitude continuous reflections occur between zones of weakly continuous reflections (particularly in the upper .5 sec); equivalent-age rocks penetrated in the nearby COST G-2 well (Figure 6) are interbedded gray silty shale, calcareous sandstone and thin chalky limestone (Scholle, Schwab, and Krivoy, 1980; Amato and Simonis, 1980).

Inferred gaps in the seismostratigraphic record for Georges Bank are more subtle than those in the record for the Baltimore Canyon trough; the rapid changes in thickness of seismic units evident there are lacking beneath Georges Bank, so that unconformities are marked by a low-angle onlap and truncation of reflections (uppermost two

arrows on right---Figure 6). An angular unconformity is present over much of the Georges Bank basin, toward the base in the sedimentary section. It appears to separate probable synrift sedimentary rocks of Early Jurassic(?) age from the postrift sequences (Figure 7). This unconformity is most pronounced at the block-faulted edge of the basin (Schlee and Fritsch, 1982) but also shows up on Figure 6 as a slight divergence of reflections at the lowest arrow.

The seismic unconformities detected on this profile (Figure 7) are numerous and appear to coincide closely with the biostratigraphic gaps detected in the Cretaceous and early Cenozoic. The unconformities older than late Hauterivian (near the top of the Neocomian) are difficult to date because fossils are scarce in rocks of Late and Middle Jurassic age and nearly absent in strata older than Middle Jurassic. Our inference as to the age of the units older than Callovian (Figure 7) comes from matching the major and minor coastal onlap shifts from the Vail curve (Vail, Mitchum, and

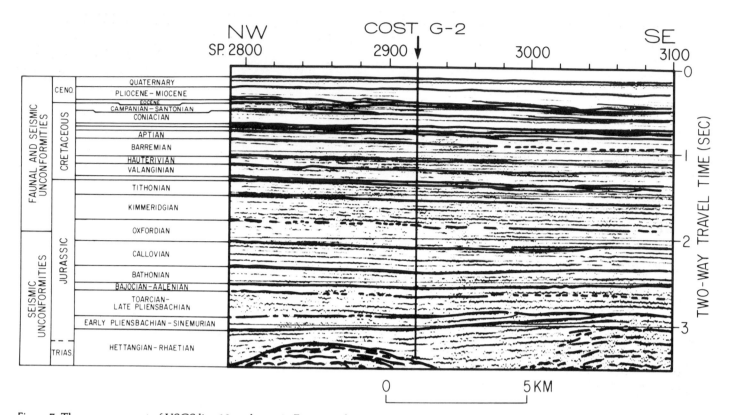

Figure 7: The same segment of USGS line 19 as shown in Figure 6, showing ages of seismic unconformities (heavy lines). The ages of the gaps older than Late Cretaceous were inferred by matching the horizons along which reflections were terminated, with sharp pullbacks in coastal onlap from the Vail curve (Vail, Mitchum, and Thompson, 1977). Sources of time scale given in "Limitations of Data."

Thompson, 1977; Figure 2) with seismic unconformities on the profile (Poag, 1982a, 1982b). If these matches are valid, then the "breakup" unconformity is of Early Jurassic age (early Pliensbachian-Sinemurian) and the rift sequence of carbonate rocks is earliest Jurassic or Late Triassic in age (as should be the salt, at the bottom of the COST G-2 well). An independent estimate was provided by Van Houten (1977; Figure 2); on the basis of a tentative correlation of several basinal sequences from both continents, he estimated that *open* marine flooding of the seaward marginal areas of eastern North America and northwest Africa took place in the Pliensbachian.

Scotian Basin

Of the wells drilled on the Scotian shelf, the Shell Mohawk B-93 is the nearest one to Georges Bank. In this well, nine biostratigraphic gaps are present (Figure 2): late Albian-middle Cenomanian; late Cenomanian; late Turonian-early Coniacian; late Coniacian-early Santonian; late Santonian; Maestrichtian-middle Paleocene; late Paleocene-early Eocene; middle Eocene-middle Miocene; late Pliocene-early Pleistocene (Poag, unpublished data). These gaps demonstrate the similarity of depositional sequences there with those of the Georges Bank basin.

Figure 8 shows a compilation of the approximate time span represented by all biostratigraphic gaps in each well. This compilation is limited to the last 100 m.y., because the biochronology is too imprecise in the older section. The range of cumulative hiatus durations is 24 to 60 m.y., being

greatest in the Georges Bank basin and least in the Baltimore Canyon trough.

The importance of these gaps, with regard to potential accumulation and maturation, can be seen by calculating the potential missing sediment column (compacted) for each well (Figure 8). This calculation is done by multiplying the average rate of sediment accumulation by the total duration of hiatuses. The possible addition of as much as 576 to 1575 m (1,890 to 5,167 ft) of section clearly would have affected the petroleum maturation in each basin.

CORRELATIONS WITH NEARBY SEDIMENTARY SEQUENCES

Studies of the U.S. Atlantic Coastal Plain (Blackwelder, 1981; Gibson, 1983; Ward, Lawrence, and Blackwelder, 1978; Ward, Lawrence, and Blackwelder 1978; Ward et al, 1979; J.E. Hazel, oral communication, 1982) show that sedimentary sequences and stratigraphic gaps similar to those offshore can be recognized in outcrops and well bores. However, the more widespread occurrence of updip nonmarine facies limits biostratigraphic resolution more onshore than offshore. Figure 9 compares the stratigraphic positions of Cenozoic and Late Cretaceous hiatuses in composite onshore and offshore sections.

The Scotian basin north of Georges Bank has been more extensively drilled than any segment of the U.S. Atlantic margin, and several important stratigraphic interpretations have been published (Jansa and Wade, 1975; Ascoli, 1976;

BORE HOLE	COST GE-1	COST B-3	COST G-2	SHELL MOHAWK B-93
MILLION YEARS REPRESENTED BY HIATUSES	41	24	60	55
MILLION YEARS REPRESENTED BY ACCUMULATION	59	76	40	45
THICKNESS OF COMPACTED SEDIMENT COLUMN (meters)	1700	1799	594	1286
AVERAGE RATE OF COMPACTED SEDIMENT ACCUMULATION (meters per m.y.)	29	24	15	29

Figure 8: Compilation of cumulative duration of hiatuses, cumulative time represented by sediments, thicknesses of compacted sediment columns, and average rates of accumulation during last 100 m.y. (Cenomanian to present) for a well in each of the four offshore basins studied. Strata older than Cenomanian are not included because of poor biostratigraphic resolution.

Given, 1977; Eliuk, 1978; Barss, Bujak, and Williams, 1979). The general scheme of lithofacies and depositional sequences appears to be nearly the same as that of the Georges Bank basin (Poag 1982a, 1982b). However, the analysis of stratigraphic gaps and seismic disconformities in the Scotian basin has been limited (King, Maclean, and Fader, 1974). A recent examination of the Mohawk B-93 well (Poag, unpublished data) shows that the biostratigraphic gaps are nearly the same as those of the COST G-2 well (Figure 2).

REGIONAL TRENDS

Current data from boreholes and seismostratigraphic interpretation show that the bulk of deposition in the Atlantic margin basins took place in the Mesozoic. Triassic strata scarcely have been recovered offshore (perhaps in the bottom of the G-1 and G-2 wells, and definitely in several wells on the Scotian shelf; Jansa and Wade, 1975; Barss, Bujak, and Williams, 1979; Poag, 1982a, 1982b). Where sampled, these Triassic strata are chiefly shallow-water carbonate deposits, evaporite deposits, and arkosic red beds (Jansa and Wade, 1975; Given, 1977). They formed chiefly in grabens and half-grabens during the rifting stage of continental breakup. As much as 8 km (5 mi) of these strata may be present in the Georges Bank basin (Mattick, Schlee, and Bayer, 1981; Schlee and Fritsch, 1982) and 9 km (5.6 mi) of strata are present in the Baltimore Canyon trough (Schlee, 1981; Figure 7).

As sea-floor spreading began in the Early Jurassic, about 185 to 190 m.y. ago, carbonate deposition dominated the outer shelves all along the eastern seaboard and a shelf-edge reef formed in a discontinuous linear trend from Florida to Nova Scotia. Carbonate deposition prevailed on the shelf until the middle of the Early Cretaceous (Hauterivian and Barremian) when terrigenous clastics began to bury the shelf-edge reefs and to spill over onto the continental slope and rise (Poag, 1980, 1982a, 1982b).

The biostratigraphic record in the Triassic to middle Cretaceous interval is insufficient to allow us to confidently distinguish biostratigraphic gaps, but a persistent series of seismic disconformities can be identified and appears to be approximately isochronous from basin to basin. These disconformities have been correlated with the scheme of Vail,

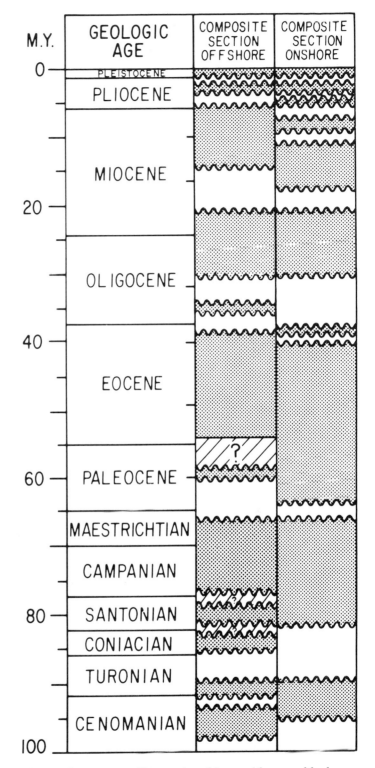

Figure 9: Comparison of biostratigraphic gaps (shown as blank segments) in composite onshore and offshore sections of U.S. Atlantic margin. Uncertainty as to presence of strata of a given age is indicated by diagonal hachures. Strata older than Cenomanian are not considered because of poor biostratigraphic resolution. Sources of time scale given in "Limitations of Data."

Mitchum, and Thompson (1977) to subdivide the undated intervals of the Georges Bank basin (Poag 1982a, 1982b; Figure 7).

The middle to Upper Cretaceous strata constitute a generally fossiliferous transgressive sequence punctuated by six gaps that appear to be of regional extent, although some have not yet been identified in all wells. Those gaps presently recognized are (Figure 2): Albian-early Cenomanian, late Cenomanian, late Turonian-early Coniacian, late Coniacian-early Santonian, late Santonian-early Campanian, and late Maestrichtian-late Paleocene.

Depositional patterns are more complicated in the Cenozoic. This greater complexity may result in part from the fact that the planktonic microfossils provide a more refined biochronology there. However, the chief reason seems to be that the major phase of thermal and isostatic subsidence terminated in the early Tertiary. This resulted in the greater prominence of local events (fluctuations in subsidence and sediment accumulation rates; changing flow pattern of the Gulf Stream) in the depositional history of the subject basins. Another factor may have been the effects of glacioeustatic sea levels. General depositional trends for the Cenozoic show a major transgressive phase in the Paleogene in which white carbonate-rich strata accumulated all along the margin, especially in the middle Eocene. The Miocene-Holocene strata are, in contrast, largely terrigenous clastics. Deltaic lobes and upwelling water masses are particularly characteristic of the Miocene; shallow-water glaciomarine and fluviomarine clastics characterize the Pleistocene strata. Five regional gaps have been identified in the Cenozoic record: late Eocene-early Oligocene; early Oligocene-late Oligocene; early Miocene; late Miocene-early Pliocene, and basal Pleistocene.

Little doubt exists that the regional gaps seen in the Atlantic margin record are equivalent to several of those discussed by Vail and his colleagues (1977 and subsequently), which are attributed to eustatic sea-level changes. The paleobathymetric record (Figure 10) also shows a series of transgressive pulses punctuated by hiatuses or regressions at the levels where the major disconformities of Vail and others are present.

A comparison of the paleoenvironmental data from the three COST wells, the Mohawk B-93, and the Vail curve of coastal onlap shows similar trends in the Jurassic and Cretaceous (Figure 10). Paleodepth reached a maximum in the Late Jurassic-Early Cretaceous. It decreased in the Valanginian, increased briefly during the Hauterivian, and decreased again in the Barremian. During the Cenomanian and Turonian it built to a maximum again. After another short-term fall in late Turonian-Coniacian, it reached a maximum for the Mesozoic in the Santonian-Campanian interval. With some fluctuations, paleodepth remained great in the Paleocene and Eocene.

Only some of the major disconformities of Vail and others have been recognized. Particularly during the Tertiary, deposition seems to have been continuous at several places despite an inferred sea-level drop as shown by the curve. Deposition seems to have been continuous at the COST GE-1 well during inferred middle Eocene lowstands of Vail, Mitchum, and Thompson (1977). At the B-2 and B-3 wells, an inferred middle Miocene lowstand has not been detected (Figure 2). However, a key unconformity is particularly noticeable in the Baltimore Canyon trough (Figure 2), where it coincides approximately with the late Oligocene sea-level

drop of approximately 400 m (1,312 ft) (Vail, Mitchum, and Thompson 1977).

The intervals of continued sedimentation point up the importance of local factors in obscuring the effects of short-term global shifts in sea level, and possibly may indicate that the ages of some of the shifts on the curve of Vail, Mitchum, and Thompson, (1977) are slightly older or younger than presently shown. For example, the effects of varying subsidence rates and fluctuating localized input of sediment may combine to overshadow the effects of a shift in sea level, particularly for a deep-water area.

CAUSES OF THE STRATIGRAPHIC GAPS

The interregional nature of some of the unconformities in the four basins studied strongly suggests that they are related to short-term fluctuations in sea level. What then is the cause of the sea-level shift? Vail, Mitchum, and Thompson's (1977) assumption that these sea-level fluctuations are glacioeustatic lacks firm documentation but has been supported, in part, by some authors' interpretations of the oxygen-isotope record in deep-sea sediments (Matthews and Poore, 1980; Matthews, this volume). In theory, ice-volume changes provide a convenient and rapid method of changing sea level, and can be reasonably documented throughout the late Paleogene to Holocene interval. However, ice-volume change fails to explain the conflicts with the record of unchanging oxygen-isotope ratios across the middle Oligocene interval, which Vail, Mitchum, and Thompson (1977) claimed was the period of greatest sea-level drop in the Mesozoic-Cenozoic interval. Clearly, other agents may have acted at various times, alone or in unison with eustatic sea level, to produce interregional unconformities. Such agents might include changing sediment accumulation rates, basin subsidence rates, paleoceanographic regimes, or local tectonism.

Changes in accumulation rates

Figure 11 illustrates minimum-accumulation-rate curves (sedimentary columns not decompacted) for three COST wells and the Mohawk B-93 well. In general, the Early and Late Jurassic (and Triassic?) rates were highest (10 to 14 cm/1,000 yrs, or 3.9 to 5.5 in/1,000 yrs). A significant reduction (3 to 5 cm/1,000 yrs, or 1.2 to 2.0 in/1,000 yrs) took place in the Early Cretaceous, and rates were generally reduced even further during the Cenozoic (0.5 to 3 cm/1,000 yrs, or 0.2 to 1.2 in/1,000 yrs). However, some notable exceptions can be seen. For example, at GE-1, middle Eocene rates jumped to 6.0 cm/1,000 yrs (2.4 in/1,000 yrs), and at B-3, the middle Miocene rate was 8 cm/1,000 yrs (3.1 in/1,000 yrs) (at B-2, the middle Miocene rate was an astounding 24.8 cm/1,000 yrs/[9.8 in/1,000 yrs]; Poag, 1980). The middle Miocene rates were high enough to maintain relatively shallow-water conditions off New Jersey and to mask the sea-level rise seen elsewhere at this time.

Changes in subsidence rates

Figure 12 shows relative subsidence rates calculated for the three COST wells, using the geohistory method of van

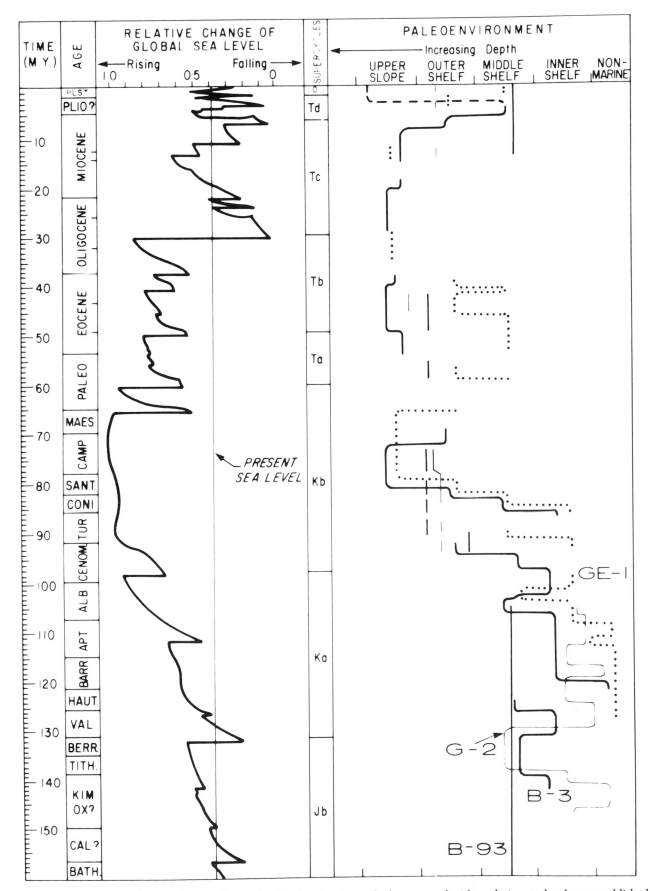

Figure 10: Paleoenvironmental curves for a well in each of the four basins studied, compared with a relative sea-level curve published by Vail, Mitchum, and Thompson (1977). Sources of time scale given in "Limitations of Data."

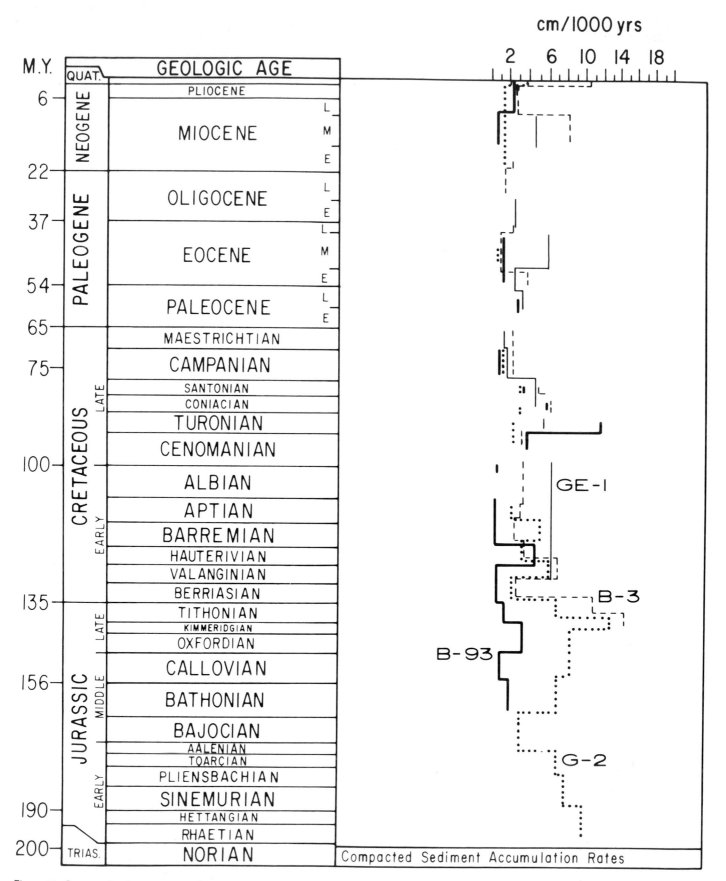

Figure 11: Curves of sediment accumulation rates (calculated for compacted sediment columns) for wells in each of the four basins studied. Sources of time scale given in "Limitations of Data."

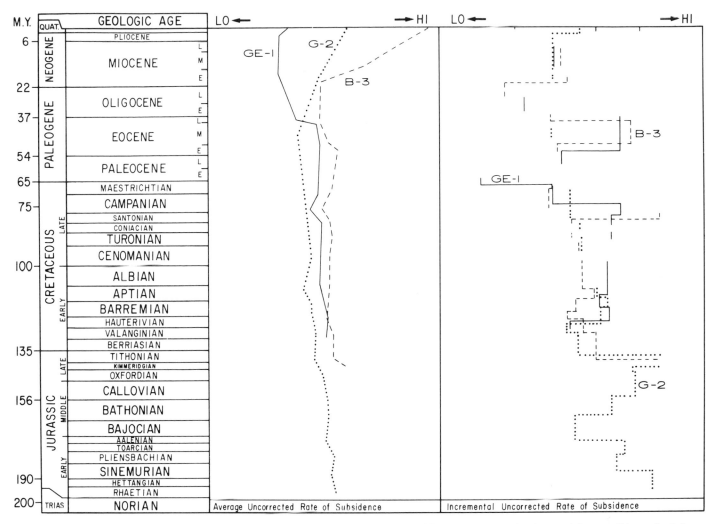

Figure 12: Curves of relative *average* and *incremental* subsidence rates (uncorrected for compaction of sediments) for a well in each of the three U.S. Atlantic offshore basins. See van Hinte (1978) and Poag and Hall (1979) for detailed explanations of the calculation methods. Sources of time scale given in "Limitations of Data."

Hinte (1978) as modified by Poag and Hall (1979). The points used to construct the rate curves represent the base of each chronostratigraphic stage (or group of stages) as represented in each well. For the *average* curves, the calculated subsidence of each stage boundary represents its total subsidence from the time of original deposition to the present (for example, base of the Cenomanian to present). The *incremental* curves differ in that the subsidence of each stage is calculated only for that increment of geologic time during which the stage (or group of stages) was being deposited (for example, base of Cenomanian to top of Cenomanian).

Because the geologic time (hiatuses) represented by the stratigraphic gaps is not included in the calculations of incremental subsidence, this incremental method yields higher or lower rates of subsidence than the averaging method. *Absolute* values for the rates vary considerably depending on the paleodepth, ancient sea level, and time-scale estimates used for the calculations. Because the *relative* values do not change, only they are illustrated in Figure 12. None of the rates have been corrected for the effect of sediment compaction.

In general, average subsidence was rapid during deposi-

tion of the oldest units penetrated in each well and declined during the Cretaceous. This subsidence history could be explained by the rapid cooling of the underlying crust following the initiation of sea-floor spreading (Watts and Steckler, 1979). However, Cenozoic units penetrated in each well have distinctly independent histories (Figure 12). The incremental curves show the same general trends, but emphasize that subsidence was markedly episodic, being characterized by rapid changes in rate and direction of motion. Such episodicity has been demonstrated by Watts and Steckler (1979), and Heller, Wentworth, and Poag (1982), who used different analytical techniques. Episodic subsidence could be expected to bring about local or regional sea-level changes and hiatuses, which in some places might accentuate or reduce the effects of eustatic sea-level changes.

Paleoceanographic changes

Investigations of biostratigraphic changes, changes in stable isotope ratios, and depositional patterns in the deep sea bring to light a record of major changes in Atlantic oceanic circulation, water-mass composition, and stratification (Berggren and Hollister, 1974; Haq, Premoli-Silva, and Lohmann, 1977; Fischer and Arthur, 1977; Tucholke and Vogt, 1979; Gradstein and Srivastava, 1980). Such events often coincided with, and may have been, in part, the consequence of, paleoclimatic changes (Frakes, 1979). These changes affected not only the deep ocean, but also marginal seas, and also could amplify or dampen the effects of eustatic sea-level changes.

Tectonic effects

Tectonism was used by Pitman (1978, 1979) to explain changes in worldwide sea-level and sedimentary patterns on the Atlantic margin. Pitman postulates that the second order shift in sea level (supercycles of Vail, Mitchum, and Thompson, 1977) could be related to changes in the spreading rate of the midocean ridge system. The spreading-rate changes are thought to alter the volume of the ridge system (slower rate, lower volume); the volume changes are sufficient to affect the *rates* of sea-level rise or fall (as in the Cenozoic) and hence cause transgressions or regressions (when combined with a changing rate of shelf subsidence). On the basis of volume estimates, Pitman (1978; Figure 4) postulated a minor regression in the Late Cretaceous, major regressions in the Paleocene and the Oligocene, and transgressions during the Eocene and early Miocene. An Eocene transgression above a conspicuous Paleocene unconformity does appear to be represented in the biostratigraphy of the COST B-3 and GE-1 wells (Figure 2). The biostratigraphic record in support of a late Oliogocene-Miocene transgression is too fragmentary to be definitive.

The phenomenon of coastal onlap by progressively younger formations was modeled by Watts (1981), who showed that it could be caused by tectonic factors. In his model, flexural strength of the basement increases with time (from initial separation of continents) and results in a progressive onlap from Jurassic to the present, of formations beneath the inner continental shelf and Atlantic Coastal Plain. The model assumes that the areas of sediment infill, in the form of the shelf, slope, and rise, remain constant during the evolution of the margin. Watts' model is appealing because it offers an explanation for the rapid onlap by units of Cretaceous age. It is less successful in explaining the *decrease* in age represented by the outcrop pattern in a seaward direction observed on some areas of the Atlantic Coastal Plain for certain units of Early Tertiary age. However, Watts thinks that the outcrop pattern may have been modified by later tilting and subaerial erosion of these areas. Watts' flexural model provides an interesting alternative to eustasy in explaining the long-term trend in sea-level changes (the supercycles of Vail, Mitchum, and Thompson, 1977); of key future importance is separating the effects of both "flexural hardening of the crust" and eustasy for adjoining areas of platform and basin development.

A combination of changes in the rate of sea-level variation and in rates of basin subsidence has been used by Sheridan (1976) to explain the sedimentary history of the offshore Atlantic basins. The rates of sea-level change are taken from Pitman (1978) and the analysis of basin subsidence is modified from a major study by Brown, Miller, and Swain (1972)

of the sedimentary units of the Atlantic Coastal Plain, North Carolina to New York. Briefly, Brown, Miller, and Swain (1972) thought that the basement beneath the Atlantic Coastal Plain was broken into distinguishable blocks. Hinge zones that bound these blocks were controlled by faults that had two main orientations (north-south and northeast-southwest) and were periodically activated by one of two stress systems. Sheridan (1974; 1976) modified the pattern of tectonic fragmentation outlined by Brown, Miller, and Swain (1972) and extended it to the submerged continental margin, where he defined 17 blocks (from Labrador to Florida) that were periodically moved by one of three stress systems.

Sheridan's (1974) history of events in evolution of the basins is important for comparison with our results. He noted that activity by these stress systems caused a regression sometime in the Barremian-Aptian, and may have caused one in the Coniacian, one in the Paleocene, and one in the Oligocene. Though some of these regressions coincide with the sedimentary gaps seen in COST wells (Coniacian, early Paleocene and Oligocene), other gaps in the record (Albian-early Cenomanian, early Miocene, and late Miocene-early Pliocene) are unexplained by his schedule of transgressions and regressions.

CONCLUSIONS

We examined trends in relative sea level and the position of gaps in the Mesozoic-Cenozoic sedimentary record in six deep stratigraphic holes drilled on the Atlantic continental margin. Records from these offshore holes have been compared to the record preserved under the Atlantic Coastal Plain and to the pattern of reflections used to define depositional sequences on an extensive grid of multichannel seismic reflection profiles.

Though stratigraphic gaps identified in the wells are not as numerous as those seen on seismic reflection profiles in areas of rapid sediment buildup, we find at least seven interregional unconformities that can be traced from the Blake Plateau basin to the Scotian shelf, and from the coastal plain to the continental rise. These gaps are in the early Cenomanian, near the Turonian-Coniacian boundary, between the late Maestrichtian and early to middle Paleocene, from the latest Eocene to the early Oligocene, within the late Oligocene, during the early Miocene, and in the late Pliocene-early Pleistocene. At least six of these interregional unconformities also coincide in stratigraphic position with major unconformities that bound depositional supercycles in the Vail model of global depositional sequences. An additional four unconformities are consistently present in two of the three basins (late Cenomanian, Coniacian-Santonian, Santonian-Campanian, early Pliocene), and better stratigraphic control may prove in the future that these also are interregional. These unconformities serve as approximately isochronous horizons that are key elements in constructing the regional stratigraphic framework and delineating the depositional sequences of this margin.

Paleoenvironmental studies of the microfauna and lithofacies from the holes indicate that relative sea level rose sporadically during the Cretaceous, and reached a maximum in the Late Cretaceous when water depths typical of the upper slope prevailed over the present outer shelf. This trend is similar to that shown by the curve of Vail, Mitchum, and Thompson (1977) Pitman (1978, 1979) and Kominz (this volume). Above an interregional unconformity, similar slope-type water depths prevailed off the northeastern U.S. shelf during the early Tertiary. However, for the remainder of the Tertiary, relative sea level fluctuated widely and interbasin variability was marked.

Lithofacies, paleobathymetric estimates, and rates of basin subsidence indicate that sedimentary sequences in each U.S. Atlantic margin basin were deposited in near-unison during the Jurassic and Cretaceous, during that part of the postrift depositional phase when cooling of the underlying crust and its consequent subsidence were most rapid.

During the Tertiary, the thickness and types of depositional sequences became more variable from basin to basin. By this time, the crust cooled to the extent that subsidence was dominated by the effect of sedimentary loading, which may have contributed to episodic subsidence. The slower rate of subsidence coupled with variable sediment input, and perhaps some local tectonism, modified the effects of eustasy and created a more complex system of depositional sequences.

We conclude that our data fit the Vail model well enough to justify its cautious use as a predictive tool in deciphering the geologic history of other parts of the U.S. margin, such as the Carolina trough, where drill-hole data are not yet available.

ACKNOWLEDGMENTS

The writers thank Peter A. Scholle and Page C. Valentine for critically reviewing the original manuscript, and P. Ascoli for providing samples from the Mohawk B-93 well.

REFERENCES CITED

Abbott, W.H., 1978, Correlation and zonation of Miocene strata along the Atlantic margin of North America using diatoms and silicoflagellates: Marine Micropaleontology, v. 3, p. 15-34.

—— 1980, Diatoms and stratigraphically significant silicoflagellates from the Atlantic margin Coring Project and other Atlantic margin sites: Micropaleontology, v. 26, p. 49-80.

Amato, R. V., and E.K. Simonis, eds., 1980, Geologic and operational summary, COST No. G-2 well, Georges Bank area, North Atlantic OCS: U.S. Geological Survey Open-File Report 80-269, 116 p.

Ascoli, P., 1976, Foraminiferal and ostracod biostratigraphy of the Mesozoic-Cenozoic, Scotian shelf, Atlantic Canada: Maritime Sediments Special Publication 1, part B, p. 653-677.

Barss, M.S., J.P. Bujak, and G.L. Williams, 1979, Palynological zonation and correlation of sixty-seven wells, eastern Canada: Geological Survey of Canada, Paper 78-24, 118 p.

Bebout, J.W., 1980, Biostratigraphy, in R.V. Amato and E.K. Simonis, eds., Geologic and operational summary, COST No. G-2 well, Georges Bank area, North Atlantic OCS: U.S. Geological Survey Open-File Report 80-269, p. 20-28.

Berggren, W.A., and C.D. Hollister, 1974, Paleogeography, paleobiogeography, and the history of circulation in the Atlantic Ocean, in W.W. Hay, ed., Studies in paleo-oceanography: Society of Economic Paleontologists and Mineralogists, Special Publication 20, p. 126-186.

——, and J.A. van Couvering, 1974, The late Neogene; biostratigraphy, biochronology, and paleoclimatology of the last 15 million years in marine and continental sediments: Palaeogeography, Palaeoclimatology, Palaeoecology, v. 16, p. 1-216.

Blackwelder, B.W., 1981, Late Cenozoic marine deposition in the United States Atlantic Coastal Plain related to tectonism and global climate: Palaeogeography, Palaeoclimatology, Palaeoecology, v. 34, p. 87-114.

Blow, W.H., 1969, Late middle Eocene to Recent planktonic foraminiferal biostratigraphy: Geneva, Proceedings 1st International Conference on Planktonic Microfossils, v. 1, p. 199-422.

——, 1979, The Cainozoic Globigerinida; a study of the morphology, taxonomy, evolutionary relationships and the stratigraphical distribution of some Globigerinida (mainly Globigerinacea): Leiden, E.J. Brill, v. 1, 752 p., v .2, 660 p., v. 3, 264 pls.

Bolli, H.M., 1966, Zonation of Cretaceous to Pliocene marine sediments based on planktonic foraminifera: Asociacion Venezuelana de Geologia, Mineralogia, y Petrologia, Boletin, Informativo, v. 9, p. 3-32.

Brown, P.M., J.A. Miller, and F.M. Swain, 1972, Structural and stratigraphic framework and spatial distribution of permeability of the Atlantic Coastal Plain, North Carolina to New York: U.S. Geological Survey Professional Paper 796, 70 p.

Bukry, D., 1971, Cenozoic calcareous nannofossils from the Pacific Ocean: San Diego Society of Natural History, Transactions, v. 16, p. 303-327.

——, 1973, Low-latitude coccolith biostratigraphic zonations, in N.T. Edgar, et al, eds., Initial reports of the deep-sea drilling project, v. 15: Washington, D.C., U.S. Government Printing Office, p. 685-703.

Dillon, W.P., et al, 1979, Structure, biostratigraphy, and seismic stratigraphy along a common-depth-point seismic profile through three drill sites on the continental margin off Jacksonville, Florida: U.S. Geological Survey Miscellaneous Field Studies Map MF 1090, 1 sheet.

Douglas, R.G., 1979, Benthic foraminiferal ecology and paleoecology; review of concepts and methods: in J.H. Lipps, et al, eds., Foraminiferal ecology and paleoecology: Society of Economic Paleontologists and Mineralogists, Short Course No. 6, p. 21-53.

——, and F. Woodruff, 1981, Deep-sea benthic foraminifera, in, C. Emiliani, ed., The oceanic lithosphere: The Sea, Wiley-Interscience, v. 7, p. 1233-1327.

Eliuk, L.S., 1978, The Abenaki Formation, Nova Scotia Shelf, Canada - a depositional and diagenetic model for a Mesozoic carbonate platform: Bulletin of Canadian Petroleum Geology, v. 26, p. 424-514.

Fischer, A.G., and M.A. Arthur, 1977, Secular variations in

the pelagic realm, in H.E. Cook, and P. Enos, eds., Deep-water carbonate environments: Society of Economic Paleontologists and Mineralogists, Special Publication No. 25, p. 19-50.

Frakes, L.A., 1979, Climates throughout geologic time: Amsterdam, Elsevier Scientific Publications, 310 p.

Gibson, T.G., 1983, Stratigraphy of Miocene through lower Pleistocene strata of the United States central Atlantic Coastal Plain in, C.E. Ray, ed., Geology and paleontology of the Lee Creek Mine, North Carolina I: Smithsonian Contributions to Paleobiology, p. 35-80.

Given, M.M., 1977, Mesozoic and early Cenozoic geology of offshore Nova Scotia: Bulletin of Canadian Petroleum Geology, v. 25, p. 63-91.

Gradstein, F.M., and S.P. Srivastava, 1980, Aspects of Cenozoic stratigraphy and paleoceanography of the Labrador Sea and Baffin Bay: Palaeogeography, Palaeoclimatology, Palaeoecology, v. 30, p. 261-295.

Grow, J.A., and R.E. Sheridan, 1981, Deep structure and evolution of the continental margin off the eastern United States: Paris, Colloquium C3, 26th International Geological Congress, Geology of Continental Margins; Oceanologica Acta, supplement, v. 4, p. 11-19.

——, R.E. Mattick, and J.S. Schlee, 1979, Multichannel depth sections and interval velocities over outer continental shelf and upper continental slope between Cape Hatteras and Cape Cod, in J.S. Watkins, L. Montadert, and P.W. Dickerson, eds., Geological and geophysical investigations of continental margins: AAPG Memoir 29, p. 65-83.

Hallam, A., 1963, Eustatic control of major cyclic changes in Jurassic sedimentation: Geological Magazine, v. 100, p. 444-450.

Halley, R.B., 1979, Petrographic summary, in P.A. Scholle, ed., Geological studies of the COST GE-1 well, United States south Atlantic outer continental shelf area: U.S. Geological Survey Circular 800, p. 42-48.

Hancock, J.M., and E.G. Kauffman, 1979, The great transgressions of the Late Cretaceous: Journal of the Geological Society of London, v. 136, p. 175-186.

Haq, B.U., 1981, Paleogene paleoceanography; Early Cenozoic oceans revisited: Paris, Colloquium C4, 26th International Geological Congress, Paleoceanography; Oceanologica Acta, supplement, v. 4, p. 71-82

——, I. Premoli-Silva, and G.P. Lohmann, 1977, Calcareous plankton paleobiogeographic evidence for major climatic fluctuations in the early Cenozoic Atlantic Ocean: Journal of Geophysical Research, v. 82, p. 3861-3876.

Hardenbol, J., and W.A. Berggren, 1978, A new Paleogene numerical time scale: AAPG Studies in Geology No. 6, p. 213-234.

——, P.R. Vail, and J. Ferrer, 1981, Interpreting paleoenvironments, subsidence history, and sea-level changes of passive margins from seismic and biostratigraphy: Paris, Colloquium C3, 26th International Geological Congress, Geology of Continental Margins; Oceanologica Acta, supplement, v. 4, p. 33-44.

Harland, W.B., A.G. Smith, and B. Wilcock, eds., 1964, The Phanerozoic time scale - a symposium: Geological Society of London Quarterly Journal, v. 120, supplement, 458 p.

Harrison, C.G.A., et al, 1981, Sea-level variations, global sedimentation rates, and the hypsographic curve: Earth and Planetary Science Letters, v. 54, p. 1-16.

Hathaway J.C., et al, 1979, U.S. Geological Survey core drilling on the U.S. Atlantic shelf: Science, v. 206, n. 4418, p. 515-527.

Hay, W.W., et al, 1967, Calcareous nannoplankton zonation of the Cenozoic of the Gulf Coast and the Caribbean - Antillean area and transoceanic correlation: Gulf Coast Association of Geological Societies, Transactions, v. 17, p. 428-480.

Hays, J.D., and W.C. Pitman III, 1973, Lithospheric plate motion, sea-level changes, and climatic and ecological consequences: Nature, v. 246, p. 18-22.

Heller, P.L., C.M. Wentworth, and C.W. Poag, 1982, Episodic post-rift subsidence of the U.S. Atlantic continental margin: Geological Society of America Bulletin, v. 93, p. 379-390.

Jansa, L.F., and J.A. Wade, 1975, Geology of the continental margin off Nova Scotia and Newfoundland: Geological Survey of Canada, Paper 74-30, p. 51-105.

King, L.H., B. Maclean, and G.B. Fader, 1974, Unconformities on the Scotian Shelf: Canadian Journal of Earth Science, v. 11, p. 89-100.

Klitgord, K.D., J.S. Schlee, and Karl Hinz, 1982, Basement structure, sedimentation, and tectonic history of the Georges Bank basin, in P.A. Scholle, and C.R. Wenkam, eds., Geological studies of the COST Nos. G-1 and G-2 wells, United States North Atlantic Outer Continental Shelf: U.S. Geological Survey Circular 861, p. 160-186.

Lachance, D.J., 1980, Lithology, in R.V. Amato, and J.W. Bebout, eds., Geologic and operational summary, COST No. G-1 well, Georges Bank area, North Atlantic OCS: U.S. Geological Survey Open-File Report 80-268, p. 16-21.

Libby-French, J., 1981, Lithostratigraphy of the Shell 272-1 and 273-1 wells; implications as to depositional history of the Baltimore Canyon trough, mid-Atlantic OCS: AAPG Bulletin, v. 65, p. 1476-1484.

Lohmann, G.P., 1978, Abyssal benthonic foraminifera as hydrographic indicators in the western South Atlantic Ocean: Journal of Foraminiferal Research, v. 8, p. 6-34.

Martini, E., 1971, Standard Tertiary and Quaternary calcareous nannoplankton zonation: Roma, Proceedings, 2nd Planktonic Conference, v. 2, p. 739-777.

Matthews, R.K., and R.Z. Poore, 1980, Tertiary ^{18}0 record and glacio-eustatic sea-level fluctuations: Geology, v. 8, p. 501-504.

Mattick, R.E., J.S. Schlee, and K. Bayer, 1981, The geology and hydrocarbon potential of the Georges Bank-Baltimore Canyon area, in J.M. Kerr and A.J. Ferguson, eds., Geology of the north Atlantic borderlands: Canadian Society of Petroleum Geologists, Memoir 7, p. 461-486.

Murray, J.W., 1973, Distribution and ecology of living benthic foraminiferids: New York, Crane, Russak, 274 p.

Ness, G., S. Levi, and R. Couch, 1980, Marine magnetic anomaly timescales for the Cenozoic and Late Cretaceous; precis, critique, and synthesis: Review of Geophysics and Space Physics, v. 18, p. 753-770.

Obradovich, J.D., and W.A. Cobban, 1975, A time-scale

for the Late Cretaceous of the western interior of North America: Geological Association of Canada Special Paper 13, p. 31-54.

Paull, C.K., and W.P. Dillon, 1980, Structure, stratigraphy, and geologic history of the Florida-Hatteras Shelf and inner Blake Plateau: AAPG Bulletin, v. 64, p. 339-358.

Phleger, F.B, 1960, Ecology and distribution of Recent foraminifera: Baltimore, Johns Hopkins Press, 297 p.

———, and F.L. Parker, 1951, Ecology of foraminifera, northwest Gulf of Mexico: Geological Society of America Memoir 46, part 1, p. 1-88; part 2, p. 1-64.

Pitman, W.C. III, 1978, Relationship between eustacy and stratigraphic sequences of passive margins: Geological Society of America Bulletin, v. 89, p. 1389-1403.

Pitman, W.C. III, 1979, The effect of eustatic sea level changes on stratigraphic sequences at Atlantic margins, in J.S. Watkins, L. Montadert, and P.W. Dickerson eds., Geological and geophysical investigations of continental margins: AAPG Memoir 29, p. 453-460.

Poag, C.W., 1977, Foraminiferal biostratigraphy, in P.A. Scholle, ed., Geological studies on the COST No. B-2 well, U.S. mid-Atlantic outer continental shelf area: U.S. Geological Survey Circular 750, p. 35-36.

———, 1978, Stratigraphy of the Atlantic continental shelf and slope of the United States: Annual Review of Earth and Planetary Science Letters, v. 6, p. 251-280.

———, 1980, Foraminiferal stratigraphy, paleoenvironments, and depositional cycles in the outer Baltimore Canyon trough, in P.A. Scholle, ed., Geological studies of the COST No. B-3 well, United States mid-Atlantic continental slope area: U.S. Geological Survey Circular 833, p. 44-65.

———, 1981, Ecologic atlas of benthic foraminifera of the Gulf of Mexico: Stroudsburg, Pennsylvania, Hutchinson Ross Publishing, 175 p.

———, 1982a, Foraminiferal and seismic stratigraphy, paleoenvironments, and depositional cycles in the Georges Bank basin, in P.A. Scholle, and C.R. Wenkam, eds., Geological studies of the COST Nos. G-1 and G-2 wells, United States north Atlantic outer continental shelf: U.S. Geological Survey Circular 861, p. 43-92.

———, 1982b, Stratigraphic reference section for Georges Bank basin -- depositional model for New England passive margin: AAPG Bulletin, v. 66, p. 1021-1041.

———, in press, Neogene stratigraphy of the submerged U.S. Atlantic margin, in J. Armentrout, ed., Cenozoic correlations of North America: Palaeoclimatology, Palaeogeography, Palaeoecology.

———, and R.E. Hall, 1979, Foraminiferal biostratigraphy, paleoecology, and sediment accumulation rates, in P.A. Scholle, ed., Geological studies of the COST GE-1 well, United States south Atlantic outer continental shelf area: U.S. Geological Survey Circular 800, p. 49-63.

Pollack, B.M., 1980a, Sandstone petrography, in P.A. Scholle, ed., Geological studies of the COST No. B-3 well, United States mid-Atlantic continental slope area: U.S. Geological Survey Circular 833, p. 24-25.

———, 1980b, Lithology, in P.A. Scholle, ed., Geological studies of the COST No. B-3 well, United States mid-Atlantic continental slope area: U.S. Geological Survey Circular 833, p. 20-23.

Rhodehamel, E.C., 1979, Lithologic descriptions, in P.A. Scholle, ed., Geological studies of the COST GE-1 well, United States south Atlantic outer continental shelf area: U.S. Geological Survey Circular 800, p. 24-36.

Schlee, J.S., 1981, Seismic stratigraphy of the Baltimore Canyon trough: AAPG Bulletin, v. 65, p. 26-53.

———, and J. Fritsch, 1982, Seismic stratigraphy of the Georges Bank basin complex, offshore New England: AAPG Memoir 34, p. 223-251.

———, W.P. Dillon, and J.A. Grow, 1979, Structure of the continental slope off the eastern United States, in L.J. Doyle, and O.H. Pilkey, eds., Geology of continental slopes: Society of Economic Paleontologists and Mineralogists, Special Paper 27, p. 95-118.

———, et al, 1976, Regional framework off northeastern United States: AAPG Bulletin, v. 60, p. 926-951.

Scholle, P.A., ed., 1977, Geological studies on the COST No. B-2 well, U.S. mid-Atlantic outer continental shelf area: U.S. Geological Survey Circular 750, 71 p.

———, ed., 1979, Geological studies of the COST GE-1 well, United States south Atlantic outer continental shelf area: U.S. Geological Survey Circular 800, 114 p.

———, ed., 1980, Geological studies of the COST No. B-3 well, United States mid-Atlantic continental slope area: U.S. Geological Survey Circular 833, 132 p.

———, and C.R. Wenkam, eds., 1982, Geological studies of the COST Nos. G-1 and G-2 wells, United States north Atlantic outer continental shelf: U.S. Geological Survey Circular 861, 193 p.

———, K.A. Schwab, and H.L. Krivoy, 1980, Summary chart of geological data from the COST No. G-2 well, U.S. north Atlantic outer continental shelf: U.S. Geological Survey Oil and Gas Investigations Chart OC-105, 1 sheet.

Sheridan, R.E., 1974, Conceptual model for block-fault origin of the North American Atlantic continental margin geosyncline: Geology, v. 2, n. 9, p. 465-468.

———, 1976, Sedimentary basins of the Atlantic margin of North America: Tectonophysics, v. 36, p. 113-132.

Sheriff, R.E., 1977, Limitations on resolution of seismic reflections and geologic detail derivable from them, in C.E. Payton, ed., Seismic stratigraphy-applications to hydrocarbon exploration: AAPG Memoir 26, p. 3-14.

Shipley, T.H., R.T. Buffler, and J.T. Watkins, 1978, Seismic stratigraphy and geologic history of Blake Plateau and adjacent western Atlantic continental margin: AAPG Bulletin, v. 62, p. 792-812.

Simonis, E.K., 1980, Lithologic description, in R.V. Amato, and E.K. Simonis, eds., Geologic and operational summary, COST No. G-2 well, Georges Bank area, north Atlantic OCS: U.S. Geological Survey Open-File Report 80-269, p. 14-19.

Sliter, W.V., and R.A. Baker, 1972, Cretaceous bathymetric distribution of benthic foraminifera: Journal of Foraminiferal Research, v. 2, p. 167-183.

Stainforth, R.M., et al, 1975, Cenozoic planktonic foraminiferal zonation and characteristics of index forms: University of Kansas Paleontology Contributions, Art. 62, p. 1-162e; Appendix p. 163-425.

Steinkraus, W.E., 1978, Biostratigraphy, in R.V. Amato, and J.W. Bebout, eds., Geological and operational summary, COST No. GE-1 well, southeast Georgia embayment area, south Atlantic OCS: U.S. Geological Survey Open-File Report 78-668, p. 29-41.

———, 1979, Biostratigraphy, *in* R.V. Amato, and E.K. Simonis, eds., Geological and operational summary, COST No. B-3 well, Baltimore Canyon trough area, mid-Atlantic OCS: U.S. Geological Survey Open-File Report 79-1159, p. 21-31.

———, 1980, Biostratigraphy, *in* R.V. Amato, and J.W. Bebout, eds., Geologic and operational summary, COST No. G-1 well, Georges Bank area, north Atlantic OCS: U.S. Geological Survey Open-File Report 80-268, p. 39-52.

Thierstein, H.R., 1976, Mesozoic calcareous nannoplankton biostratigraphy of marine sediments: Marine Micropaleontology, v. 1, p. 325-362.

Tucholke, B.E., and P.R. Vogt, 1979, Western north Atlantic; sedimentary evolution and aspects of tectonic history, *in* B.E. Tucholke, et al, Initial reports of the deep sea drilling project: Washington, D.C., U.S. Government Printing Office, v. 43, p. 791-825.

———, R.E. Houtz, and W.J.H. Ludwig, 1982, Isopach map of sediments in the Western North Atlantic Ocean: AAPG Map Series.

Vail, P.R., and J. Hardenbol, 1979, Sea-level changes during the Tertiary: Oceanus, v. 22, p. 71-79.

———, and R.M. Mitchum, Jr., 1979, Global cycles of relative changes of sea level from seismic stratigraphy, *in* J.S. Watkins, L. Montadert, and P.W. Dickerson, eds., Geological and geophysical investigations of continental margins: AAPG Memoir 29, p. 469-472.

———, and R.G. Todd, 1981, Northern North Sea Jurassic unconformities, chronostratigraphy, and sea-level changes from seismic stratigraphy, *in*, L.V. Illing and G.D. Hobson, eds., Petroluem geology of the continental shelf of northwest Europe: London, Institute of Petroleum, p. 216-235.

———, R.M. Mitchum, Jr., and S. Thompson III, 1977, Seismic stratigraphy and global changes of sea level, Part 4; global cycles of relative changes of sea level, *in* C.E. Payton, ed., Seismic stratigraphy-applications to hydrocarbon exploration: AAPG Memoir 26, p. 83-97.

Valentine, J.W., 1973, Evolutionary paleoecology of the marine biosphere: Englewood Cliffs, New Jersey, Prentice-Hall, 511 p.

Valentine, P.C., 1977, Nannofossil biostratigraphy, *in* P.A. Scholle, ed., Geological studies on the COST No. B-2 well, U.S. mid-Atlantic outer continental shelf area: U.S. Geological Survey Circular 750, p. 37-40.

———, 1979, Calcareous nannofossil biostratigraphy and paleoenvironmental interpretation, *in* P.A. Scholle ed., Geological studies of the COST GE-1 well, United States south Atlantic outer continental shelf area: U.S. Geological Survey Circular 800, p. 64-70.

———, 1980, Calcareous nannofossil biostratigraphy, paleoenvironments, and post-Jurassic continental margin development, *in* P.A. Scholle, ed., Geological studies of the COST No. B-3 well, United States mid-Atlantic continental slope area: U.S. Geological Survey Circular 833, p. 67-83.

———, 1982, Calcareous nannofossil biostratigraphy and paleoenvironment of two deep stratigraphic wells in the Georges Bank basin, *in* P.A. Scholle and C.R. Wenkam, eds., Geological studies of the COST Nos. G-1 and G-2 wells, United States north Atlantic outer continental shelf: U.S. Geological Survey Circular 861, p. 34-42.

van Hinte, J.E., 1976a, A Jurassic time scale: AAPG Bulletin, v. 60, p. 489-497.

———, 1976b, A Cretaceous time scale: AAPG Bulletin, v. 60, p. 498- 516.

———, 1978, Geohistory analysis-application of micropaleontology in exploration geology: AAPG Bulletin, v. 62, p. 201-222.

Van Houten, F.B., 1977, Triassic-Liassic deposits of Morocco and eastern North America—a comparison: AAPG Bulletin, v. 61, p. 79-99.

Ward, L.W., et al, 1979, Stratigraphic revision of Eocene, Oligocene, and lower Miocene formations of South Carolina: Geological Notes, v. 23, p. 1-32.

———, D.R. Lawrence, and B.W. Blackwelder, 1978, Stratigraphic revision of the middle Eocene, Oligocene and lower Miocene -- Atlantic coastal plain of North Carolina: U.S. Geological Survey Bulletin, 1457-F, p. F1—F23.

Watts, A.B., 1981, The U.S. Atlantic continental margin; subsidence history, crustal structure and thermal evolution, *in* Geology of passive margins; history, structure and sedimentologic record (with special emphasis on the Atlantic margin): AAPG Course Notes Series 19, p. 2-1 — 2-75.

———, and M.S. Steckler, 1979, Subsidence and eustacy at the continental margin of eastern North America, *in* M. Talwani, W.W. Hay, and W.B.F. Ryan, eds., Deep drilling results in the Atlantic Ocean; continental margins and paleoenvironments: American Geophysical Union, Maurice Ewing Symposium, Series 3, p. 273-310.

Index ———————————————————————

A reference is indexed according to its important or "key," words.

Three columns are to the left of a keyword entry. The first column, a letter entry, represents the AAPG book series from which the reference originated. In this case, M stands for Memoir Series. Every five years, AAPG will merge all its indexes together, and the letter M will differentiate this reference from those of the AAPG Studies in Geology Series (S) or the AAPG Bulletin (B).

The following number is the series number. In this case, 36 represents a reference from AAPG Memoir 36. The third column lists the page number of this volume on which the reference can be found.